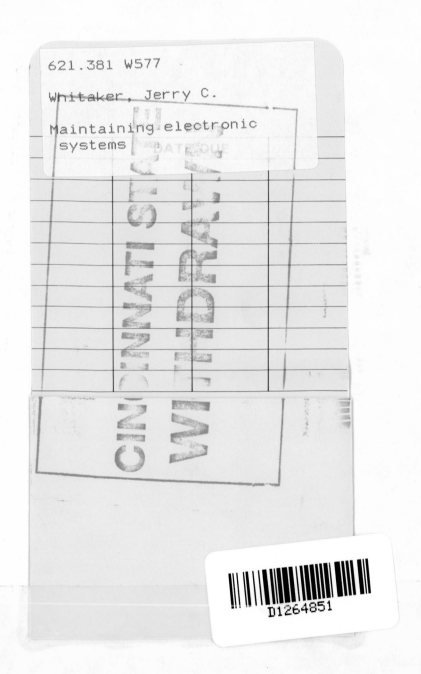

MAINTAINING ELECTRONIC SYSTEMS

MAINTAINING ELECTRONIC SYSTEMS

Jerry C. Whitaker

CRC Press
Boca Raton   Ann Arbor   Boston

Library of Congress Catalog Card Number 90-85586

Direct all inquiries to CRC Press, Inc., 2000 Corporate Blvd., N.W., Boca Raton, Florida 33431.

© 1991 by Multiscience Press, Inc.

International Standard Book Number 0-8493-7411-1
Printed in the United States of America

*This book is dedicated to the memory of
Helen and Osborne Gibson*

CONTRIBUTORS

The author wishes to express appreciation to the following individuals who provided input to this publication.

Tom Allen, John Fluke Manufacturing Company, Everett, Wash.

Carl A. Bentz, Intertec Publishing Corp., Overland Park, Kan.

Gregory Carey, Sencore, Sioux Falls, S. Dak.

Michael Dahlgren, Novell Service Division, Provo, Utah.

Bradley Dick, Intertec Publishing Corp., Overland Park, Kan.

Donald Markley, P.E., Markley and Associates Consultants, Peoria, Ill.

Conrad Persson, *Electronic Servicing & Technology* magazine, Overland Park, Kan.

Technical Staff, Eimac Division of Varian Associates, San Carlos, Calif.

CONTENTS

PREFACE

Technology is a moving target. Continuing advancements in hardware and software provide new features and increased performance for industrial customers. Those same advancements, however, place new demands on the engineering/maintenance department of a facility. Today, more than ever, the reliability of a system can have a direct and immediate impact on the profitability of an operation.

The days of troubleshooting a piece of gear armed only with a scope, voltmeter, and a general idea of how the hardware works are gone forever. Today, unless you have a detailed maintenance manual and the right test equipment, you are out of luck. The testbench of the 1980s—stocked with a VTVM, oscilloscope, signal generator, and signal analyzer—is a relic of the past. The workbench of today more closely resembles a small computer repair center than anything else.

It is true that some equipment problems still can be located with little more than a digital multimeter (DMM) and oscilloscope, given enough time and effort. But time costs money. Few technical managers are willing to make the trade. With current-technology equipment, the proper test equipment is a must.

Two of the more important pieces of equipment for a maintenance technician servicing current-technology products are good lighting and a whopping big magnifier! Although that is certainly an exaggeration, it points up a significant problem in equipment maintenance today: There are many tiny components, most of them jammed into a small amount of circuit board real estate. Tight component packaging makes printed wiring boards (PWBs) difficult to repair, at best. When complex and interrelated circuitry is added to the servicing equation, repair down to the component level may be virtually impossible. The equipment is just too complex, electrically and mechanically. The sophistication of hardware today has ushered in a new era in equipment maintenance—that of *repair by replacement.*

Some equipment manufacturers have built sophisticated test and diagnostic routines into their products. This trend is welcomed, and is likely to accelerate as the *maintainability* of products becomes an important selling point. Still, however, specialized test equipment is often necessary to trace a problem to the board level.

ANALYTICAL APPROACH TO MAINTENANCE

Because of the requirement for maximum uptime and top performance, a *comprehensive preventive maintenance* (CPM) program should be considered for any facility. Priority-based considerations of reliability and economics are applied to identify the applicable and appropriate preventive maintenance tasks to be performed. CPM involves a realistic assessment of the vulnerable sections or components within the system, and a cause-and-effect analysis of the consequences of component failure. Basic to this analysis is the importance of keeping the system up and running at all times. Obvious applications of CPM include the stocking of critical spare parts used in stages of the equipment exposed to high temperatures and/or high voltages, or the installation of standby power/transient overvoltage protection equipment at the ac input point of critical equipment. Usually, the sections of a system most vulnerable to failure are those exposed to the outside world.

The primary goals of any CPM program are to prevent equipment deterioration and/or failure, and to detect impending failures. There are, logically, three broad categories into which preventive maintenance work can be classified:

- *Time-directed:* Tasks performed based upon a timetable established by the system manufacturer or user.
- *Condition-directed:* Maintenance functions undertaken because of feedback from the equipment itself (such as declining power output from a vacuum tube).
- *Failure-directed:* Maintenance performed first to return the system to operation, and second to prevent future failures through the addition of protection devices or component upgrades recommended by the manufacturer.

Regardless of whether such work is described as CPM or just plain common sense, the result is the same. Preventive maintenance is a requirement for reliability.

TRAINING OF MAINTENANCE PERSONNEL

The increasingly complex hardware used in industry today requires competent technical personnel to keep it running. The need for well-trained engineers has never been greater. Proper maintenance procedures are vital to top performance. A comprehensive training program can prevent equipment failures that affect productivity, worker morale, and income. Good maintenance is good business.

Maintenance personnel today must think in a "systems mode" to troubleshoot much of the hardware now in the field. New technologies and changing economic conditions have reshaped the way maintenance professionals view their jobs. As technology drives equipment design forward, maintenance difficulties will continue to increase. Such problems can be met only through improved test equipment and increased technician training.

The goal of every maintenance engineer is to ensure top-quality performance from each piece of hardware. These objectives do not just happen. They are the result of a carefully planned maintenance effort.

It is easy to get into a rut and conclude that the old ways, tried and true, are best. Change for the sake of change does not make sense, but the electronics industry has gone through a revolution within the past 10 years. Every facility should re-evaluate its inventory of tools, supplies, and procedures.

Technology has altered the way electronic products are designed and constructed. The service bench needs to keep up as well. That is the goal of this book.

Jerry C. Whitaker

1

RELIABILITY ENGINEERING

1.1 INTRODUCTION

Before the advent of semiconductors, reliability of electronic equipment was an elusive goal. Rackfuls of vacuum tubes did not lend themselves to long-term stability and dependable operation. In actual circuit complexity, the systems generally were simple, compared with today's computer-based hardware. However, the primary active devices that made the equipment of 30 years ago operate—tubes—were fragile components with a more-or-less limited lifetime.

Enter solid-state electronics and the wonders that it has provided. We have been blessed with increased reliability and performance, reduced space requirements and heat generation, and practical as well as affordable systems that were little more than dreams 30, or even 20, years ago. However, as technology marches on, engineers are encountering new problems in the areas of preventive maintenance and troubleshooting.

The electronics industry has grown up in a technical sense. And the sophistication of the equipment requires new approaches to *reliability* and *maintainability*. Reliability is the primary concern of any equipment user. The best computer system in the world is useless if it works only part of the time. The most powerful and sophisticated transmitter is of no value if it won't stay on the air. Maintainability ranks right behind reliability in concern to professional users. When equipment fails, the user must be able to return it to service within a reasonable length of time.

The science of reliability and maintainability is not just an esoteric concept developed by the aerospace industry to satisfy governmental dictates. It is a science that has fostered continued improvements in the components and systems that we enjoy today.

1.1.1 Objectives

The ultimate goal of any maintenance department is zero downtime. This is an imposing objective, but one that can be approximated by examining the vulnerable areas of plant operation and taking steps to prevent any sequence of events that could result in system failure. In cases where failure prevention is not practical, a reliability assessment should encompass the stocking of spare parts, circuit boards, or even entire systems. A large facility often can cost-justify the purchase of backup gear to be used as spares for the entire complex. Backup hardware is expensive, but so is downtime.

Failures can, and do, occur in electronic systems. The goal of product quality assurance at every step in the manufacturing and operating chain is to ensure that failures do not produce a systematic or repeatable pattern. The ideal is to eliminate failures altogether. Short of that, the secondary goal is to end up with a random distribution of failure modes. This indicates that the design of the system is fundamentally correct and that failures are caused by random events that cannot be predicted. In an imperfect world, it is often the best we can hope for. Reliability and maintainability must be *built into* products or systems at every step in the design, construction, and maintenance process. It cannot be treated as an afterthought.

1.1.2 Terminology

To implement the principles of reliability engineering, it is essential to understand the following basic terms:

- Availability: The probability that a system that is subject to repair will operate satisfactorily on demand.
- Average life: The mean value for a normal distribution of product or component lives. (Generally applied to mechanical failures resulting from "wearout.")
- Burn-in: Initially high failure rate encountered when a component is first placed on test. Burn-in failures usually are associated with manufacturing defects and the debugging phase of early service.
- Defect: Any deviation of a unit or product from specified requirements. A unit or product may contain more than one defect.
- Degradation failure: A failure that results from a gradual change, over time, in the performance characteristics of a system or part.
- Downtime: Time during which equipment is not capable of doing useful work because of malfunction. This does not include preventive maintenance time. In other words, downtime is measured from the occurrence of a malfunction to its correction.

- Failure: A detected cessation of ability to perform a specified function or functions within previously established limits. It is beyond adjustment by the operator by means of controls normally accessible during routine operation of the system. (This requires that measurable limits be established to define "satisfactory performance.")
- Failure mode and effects analysis (FMEA): An iterative documented process performed to identify basic faults at the component level and to determine their effects at higher levels of assembly.
- Failure rate: The rate at which failure occurs during an interval of time as a function of the total interval length.
- Fault-tree analysis (FTA): An iterative documented process of a systematic nature that is performed to identify basic faults, determine their causes and effects, and establish their probabilities of occurrence.
- Lot size: A specific quantity of similar material or a collection of similar units from a common source; in inspection work, the quantity offered for inspection and acceptance at any one time. It may be a collection of raw material, parts, subassemblies inspected during production, or a consignment of finished products to be sent out for service.
- Maintainability: The probability that a failure will be repaired within a specified time after the failure occurs.
- Mean Time Between Failure (MTBF): The measured operating time of a single piece of equipment divided by the total number of failures during the measured period of time. This measurement normally is made during the period between early life and wearout failures.
- Mean Time to Repair (MTTR): The measured repair time divided by the total number of failures of the equipment.
- Mode of failure: The physical description of the manner in which a failure occurs and the operating condition of the equipment or part at the time of the failure.
- Part failure rate: The rate at which a part fails to perform its intended function.
- Quality assurance (QA): All those activities, including surveillance, inspection, control, and documentation, undertaken to ensure that the product will meet its performance specifications.
- Reliability: The probability that an item will perform satisfactorily for a specified period of time under a stated set of use conditions.
- Reliability growth: Action taken to move a hardware item toward its reliability potential, during development, manufacturing, or operation.
- Reliability predictions: Compiled failure rates for parts, components, subassemblies, assemblies, and systems. These generic failure rates are used as basic data to predict a value for reliability.
- Sample: One or more units selected at random from a quantity of product to represent that product for inspection purposes.

- Sequential sampling: Sampling inspection in which, after each unit is inspected, the decision is made to accept, reject, or inspect another unit. (Note: Sequential sampling as defined here is sometimes called *unit sequential sampling* or *multiple sampling.*)
- System: A combination of parts, assemblies, and sets joined together to perform a specific operational function or functions.
- Test to failure: Testing conducted on one or more items until a predetermined number of failures have been observed. Failures are induced by increasing electrical, mechanical, and/or environmental stress levels, usually in contrast to life tests, in which failures occur after extended exposure to predetermined stress levels. A life test can be considered a test to failure using age as the stress.

1.2   QUALITY ASSURANCE

Electronic components and systems manufacturers design and implement quality assurance procedures for one fundamental reason: Nothing is perfect. The goal of a QA program is to ensure, for both the manufacturer and the customer, that all but some small, mutually acceptable percentage of devices or systems shipped will be as close to perfection as economics and the state of the art allow. There are tradeoffs in this process. It would be unrealistic, for example, to perform extensive testing to identify potential failures if the cost of that testing exceeded the cost savings that would be generated by not having to replace the devices later in the field. The focal points of any QA effort are *quality* and *reliability*. These terms are not synonymous. Although they are related, they do not provide the same measure of a product: Quality is the measure of a product's performance relative to some established criteria. Reliability is the measure of a product's life expectancy. Stated from a different perspective, quality answers the question of whether the product meets applicable specifications *now*; reliability answers the question of *how long* the product will continue to meet its specifications.

1.2.1   Inspection Process

Quality assurance for components normally is performed through sampling rather than through 100 percent inspection. The primary means used by QA departments for controlling product quality at the various processing steps include:

- Gates method: Every lot passing through a critical production stage is sampled mandatorily. Material cannot move on to the next operation until QA has inspected and accepted the lot.

- Monitor points method: Some attribute of the component is sampled periodically. QA personnel sample devices at a predetermined frequency to verify that machines and operators are producing material that meets preestablished criteria.
- Quality audit: A separate group within the QA department carries out an audit to ensure that all production steps throughout the manufacturer's facility are in accordance with current specifications.
- Statistical quality control: Through this technique, based on computer modeling, data accumulated at each gate and monitor point is incorporated to construct statistical profiles for each product, operation, and piece of equipment within the plant. Analysis of this data over time allows quality engineers to assess trends in product performance and failure rates.

Quality assurance for a finished subassembly or system may range from a simple go-no test to a thorough operational checkout that may take days, or even weeks, to complete. Most manufacturers test individual printed wiring boards (PWBs) before integrating them into a larger system. Test fixtures of various types provide manual or automated testing of the cards and alignment of circuits, if required.

1.2.2   Reliability Evaluation

Reliability prediction is the process of quantitatively assessing the reliability of a component or system during development, before large-scale fabrication and field operation. During product development, predictions serve as a guide by which design alternatives may be judged for reliability. To be effective, the prediction technique must relate engineering variables to reliability variables.

A prediction of reliability is obtained by determining the reliability of each critical item at the lowest system level, then proceeding through intermediate levels until an estimate of overall reliability is obtained. This prediction method depends on the availability of accurate evaluation models that reflect the reliability of lower-level components. Various formal prediction procedures are used, based on theoretical and statistical concepts.

Parts-Count Method. The parts-count approach to reliability prediction provides an estimate of reliability based on a count by part type, such as integrated circuits (ICs), transistors, resistors, capacitors, and other components. This method is useful during the early design stage of a product, when the amount of available detail is limited. It involves counting the number of components of each type, multiplying that number by a generic failure rate for each part type, and summing the products to obtain the failure rate of each functional circuit, subassembly assembly, and/or block depicted in the system block diagram. This method is useful in the design phase because it allows rapid estimates of reliability, permitting assessment of the feasibility of a given concept.

Stress-Analysis Method. The stress-analysis technique is similar to the parts-count method, but utilizes a detailed parts model, plus calculation of circuit stress values for each part, before determining the failure rate. Each part is evaluated in its electric-circuit and mechanical-assembly application based on an electrical and thermal stress analysis. After part failure rates have been established, a combined failure rate for each functional block is determined.

1.2.3   Failure Analysis

Failure mode and effects analysis can be performed with data taken from actual failure modes observed in the field, or from hypothetical failure modes derived from one of the following:

- Design analysis
- Reliability-prediction activities
- Experience with how specific parts fail

In the most complete form of FMEA, failure modes are identified at the component level. Failures are analytically induced into each component, and failure effects are evaluated and noted, including the severity and frequency (or probability) of occurrence. Using this approach, the probability of various system failures can be calculated, based on the probability of lower-level failure modes. Fault-tree analysis is a tool commonly used to analyze failure modes found during design, factory test, or field operations. The approach involves several steps, including the development of a detailed logic diagram, which depicts basic faults and events that can lead to system failure and/or safety hazards. This data is used to formulate corrective suggestions which, when implemented, will eliminate or minimize faults considered critical. An example FTA chart is shown in Figure 1.1. *Failure mode effects and criticality analysis* (FMECA) is an extension of FMEA. The purpose of FMECA is to provide a systematic, critical examination of potential failure modes of systems/facilities and their causes for the following purposes:

- Assess the safety of various systems or components
- Analyze the effects of each failure mode on system operation
- Identify corrective actions (such as design modifications)

The FMECA process is important for reliability. It is also useful in maintainability, in maintenance plan analysis, and for failure detection and subsystem design. In its general form, FMECA is an analysis procedure that:

- Documents all probable failure modes in a system within specified ground rules.
- Identifies single failure points.

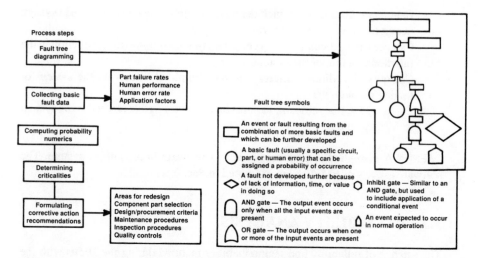

Process steps

Fault tree diagramming

Collecting basic fault data → Part failure rates / Human performance / Human error rate / Application factors

Computing probability numerics

Determining criticalities

Formulating corrective action recommendations → Areas for redesign / Component part selection / Design/procurement criteria / Maintenance procedures / Inspection procedures / Quality controls

Fault tree symbols

☐ An event or fault resulting from the combination of more basic faults and which can be further developed

○ A basic fault (usually a specific circuit, part, or human error) that can be assigned a probability of occurrence

◇ A fault not developed further because of lack of information, time, or value in doing so

AND gate — The output event occurs only when all the input events are present

OR gate — The output occurs when one or more of the input events are present

○ Inhibit gate — Similar to an AND gate, but used to include application of a conditional event

An event expected to occur in normal operation

Figure 1.1 Example fault-tree analysis diagram. (Data from: D. Fink and D. Christiansen, *Electronics Engineers' Handbook*, 3rd ed., McGraw-Hill, New York, 1989)

- Ranks each failure in accordance with a severity classification grid.
- Determines by failure mode analysis the effect of each failure on system operation.

The primary steps in the FMECA process are:

1. Identify all potential component and interface failure modes, and analyze their effects on the task or mission to be accomplished.
2. Evaluate each failure mode in terms of the worst potential consequences as a function of the severity of failure.
3. Establish a *criticality factor* to rank the failure modes for corrective action.
4. Identify corrective designs required to eliminate the failure or to control the risk.
5. Identify the effects of the corrective actions.
6. Document the analysis procedure and summarize problems that could not be corrected by changes in design.

Four categories of severity are identified under FMECA to provide a qualitative measure of the worst potential consequences resulting from a design error or a component failure:

- Category I: Catastrophic. Includes a loss of life and severe reduction in system output to an unacceptable level.
- Category II: Critical. Includes personal injury and significant reduction in the functional output of the system.
- Category III: Minor. Failures that have minimal effects on the system or facility output level.
- Category IV: Insignificant. Failures that have negligible effects on the system or facility output.

FMECA is applied to both hardware and software in a complex system. The software aspects of FMECA are examined in Sec. 1.5.1.

1.3   RELIABILITY ANALYSIS

The science of reliability and maintainability matured during the 1960s with the development of sophisticated computer systems and complex military and space-craft electronics. Components and systems never fail without a reason. That reason may be difficult to find, but determination of failure modes and weak areas in system design or installation is fundamental to increasing the reliability of any component or system, whether it is an integrated circuit, aircraft autopilot, or broadcast transmitter.

All equipment failures are logical; some are predictable. A system failure usually is related to poor-quality components or to abuse of the system or a part within, either because of underrating or environmental stress. Even the best-designed components can be manufactured poorly. A process may go awry, or a step involving operator intervention may result in an occasional device that is substandard or likely to fail under normal stress. Hence, the process of screening and/or *burn-in* to weed out problem parts is a universally accepted quality control tool for achieving high reliability.

1.3.1   Statistical Reliability

Figure 1.2 illustrates what is commonly known as the *bathtub curve*. It divides the expected lifetime of a class of parts into three segments: infant mortality, useful life, and wearout. A typical burn-in procedure consists of the following steps:

1. The parts are electrically biased and loaded (that is, they are connected in a circuit representing a typical application).
2. The parts are cycled on and off (power is applied, then removed) a predetermined number of times. The number of cycles may range from ten to several thousand during the burn-in period, depending on the component under test.

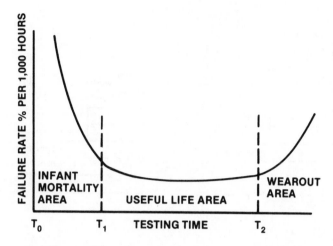

Figure 1.2 The statistical distribution of equipment or component failures vs. time for electronic systems and devices.

3. The components under load are exposed to a high temperature (typically 125 to 150°C) for a selected time (typically 72 to 168 hours). This represents an accelerated life test for the part.

An alternative approach involves temperature shock testing, in which the component product is subjected to temperature extremes, with rapid changes between the hot-soak and cold-soak conditions. After the stress period, the components are tested for adherence to specifications. Parts meeting the established specifications are accepted for shipment to customers. Parts that fail to meet specifications are discarded.

Figure 1.3 illustrates the benefits of temperature cycling to product reliability. The charts (from Ref. 1) show the distribution of component failures identified through steady-state high-temperature burn-in compared with those identified through temperature cycling. Note that cycling screened out a significant number of IC failures. The distribution of failures under temperature cycling usually resembles the distribution of field failures. Temperature cycling more closely simulates real-world conditions than steady-state burn-in. The goal of burn-in testing is to ensure that the component lot is advanced past the infant mortality stage (T-1 on the bathtub curve). This process is used not only for individual components, but for entire systems.

Such a systems approach to reliability is effective, but not foolproof. The burn-in period is a function of statistical analysis; there are no guarantees. The natural enemies of electronic parts are heat, vibration, and excessive voltage. Figure 1.4 documents failures vs. hours in the field for a piece of avionics equipment. The

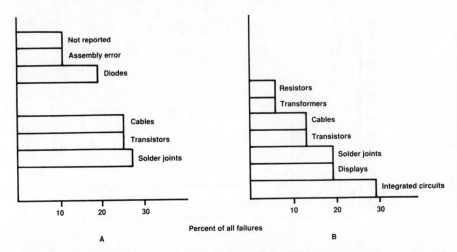

Figure 1.3   Distribution of component failures identified through burn-in testing. (a) Steady-state high-temperature burn-in. (b) Temperature cycling. (Data from: Richard Powell, "Temperature Cycling Vs. Steady-State Burn-In," *Circuits Manufacturing*, Benwill Publishing, September 1976)

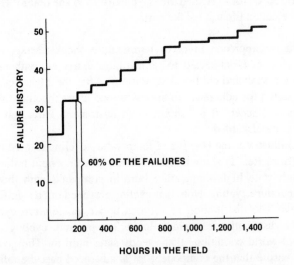

Figure 1.4   The failure history of a piece of avionics equipment vs. time. Note that 60 percent of the failures occurred within the first 200 hours of service. (Data from: J. Capitano and J. Feinstein, "Environmental Stress Screening Demonstrates Its Value in the Field," *Proceedings of the IEEE Reliability and Maintainability Symposium*, IEEE, New York, 1986)

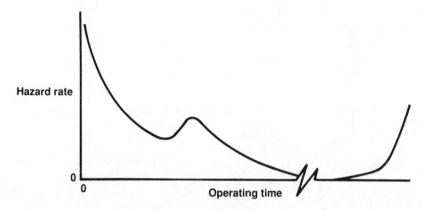

Figure 1.5   The roller-coaster hazard rate curve for electronic systems. (Data from: Kam L. Wong, "Demonstrating Reliability and Reliability Growth with Environmental Stress Screening Data," *Proceedings of the IEEE Reliability and Maintainability Symposium,* IEEE, New York, 1990)

conclusion is made that a burn-in period of 200 hours or more will eliminate 60 percent of the expected failures. However, the burn-in period for another system using different components may well be a different number of hours.

What does this mean to the end-user? Simply that infant mortality is a statistical fact of life in the operation of any piece of hardware. Most engineers can relate to the problems of "working the bugs out" of a new piece of equipment.

The goal of burn-in testing is to catch system problems and potential faults before the device or unit leaves the manufacturer. Therefore, the longer the burn-in period, the greater the likelihood of catching additional failures. Extended burn-in, however, means more time and money. Longer burn-in translates to longer delivery delays and additional costs for the equipment manufacturer, which are likely to be passed on to the user. The point at which a product is shipped is determined largely through experience with similar components or systems and the financial requirement to move products to users.

Roller-Coaster Hazard Rate. The bathtub curve has been used for decades to represent the failure rate of an electronic system. Recent data, however, has raised questions regarding the accuracy of the curve shape. A growing number of reliability scientists now say that the probability of failure, known in the trade as the *hazard rate*, is represented more accurately as a roller-coaster track, as illustrated in Figure 1.5. Hazard-rate calculations require analysis of the number of failures of the system under test, as well as the number of survivors. Experts say that previous estimating processes and averaging tended to smooth the roller-

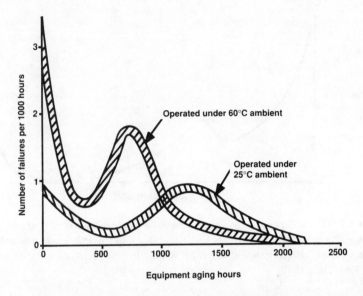

Figure 1.6   The effects of environmental conditions on the roller-coaster hazard rate curve. (Data from: Kam L. Wong, "Demonstrating Reliability and Reliability Growth with Environmental Stress Screening Data," *Proceedings of the IEEE Reliability and Maintainability Symposium,* IEEE, New York, 1990)

coaster curve so that the humps were less pronounced, leading to an incorrect conclusion insofar as the hazard rate was concerned. The testing environment also has a significant effect on the shape of the hazard curve, as illustrated in Figure 1.6. Note that with higher operating temperatures (greater environmental stress), the roller-coaster hump moves to an earlier age.

1.3.2   Environmental Stress Screening

The science of reliability analysis is rooted in the understanding that there is no such thing as a random failure; every failure has a cause. For reasonably designed and constructed electronic equipment, failures result from built-in flaws or *latent defects.* Because different flaws are sensitive to different stresses, a variety of environmental forces must be applied to a unit under test to identify any latent defects. This is the underlying concept behind *environmental stress screening* (ESS).

ESS, which has come into widespread use for aeronautics and military products, takes the burn-in process a step further by combining two of the major environmental factors that cause parts or units to fail: heat and vibration. Qualification

Table 1.1   Comparison of Conventional Reliability Testing and Environmental Stress Screening. Data from: Kam L. Wong, "Demonstrating Reliability and Reliability Growth with Environmental Stress Screening Data," *Proceedings of the IEEE Reliability and Maintainability Symposium,* IEEE, New York, 1990.

Parameter	Conventional testing	Environmental stress screening
Test condition	Simulates operational equipment profile	Accelerated stress condition
Test sample size	Small	100% of production
Total test time	Limited	Very large
Number of failures	Small	Large
Reliability growth	Potential for gathering useful data small	Potential for gathering useful data very good

testing for products at a factory practicing ESS involves a carefully planned series of checks for each unit off the assembly line. Units are subjected to random vibration and temperature cycling during production (for subassemblies and discrete components) and upon completion (for systems). The goal is to catch product defects at the earliest possible stage of production. ESS also can lead to product improvements in design and manufacture if feedback from the qualification stage to the design and manufacturing stages is implemented. Figure 1.7 illustrates the improvement in reliability that typically can be achieved through ESS over simple burn-in screening, and through ESS with feedback to earlier production stages. Significant reductions in equipment failures in the field can be gained. Table 1.1 compares the merits of conventional reliability testing and ESS. Designing an ESS procedure for a given product is no easy task. The environmental stresses imposed on the product must be great enough to cause fallout of marginal components during qualification testing. The stresses must not be so great, however, as to cause failures in good products. Any unit that is stressed beyond its design limits eventually will fail. The proper selection of stress parameters—generally, random vibration on a specially designed vibration generator, and temperature cycling in an environmental chamber—can, in minutes, lead to the discovery of product flaws that might take weeks or months to manifest themselves in the field. The result is greater product reliability for the user.

The ESS concept requires that every single product undergo qualification testing before it is implemented into a larger system for shipment to an end-user. The flaws

Table 1.2   System Failure Modes Typically Uncovered by Environmental Stress Screening. Data from: Wayne Tustin, "Recipe for Reliability: Shake and Bake," *IEEE Spectrum*, IEEE, New York, December 1986.

	FAILURE REVEALED THROUGH	
TYPE OF DEFECT	THERMAL SCREENING	VIBRATION SCREENING
COMPONENT DEFECTS	X	X
IMPROPERLY INSTALLED PART	X	
FAULTY SOLDER CONNECTION	X	X
FAULTY ETCHING	X	X
LOOSE CONTACT		X
BAD WIRE INSULATION	X	
LOOSE WIRE TERMINATION	X	X
IMPROPER CRIMP	X	
CONTAMINATION	X	
DEBRIS WITHIN ASSEMBLY		X
LOOSE HARDWARE		X
MECHANICAL FLAW		X

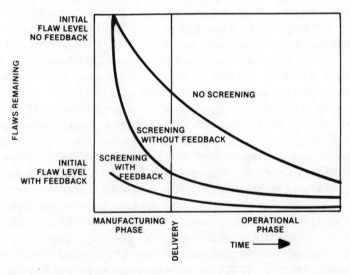

Figure 1.7   The effects of environmental stress screening on the reliability bathtub curve. (Data from: Wayne Tustin, "Recipe for Reliability: Shake and Bake," *IEEE Spectrum*, IEEE, New York, December 1986)

Table 1.3   Integrated Circuit Screening Processes. Data from: Gregg K. Hobbs, "Development of Stress Screens," *Proceedings of the IEEE Reliability and Maintainability Symposium*, IEEE, New York, 1987.

Screening test	Substrate mounting defects	Bulk silicon defects	Substrate defects	Substrate surface defects	Bonding and wire defects	Particle contamination, extraneous material	Seal defects	Package defects	External lead defects	Thermal mismatch	Electrical stability
Internal visual exam	●			●	●	●		●			
External visual exam								●	●	●	
Stabilization bake		●	●	●							●
Thermal cycling	●			●	●			●	●	●	
Thermal shock	●			●	●			●	●	●	
Centrifuge	●				●	●		●			
Shock	●				●	●		●			
Vibration	●				●	●		●			
X-ray	●				●	●	●	●			
Burn-in	●	●	●	●							●
Leakage tests							●				

(Failure mechanism)

uncovered by ESS vary from one unit to the next, but types of failures tend to respond to particular environmental stresses, as illustrated in Table 1.2. Table 1.3 lists the effects of environmental stresses on integrated circuit elements and parameters. This data clearly demonstrates that if the burn-in screens do not match the flaws sought, the flaws probably will not be found.

The concept of flaw-stimulus relationships also can be shown in Venn diagram form. Figure 1.8 shows a Venn diagram for a hypothetical, but specific, product. The required screen would be different for a different product. For clarity, not all stimuli are shown. Note that there are many latent defects that will not be uncovered by any one stimulus. For example, a solder splash that is just barely clinging to a circuit board probably would not be broken loose by high-temperature burn-in or voltage cycling. Vibration or thermal cycling, however, probably would break the particle loose. Remember also that the defect may be observable only during a stimulation, and not during a static bench test.

The levels of stress imposed on a product during ESS should be greater than the stress to which the product will be subjected during its operational lifetime, but still be below the maximum design parameters. This rule of thumb is pushed to the limits under an *enhanced screening* process. Enhanced screening places the component or system at well above the expected field environmental levels. This process has been found to be useful and cost-effective on many programs and

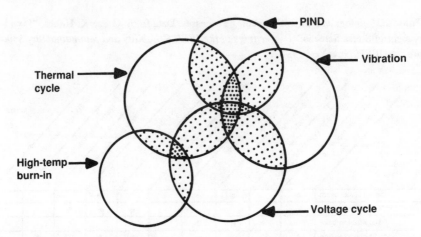

Figure 1.8   Venn diagram representation of the relationship between flaw precipitation and applied environmental stress. (Data from: Gregg K. Hobbs, "Development of Stress Screens," *Proceedings of the IEEE Reliability and Maintainability Symposium,* IEEE, New York, 1987)

products. Enhanced screening, however, requires the development of screens that are identified carefully during product design development, so that the product can survive the qualification tests. Enhanced-screening techniques often are required for cost-effective products on a cradle-to-grave basis. That is, early design changes for screenability save tremendous costs over the lifetime of the product.

The types of products that can be checked economically through ESS fall into two categories: high-dollar items and mass-produced items. Units that are physically large, such as RF generators or mainframe computers, are difficult to test in the finished state. Still, qualification tests incorporating more primitive methods, such as the use of cam-driven truck-bed shakers, are practical. Because most large systems generate a great deal of heat, subjecting the equipment to temperature extremes also may be accomplished. Sophisticated ESS for large systems, however, must rely on qualification testing at the subassembly stage.

The basic hardware complement for an ESS test station includes a thermal chamber shaker and controller/monitor. A typical test sequence includes 10 minutes of exposure to random vibration, followed by 10 cycles between temperature minimum and maximum. To save time, the two tests may be performed simultaneously.

1.3.3   Latent Defects

The cumulative failure rate observed during the early life of an electronic system is dominated by the latent defect content of the product, not its inherent failure rate.

Product design is the major determinant of inherent failure rate. A product design will show a higher-than-expected inherent rate if the system contains components that are marginally overstressed, have inadequate functional margin, or contain a subpopulation of components that experience a wearout life shorter than the useful life of the product. Industry has grown to expect the high instantaneous failure rate observed when a product is first placed into service. The burn-in process, whether ESS or conventional, is aimed at shielding customers from the detrimental effects of infant mortality. The key to reducing failure early in the life of a product is to reduce the number of latent defects.

A latent defect is some abnormal characteristic of the product or its parts that is likely to result in failure at some point, depending on:

• The degree of abnormality
• The magnitude of applied stress
• The duration of applied stress

For example, consider a solder joint on a double-sided printed wiring board. One characteristic of the joint is the degree to which the plated through-hole is filled with solder, characterized as "percent fill." All other characteristics being acceptable, a joint that is 100 percent filled offers the maximum mechanical strength, minimum resistance, and greatest current-carrying capacity. Conversely, a joint that is 0 percent filled has no mechanical strength, and only if the lead is touching the barrel or pad does it have any significant electrical properties. Between these two extreme cases are degrees of abnormality. For a fixed magnitude of applied stress: A grossly abnormal solder joint probably will fail in a short time. A moderately abnormal solder joint probably will fail, but after a longer period of time than a grossly abnormal joint. A mildly abnormal solder joint probably will fail, but after a much longer period of time than in either of the preceding conditions. Figure 1.9 illustrates this concept. A similar time-stress relationship holds for a fixed degree of abnormality and variable applied stress.

A latent defect eventually will become a *patent defect* through exposure to environmental, or other, stimuli. A patent defect is a flaw that has advanced to the point at which an abnormality actually exists. To continue with the solder example, a cold solder joint represents a flaw (latent defect). After vibration and/or thermal cycling, the joint will (it is assumed) crack. The joint now has become a detectable (patent) defect. Some latent defects can be stimulated into patent defects by thermal cycling, some by vibration, and some by voltage cycling. Not all flaws respond to all stimuli. The total number of physical and functional defects found per unit of product during the manufacturing process correlates strongly with the average latent defect content of shipping product. Product- and process-design changes aimed at reducing latent defects not only improve the reliability of shipping product, but also result in substantial manufacturing cost savings.

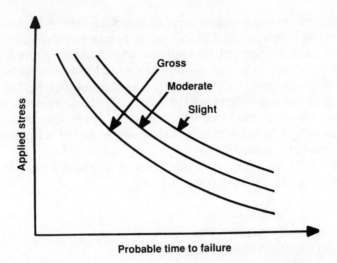

Figure 1.9   Estimation of the probable time to failure caused by an abnormal solder joint. (Data from: William B. Smith, "Integrated Product and Process Design to Achieve High Reliability in Both Early and Useful Life of the Product," *Proceedings of the IEEE Reliability and Maintainability Symposium*, IEEE, New York, 1987)

1.3.4   Field Experience

A wealth of research has been conducted by the aerospace industry into what types of parts are likely to fail in a given application. This data provides an important, although at times contradictory, perspective on what to look for when a system fails.

U.S. Navy Research. In Ref. 2, research conducted by the U.S. Navy cites connectors and cabling as the main culprits in system failures (see Figure 1.10). The study concludes that:

- Approximately 60 percent of all avionics failures are the result of high-level components, primarily connectors and cabling.
- Twenty-five percent of all failures are caused by maintenance and test procedures.
- Thirteen percent are the result of overstress and abuse of system components.
- Less than 2 percent are the result of IC failures.

The high percentage of connector- and cable-related problems will come as no surprise to most maintenance technicians. Wiring harnesses may be subjected to significant physical abuse under the best operating conditions.

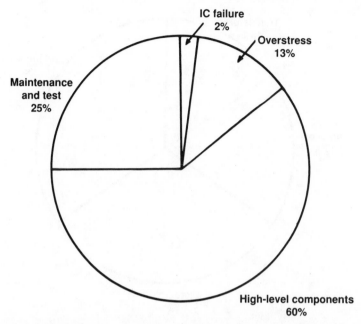

Figure 1.10   The distribution of failure modes in the field for an avionics system. (Data from: Richard J. Clark, "Electronic Packaging and Interconnect Technology Working Group Report (IDA/OSD R&M Study)," IDA Record Document D-39, August 1983)

Collins Avionics Research. In Ref. 3, research conducted by Collins Avionics into removal rates for components indicates that most of the failures were caused by semiconductors (see Figure 1.11). Specifically, the study shows:

- Microcircuit failures account for 32 percent of device removals.
- Capacitors represent 19 percent of device removals.
- Transistor failures account for 15 percent of device removals.
- All other parts account for 34 percent of device removals.

On the face of it, this data would seem to contradict the Navy research. Several important points, however, must be considered: Data based on "parts removed" may or may not provide a true picture of failure rates. In the course of servicing a piece of equipment, the maintenance technician may replace several parts before the system is returned to service. Simply because a component is replaced does not mean that it was defective. Data contained in various technical papers reflects the parts mix of the system being studied. For example, a system consisting of several chassis, each including a large number of PWBs, will be more vulnerable to

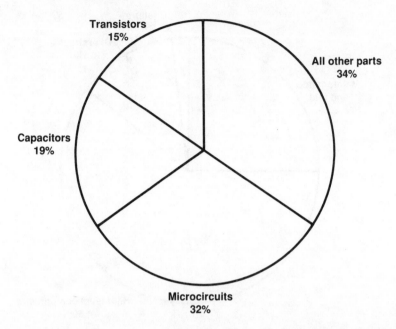

Figure 1.11 Component replacement distribution for avionics hardware. (Data from: M. F. Wilson and M. L. Woodruff, "Economic Benefits of Parts Quality Knowledge," *Proceedings of the IEEE Reliability and Maintainability Symposium,* IEEE, New York, 1985)

connector-related failures than a self-contained system with only a few input/output (I/O) connectors and a single PWB. The number of adjustable components may change the failure rates for a class of devices. For example, a piece of analog equipment with a large number of trimmer pots probably will experience a greater number of "resistor" failures than a digital system with no trimmer pots. The scope of the research also must be considered. When a manufacturer examines failure rates for a given piece of equipment, the research usually does not include cables or assemblies exterior to the manufacturer's box.

General Electric Research. In Ref. 4, research conducted by General Electric into field failures cites PWB-related problems as the major cause of system failures (see Figure 1.12). Conclusions from the research are:

- PWBs and solder joints account for 38 percent of failures.
- Integrated circuits account for 23 percent of failures.
- Resistors account for 11 percent of failures.
- Semiconductors (other than ICs) account for 10 percent of failures.
- All other parts account for 18 percent of failures.

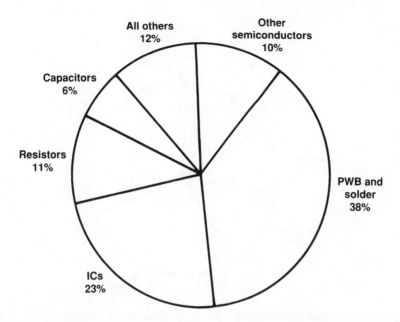

Figure 1.12   Failure rate analysis from a General Electric reliability study. (Data from: P. Griffin, "Analysis of the F/A-18 Hornet Flight Control Computer Field Mean Time Between Failure," *Proceedings of the IEEE Reliability and Maintainability Symposium,* IEEE, New York, 1985)

The high failure rate for PWBs was, according to the report, the result of a smear problem in the production process of the system being examined. A failure rate study of the system after the production process was corrected would, presumably, result in a different distribution of failures.

Hughes Aircraft Research. In Ref. 5, research conducted by Hughes Aircraft shows that field failures are caused primarily by semiconductors, including discrete diodes, discrete transistors, ICs, and hybrid microcircuits. Figure 1.13 shows the failure distribution. General conclusions of the study are:

- Integrated circuits account for 30 percent of failures.
- Resistors account for 28 percent of failures.
- Semiconductors (other than ICs) account for 20 percent of failures.
- Capacitors account for 13 percent of failures.
- All other components account for 9 percent of failures.

The Hughes report states that solder joints and PWBs are thought to be involved in a significant number of failures, although hard data was not available to verify the assumption.

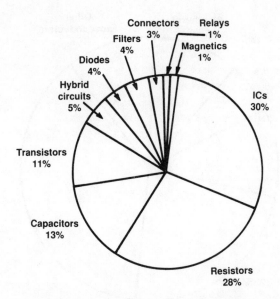

Figure 1.13 Avionics hardware parts replacement distribution. (Data from: K. L. Wong, I. Quart, J. Kallis, and A. H. Burkhard, "Culprits Causing Avionics Equipment Failures," *Proceedings of the IEEE Reliability and Maintainability Symposium,* IEEE, New York, 1987)

Parts-Count Design Tradeoffs. Designing for reliability and circuit efficiency must take place early in the development stage if parts-count reductions are to result in any significant decrease in system failure rates.

For example, in Ref. 6, the authors report that if very large-scale integrated (VLSI) chips were used in the F-16 airborne radar to implement the existing logic without reducing the number of circuit boards, the parts count would be reduced by as much as 85 percent, but reliability would increase by a factor of only 2 percent. However, if the design were optimized by the insertion of high-density VLSI chips into a logic implementation reconfigured to minimize the board count and, correspondingly, the connector count, reliability would be increased by a factor of 7 percent over the baseline.

Semicustom ICs readily can achieve significant reductions in component counts. Reducing the parts count is important because of fewer *explicit components* in the design—ICs, resistors, capacitors, and inductors—as well as the smaller number of *implicit components*—wires, solder joints, and connectors. Every resistor or capacitor adds two solder joints to a circuit; ICs may add as few as 8 or as many as 200. These joints are prone to manufacturing defects, corrosion, and failure from thermal mismatch.

Consider the following example: A semiconductor manufacturer recently developed a 2-device chip set that replaces more than 80 individual pieces of "glue" logic on the main circuit board of an IBM personal computer. The payoff is significant with respect to reliability. Assuming the reliability of each component stays the same, a design with a total of 115 components, each with a 99.9 percent probability of working properly for a year, will have an 89.1 percent chance of operation without failure for one year. A similar system with 35 components of the same reliability would have a chance of better than 96 percent.

1.3.5   Operating Environment

The operating environment of an electronic system, either because of external environmental conditions or unintentional component underrating, may be significantly more stressful than the system manufacturer or component supplier might expect. Unintentional component underrating represents a design fault, but unexpected environmental conditions are possible for many applications, particularly in remote locations.

Conditions of extreme low or high temperatures, high humidity, and vibration during transportation may significantly affect the long-term reliability of the system. For example, it is possible—and, more than likely, probable—that the vibration stress of the truck ride to a remote transmitting site will be the worst-case vibration exposure of that transmitter and all the components within during the lifetime of the system.

Manufacturers report that most of the significant vibration and shock problems for land-based products arise from the shipping and handling environment. Shipping conditions tend to be an order of magnitude more severe than the operating environment, with respect to vibration and shock. Early testing for these problems involved simulation of actual shipping and handling events, such as end-drops, truck trips, side impacts, and rolling over curbs and cobblestones. Although unsophisticated by today's standards, these tests were capable of improving product resistance to shipping-induced damage. End-users must realize that *the field* is the final test environment and burn-in chamber for the equipment.

Case Histories. The ESS approach to quality assurance produces tangible improvements in product quality. Consider the following examples (from Ref. 7):

- IBM 4381 Processor Power Supply. Computer power-supply units in the field were experiencing early-life failures. Defect analysis of supplies returned from customers was difficult because many were examined and pronounced "no defect found" (NDF). Vibration in the operating environment was suspected. Low-level (0.1 g rms, 10 to 100 Hz) random vibration tests were conducted, and the NDF supplies began to fail immediately. It was determined that the failures were the result of a pin joint contact that became intermittent when subjected to vibration. The root cause of the failure was incorrect handling in the manufacturing process. A random vibration screen was developed by the company to check all existing stock, and the manufacturing process was corrected.

- IBM 4234 Dot Band Printer. Random vibration was used as a step stress test tool from the early stages of development and was instrumental in design improvements to eliminate problems of loose fasteners and connectors. Random vibration profiles were applied as a production test tool, along with thermal cycling. Field reliability of the printer was excellent.

1.4   MAINTENANCE ACTIONS

The goal of preventive maintenance is to improve reliability. The goal of maintenance is to return a failed system to operation. Unfortunately, maintenance activities also carry the probability of introducing failures into electronic systems. In fact, studies have shown that maintenance-induced failures are a major factor contributing to the maintenance burden of man-hours, materials, and support costs. One analysis (Ref. 8) showed that on certain types of electronic equipment, maintenance-induced failures constitute an average of 20 to 30 percent of all failures, and could run as high as 48 percent. Another analysis (Ref. 9) attributed approximately 30 percent of the failed parts in an aircraft weapons assembly to electrical overstress or other abuse. Electrical overstress was defined in the report as the inadvertent application of an overvoltage during maintenance or through an improper troubleshooting procedure.

Reliability has an inverse relationship with frequency of maintenance actions. If the MTBF increases, maintenance actions, including preventive tasks, will be reduced. If reliability can be improved without significantly increasing the complexity or difficulty of performing maintenance, the overall maintenance support effort will be enhanced.

1.4.1   Maintenance Procedures

Simplification of maintenance procedures has been a goal of both commercial and military hardware for decades. Many published papers attest to the improvements in maintainability achieved in modern systems. Unfortunately, any shortcomings in design must be compensated for by maintenance technicians. Modern equipment packs more parts of greater complexity into smaller packages. Most problems related to maintainability are discovered after the product is in the field and, because product improvement programs are expensive, technicians usually must live with the inherent faults of the hardware. Furthermore, most product enhancement work is designed to improve performance, not maintainability.

Case in Point. Examples abound of systems designed with little regard to maintainability. The best-documented cases can be found in the realm of military hardware. One example (Ref. 10) is particularly interesting:

Each engine on the B-1B bomber has an accessory drive gearbox (ADG). A hinged access door with four thumb latches is provided on each compartment panel for servicing. The access door permits personnel to check the ADG oil without removing the compartment panel. However, the oil-level sight gauge requires line-of-sight reading. The way it is installed, the gauge cannot be read through the access door, even with an inspection mirror. The entire compartment panel, secured with 63 fasteners, must be removed just to see whether oil servicing is needed.

It is clear that maintainability must play a larger role in the design formula of both military and commercial equipment. With the additional complexity of hardware, designers must consider, up front, the need to fix a piece of equipment after it has left the factory.

Impact of New Technologies. Complex ICs will dominate the next generation of PWBs, and common repair techniques will be inadequate to troubleshoot and repair such boards. New design features, however, promise to render fault isolation, testing, and component replacement less time-consuming. Features that will have a favorable impact on the repair of advanced-technology PWBs include:

- Design for testability
- Built-in test
- New manufacturing processes, such as pin-grid array components, surface-mounted devices, and socketed parts

The use of advanced devices offers tremendous advantages for circuit card assembly (CCA) design, but it significantly complicates the ability to test, isolate a fault, and repair the PWB. The test problems associated with the use of advanced technologies have begun to result in components and circuits that are designed for testing by an external analyzer and/or self-test features. This work is welcomed by technicians. Advanced technologies now in the marketplace or in the immediate future include:

- Surface-mount devices (SMD) with input/output counts of 64 and more, on center spacings of less than 0.05 in (50 mil)
- Pin- and pad-grid arrays (PGAs) with I/O counts of 100 or more
- New-technology mounting schemes, including: chip on board (COB), tape automated bonding (TAB), flip-chip, multichip packaging, and wafer-scale integration.

Unless a PWB is designed to allow for testing and component removal, successful repair of cards containing these advanced devices is highly unlikely.

1.4.2   Reliability-Centered Maintenance

Reliability-centered maintenance (RCM) is a tool used by reliability, safety, and/or maintenance engineers for developing an optimum maintenance program. RCM defines the requirements and tasks to be performed in achieving, restoring, or maintaining the operational capability of an individual piece of equipment or an entire system. The RCM concept was developed during the 1970s by the commercial airline industry. Although originally structured to meet a set of specific needs, RCM may be adapted to a wide range of maintenance-important operations, including avionics hardware, transportation control systems, and other equipment whose failure could result in significant property damage or injury.

Implementing RCM requires analysis of failure mode, rate, and criticality data to determine the most effective maintenance procedures for a given situation. The RCM process initially is applied during the design/development phase and is reapplied, as necessary, during deployment to sustain an optimum maintenance program based on field experience. RCM identifies two basic types of components:

1. *Nonsafety-critical components.* Scheduled maintenance tasks are performed only when the tasks will reduce the life-cycle cost of ownership.
2. *Safety-critical components.* Scheduled maintenance tasks are performed only when the tasks will prevent a decrease in reliability and/or deterioration of safety to unacceptable levels.

RCM is based on the premise that reliability is a design characteristic to be realized and preserved during the operational life of the system. This concept is further extended to assert that efficient and cost-effective lifetime maintenance and logistical support programs can be developed using decision logic that focuses on the consequences of failure. Such a program provides these benefits:

- The probability of failure is reduced.
- Incipient failures are detected and corrected either before they occur or before they develop into major defects.
- Greater cost-effectiveness of the maintenance program is achieved.

Maintenance Classifications. Under RCM, maintenance tasks are classified as follows:

- Hard-time maintenance: Applies to failure modes that require scheduled maintenance at predetermined, fixed intervals of age or usage.
- On-condition maintenance: Applies to failure modes that require scheduled inspections or tests designed to measure deterioration of an item so that corrective maintenance can be performed.

- Condition monitoring: Applies to failure modes that require unscheduled tests or inspection of components where failure can be tolerated during operation of the system, or where impending failure can be detected through routine monitoring during normal operations.

Each step in the RCM process is critical. When executed properly, RCM results in a preventive maintenance program that ensures the system will remain in good working condition. RCM analysis involves these key steps:

1. Determine the maintenance-important items in the system.
2. Acquire failure data on the system.
3. Develop fault-tree analysis data.
4. Apply decision logic to critical failure modes.
5. Compile and record maintenance classifications.
6. Implement the RCM program.
7. Apply sustaining engineering based on field experience.

The RCM process has a life-cycle perspective. The goal is to reduce the scheduled maintenance burden and support cost while maintaining the necessary readiness state. As with other reliability tasks, the RCM process is reapplied as available data moves from a predicted state to operational experience. (For more information on RCM, see Ref. 11.)

1.5   SOFTWARE RELIABILITY

Until recently, software was considered a mystical subject, with a language and acronyms all its own. Hardware-design engineers did not understand the world of software, and the customer understood even less. The result was an unstructured development process that invited problems. The importance of software in computer-based systems has led to the application of reliability engineering to software, as well as to hardware.

The cost of software in most sophisticated products now exceeds the cost of the hardware on which it is designed to operate. Figure 1.14 illustrates how the relative costs of hardware and software have changed since 1955 for military electronics. Software costs are generated by:

- Identifying the requirements of the program.
- Writing the program.
- Debugging the program before delivery.
- Providing follow-up maintenance and program updates.
- Supplying customer technical support.

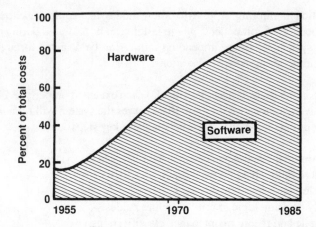

Figure 1.14   Hardware costs vs. software costs for military electronics equipment. (Data from: W. Yates and D. Shaller, "Reliability Engineering as Applied to Software," *Proceedings of the IEEE Reliability and Maintainability Symposium*, IEEE, New York, 1990)

Any discussion of software reliability must include a definition of what constitutes a failure. Software is reliable if it does what the customer intended it to do whenever it is called upon to perform. The software fails if it doesn't fulfill this promise. Statistics developed by the U.S. government (Ref. 12) show that, for military systems, the main problem in writing software is identifying exactly what the software is supposed to do. As shown in Figure 1.15, "requirements analysis" constitutes more than half of the software errors that occur during the development cycle of a major project. It follows that the up-front design stage offers the greatest opportunity to affect software reliability. The areas of software development that are most problematic can be divided into the nine categories shown in Figure 1.16. Here again, translation of customer requirements is the source of most problems.

Even though the major software errors are made before integration testing, statistics from the Air Force (Ref. 13) show that at least 76 percent of those errors are not discovered and corrected until integration and beyond. Figure 1.17 illustrates the impact of this delayed discovery on project costs. The relative cost of error detection and resolution during system integration and beyond can be 10 to 15 times greater than the cost of early resolution.

1.5.1   Hardware Vs. Software

The distinctions between hardware and software are numerous and obvious. The differences arise from the fact that a software program is, essentially, an abstraction; it is immune to the mechanical defects, electrical noise, and physical degra-

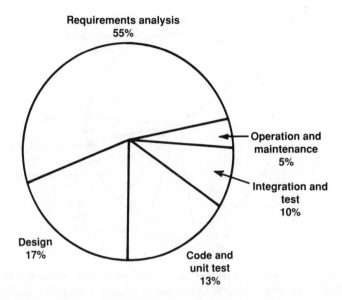

Figure 1.15 An estimation of the percentage of software errors that occur in each development cycle for military electronics systems. (Data from: W. Yates and D. Shaller, "Reliability Engineering as Applied to Software," *Proceedings of the IEEE Reliability and Maintainability Symposium*, IEEE, New York, 1990)

dation that affect hardware. As both technologies advance, however, the distinctions between hardware and software are lessening. For example, system designers freely trade hardware complexity for software simplicity, and vice versa.

Before the advent of low-cost semiconductor memory chips, complex electronic systems were highly dependent on hardware for the implementation of complex logic. Such systems were subject to the same types of malfunctions that currently arise in software, except that the overall systems were, by necessity, significantly less complex than current technology allows. With the evolution of modern semiconductors, memory costs have dropped to the point that logical complexity can be almost eliminated from hardware, or it can be incorporated into standardized ICs from which logical problems eventually are eliminated after testing.

Hardware reliability is concerned with latent defects, physical design, and the operating environment. Software reliability is pure logic; it has no problems analogous to wearout. Software also is used to mask hardware failures. *Fault-tolerant* systems are charged with maintaining system integrity in the face of a hardware malfunction. Software failures for fault-tolerant systems can be categorized as follows:

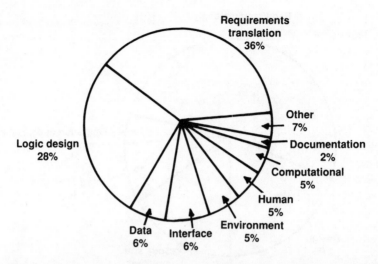

Figure 1.16 An estimation of the root causes of software errors discovered in military electronics systems. (Data from: W. Yates and D. Shaller, "Reliability Engineering as Applied to Software," *Proceedings of the IEEE Reliability and Maintainability Symposium*, IEEE, New York, 1990)

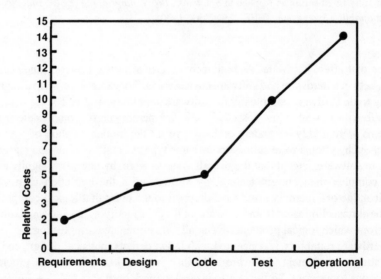

Figure 1.17 The costs of correcting software errors at key points in the development cycle. (Data from: W. Yates and D. Shaller, "Reliability Engineering as Applied to Software," *Proceedings of the IEEE Reliability and Maintainability Symposium*, IEEE, New York, 1990)

- Failure to perform functions required under normal operating conditions.
- Failure to perform functions required under abnormal conditions. Such conditions could include overload, invalid input, or another abnormal operating state.
- Failure to recover from a hardware failure.
- Failure to diagnose a hardware failure.

Fault-tolerant control systems with duplicate critical parts are expected to function despite any single hardware problem. Enumeration of all possible faults is a staggering task. Systems that operate in real time, and are subject to quasi-random inputs from tens or hundreds of ports, can be driven into internal states beyond classification. Software design and reliability testing for a fault-tolerant system is no easy matter.

Because software failures vary greatly in levels of importance, it is necessary to grade them in steps, ranging from fatal flaws to cosmetic problems. The density of software bugs also must be related to the rate at which the defects are likely to create actual failures under field conditions. A quality standard for software balances (1) the engineer's desire to remove all possible defects, with (2) the need to deliver products within a reasonable time frame and at a reasonable price point. Under conditions in which the test environment is a good model of the actual field environment, and the software code is exercised completely, the "find-and-fix" process can be used as a reliable guide to software quality. Under these conditions, the rate at which bugs are found is roughly proportional to the density of bugs in the code.

A frequently used method of checking software run on critical systems is to purposely overload the system. Reason tells us that problems can be expected to occur under the stresses that are present when queues reach their limits, registers go into overflow, and transactions are delayed. Such abnormal behavior is likely to exercise "contingency software" not used under normal conditions.

Identifying Software Failures. Distinguishing between hardware and software problems is often a difficult task. Problems manifest themselves at the system level and, sometimes, they may be remedied by changes in either the software or the hardware. Because software is more flexible, the easiest and most economical way to fix a problem is usually through a software change. In light of this, it is possible that the perceived contribution of software to system unreliability may be artificially high.

Causal Analysis. In the general application, causal analysis attempts to determine the cause of some observed phenomenon. In software development, causal analysis seeks to explain how faults have been injected into the program, and how they have led to system or software failures. Once a fault is identified, programmers must determine at what point in the process the fault was injected, and how it can be prevented in future program development.

Failure Mode, Effect, and Criticality Analysis (FMECA). FMECA is a tool for examining the programming code from a contingent perspective to determine how the software will react to unusual input or environmental conditions. Examination of the software's operation under errant conditions can lead to greatly enhanced robustness of the code. This process also leads the designer to mitigate the possible effects of failure that the system reasonably can be expected to encounter. FMECA must be applied at the beginning of a software program design effort. Considerations include the possibility of operator errors or hardware errors, such as the failure of a communications system if the device is linked to other hardware.

Fault injection is the physical realization of FMECA. Faults are introduced purposely into a system to see how it reacts and handles the condition. This stresses the fault-tolerant and fault-recovery aspects of the design. The faults can be put into either the software or the hardware. This process forces designers to look at how their systems will behave under harsh environmental conditions.

1.5.2 Improving Software Reliability

Historically, the way to improve software reliability has been to extensively test the program before delivery to the customer. Errors found during the testing process are then corrected. There is, however, a better way. The key lesson learned from the science of hardware reliability also is applicable to software: Reliability must be *designed into* the product. It is not cost-effective to test reliability into a program. Every time an error is corrected during testing, the probability exists for introducing new errors into the program. In addition, it is much more difficult to find, correct, and document errors in later stages of the development cycle.

Mathematicians, utilizing probability and statistics tools, have attempted to predict software reliability. However, most models as of this writing have fallen short of the goal of developing a software equivalent of common hardware models. Still, the basic concepts employed in hardware reliability assurance can be applied successfully to software. Reliability engineering concepts and tools may, if applied within the software development process, directly improve the reliability of the end-product.

It must be understood that in an engineering environment, there is a distinct difference between "reliability" and "quality." In the software development community, quality and reliability are synonyms. In engineering, they are two distinct concepts. "Quality" is a measure of how well a process delivers products to the customer with consistency. "Reliability" is a measure of the inherent design of the product. It follows, therefore, that reliability must be addressed up front in the design process, just as in hardware. Testing and documentation are subsets of quality. To make software more reliable, error prevention through design—not error discovery through testing—must be the thrust.

Software Maintainability. The maintainability of software is determined by those characteristics of the code and computer support resources that affect the ability of software programmers and analysts to change instructions. Such changes are made to:

- Correct program errors.
- Expand system capabilities.
- Remove specific features from the program.
- Modify software to make it compatible with hardware changes.

Error-free software does not exist in real-world programs. It is estimated that the average application program contains two to three defects per 1000 lines of source code (Ref. 14) when it is released to customers.

Modern software programs usually are constructed of modules that perform specific functions. Modular construction is achieved by dividing total program functions into a series of elements, each having a single input (entry point) and a single output (exit point). The development of a large system may permit the use of a given module a number of times, simplifying software design and debugging. Modules performing a specific function, such as I/O management or display control, also may be used on other product lines.

The top-level modules of a software program provide functional control over the lower-level modules on which they call. Therefore, the top-level software is exercised more frequently than lower-level modules. Various estimates of run time distributions show that just 4 to 10 percent of the total software package operates for 50 to 90 percent of the time in a real-time program.

1.5.3   Software System Safety

Software is not intrinsically hazardous. However, when software is integrated with hardware that performs critical functions, the potential exists for catastrophic effects. For a hazard to exist, there must be the possibility of some undesirable or unexpected action from the system. A failure of the hardware or hardware/software interaction may cause the hazard to occur. Normally, only the software that exercises direct control over the hardware, or the software that monitors the hardware, is considered *safety critical*. The following conditions constitute software hazards:

- An inadvertent or unwanted event occurs.
- An out-of-sequence event occurs.
- An event fails to occur.
- The magnitude or direction of the event is lacking or nonexistent.

Reliability and safety engineering play a major role in the system design phase of critical software. Fault-tree analysis is used to ensure that failure modes are identified and are designed out of the hardware/software. Fault-tree analysis also may be used to assess potential system hazards. Even under the most thorough testing program, it is impossible to foresee and test every conceivable software condition in a major system. The goal of software safety, therefore, is to identify all potential hazards and ensure that they have been eliminated or reduced to an acceptable level of risk.

1.6   FAULT-TOLERANT SYSTEMS

Increased reliance on centralized computing systems has led to growing concerns regarding reliability. Single-point faults that can bring an entire system down must be kept to an absolute minimum, especially when the computer performs critical functions. The failure of a computer that operates a chemical plant, for example, could have disastrous consequences. Computer manufacturers, therefore, have developed fault-tolerant systems for critical applications. A fault-tolerant system must deal with the following forces:

- Environmental conditions
- Operational errors
- Maintenance functions
- Software flaws
- Hardware flaws

If a computer system has perfect hardware and software, but is difficult to operate, difficult to maintain, or has only minimal protection against power failures, frequent outages will occur. Reliability, therefore, must address all areas of operation.

The most obvious threat to a computer system is the environment. This parameter typically is assumed to include:

- Utility company ac power
- Heat and humidity
- Common-carrier data communications lines
- The physical facility in which the computer is located
- External threats such as weather, fire, and earthquake

A fault-tolerant system is one that has the ability to tolerate or mask faults. Typically, systems are designed and rated as *n-fault* tolerant for some number *n*. A single fault-tolerant system (n = 1) is the most common. Such configurations are designed to mask most single faults within a specified repair window. If two faults occur within the repair window, the fault might not be tolerated by the system.

1.6.1   Field Experience

Research conducted by a manufacturer of fault-tolerant computer systems (Ref. 15) illustrates the improvement in reliability achieved by computer manufacturers. For the purposes of the study, failure reports compiled by the firm were examined to determine common threads and to identify failure trends. A typical system analyzed in the project consisted of 4 processors, 12 disk drives, a few hundred terminals, and related communications hardware. Terminals could include any of the following:

* Robots in an automated manufacturing plant
* Bar code readers
* Form-processing terminals
* Gas pumps or other point-of-sale devices

The study covered a 5-year period, from 1985 to 1990. Data collected demonstrated that the number of outages per system decreased dramatically over the measurement period. The *mean-time-between-reported-outage* (MTBO) increased from 8 operating years in 1985 to 20 years in 1990. Figure 1.18 plots the faults by

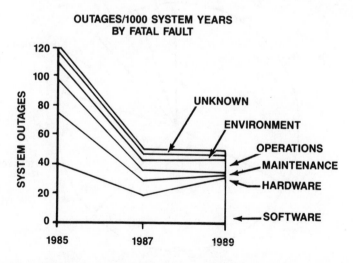

Figure 1.18   Failure rates for fault-tolerant computer systems. (Data from: Jim Gray, "A Census of Tandem System Availability Between 1985 and 1990," *IEEE Transactions on Reliability,* vol. 39, no. 4, IEEE, New York, October 1990)

category vs. time. It can be observed that significant improvements were made in overall system reliability. The categories listed are defined as follows:

- *Environment.*. Includes air conditioning; the threat of fire, lightning and other storms; utility company ac power; earthquake; sabotage; flood; and communications line failure.
- *Operations.* Includes hardware/software configuration and installation, system relocation, operational procedures, system upgrading, and overflow of system limits (*files full* conditions).
- *Maintenance.* Includes preventive and troubleshooting work performed on communications controllers and lines, disk drives, printers, facilities (power, cooling, and lights), processors, memory, and power supplies.
- *Hardware.* Includes communications controllers and lines, disk drives, tape drives, processors, memory, power supplies, terminals, printers, workstations, and cabling.
- *Software.* Includes customer-specific applications, communications, database, languages and tools, microcode, and the operating system.

Figures 1.19 and 1.20 plot the percentages of outages as a function of faults. It can be observed that most of the reliability increase came from improvements in

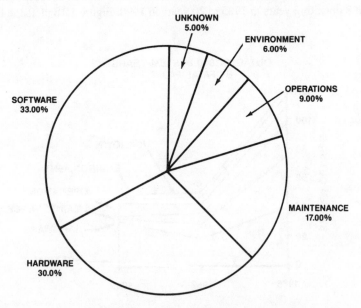

Figure 1.19 Percentage of outages in 1985 for fault-tolerant computer systems. (Data from: Jim Gray, "A Census of Tandem System Availability Between 1985 and 1990," *IEEE Transactions on Reliability,* vol. 39, no. 4, IEEE, New York, October 1990)

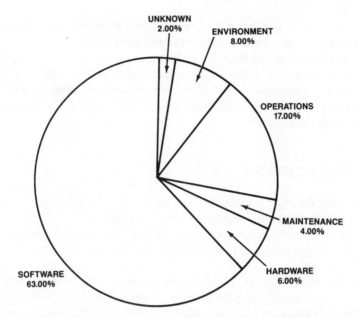

Figure 1.20 Percentage of outages in 1990 for fault-tolerant computer systems. (Data from: Jim Gray, "A Census of Tandem System Availability Between 1985 and 1990," *IEEE Transactions on Reliability*, vol. 39, no. 4, IEEE, New York, October 1990)

hardware and maintenance, which together decreased from 50 percent of the outages in 1985 to less than 10 percent in 1990. In contrast, the portion of outages caused by software increased from 33 percent to more than 60 percent of the outages reported during the same time period. Further analysis of the collected data is necessary, however, to gain a complete picture of the failure trends. The dramatic percentage increase in software-related outages was principally the result of the reduced number of failures overall. When software was analyzed by itself, outages remained relatively constant, while other failure categories improved.

The reduction in maintenance-related outages was due primarily to improved technology and improved design. High-density disk drives provide a good example of the interplay of both forces. In 1985 disk drives typically required service once a year. This process involved powering down the unit, replacing an air filter, adjusting the power systems, and (if needed) adjusting head alignment. This work provided many opportunities for maintenance-related mistakes. In addition, the disk cabinets and connectors were not designed for maintenance tasks. Special tools often were needed to perform the work. By 1990, disk drives required no scheduled maintenance, and if work was needed, no special tools were necessary. Cabling and connectors were reduced by as much as 95 percent through the application of fiber-optic lines. Disk drive MTFB increased from about 8000 hours to more than 100,000 hours in the period between 1985 and 1990.

It is not surprising that the software fault rate per system saw no improvement over the study period. Although better software is written today than in 1985, programs are more complex, and many are interrelated. As other components of a computer system become increasingly reliable, software necessarily becomes the dominant cause of outages.

1.7   REFERENCES

[1] Richard Powell, "Temperature Cycling Vs. Steady-State Burn-In," *Circuits Manufacturing*, Benwill Publishing, September 1976.

[2] Richard J. Clark, "Electronic Packaging and Interconnect Technology Working Group Report (IDA/OSD R&M Study)," IDA Record Document D-39, August 1983.

[3] M. F. Wilson and M. L. Woodruff, "Economic Benefits of Parts Quality Knowledge," *Proceedings of the IEEE Reliability and Maintainability Symposium*, IEEE, New York, 1985.

[4] P. Griffin, "Analysis of the F/A-18 Hornet Flight Control Computer Field Mean Time Between Failure," *Proceedings of the IEEE Reliability and Maintainability Symposium*, IEEE, New York, 1985.

[5] K. L. Wong, I. Quart, L. Kallis, and A. H. Burkhard, "Culprits Causing Avionics Equipment Failures," *Proceedings of the IEEE Reliability and Maintainability Symposium*, IEEE, New York, 1987.

[6] R. Horn and F. Hall, "Maintenance Centered Reliability," *Proceedings of the IEEE Reliability and Maintainability Symposium*, IEEE, New York, 1983.

[7] R. Frey, G. Ratchford, and B. Wendling, "Vibration and Shock Testing for Computers," *Proceedings of the IEEE Reliability and Maintainability Symposium*, IEEE, New York, 1990.

[8] Eqbert Maynard (OUSDRE Working Group Chairman), VHSIC Technology Working Group Report (IDA/OSD R&M Study), Document D-42, Institute of Defense Analysis, November 1983.

[9] D. Robinson and S. Sauve, "Analysis of Failed Parts on Naval Avionic Systems," Report no. D180-22840-1, Boeing Company, October 1977.

[10] Charles M. Worm, "The Real World: A Maintainer's View," *Proceedings of the IEEE Reliability and Maintainability Symposium*, IEEE, New York, 1987.

[11] D. C. Brauer and G. D. Brauer, "Reliability-Centered Maintenance," *Proceedings of the IEEE Reliability and Maintainability Symposium*, IEEE, New York, 1987.

[12] Philip S. Babel, "Software Development Integrity Program," briefing paper for the Aeronautical Systems Division, Air Force Systems Command. From Yates,

W., and D. Shaller: "Reliability Engineering as Applied to Software," *Proceedings of the IEEE Reliability and Maintainability Symposium*, IEEE, New York, 1990.

[13]Major Sue E. Hermason, USAF: Letter dated Dec. 2, 1988. From W. Yates and D. Shaller, "Reliability Engineering as Applied to Software," *Proceedings of the IEEE Reliability and Maintainability Symposium*, IEEE, New York, 1990.

[14]F. Hall, R. A. Paul, and W. E. Snow, "R&M Engineering for Off-the-Shelf Critical Software," *Proceedings of the IEEE Reliability and Maintainability Symposium*, IEEE, New York, 1988.

[15]Gray, Jim, "A Census of Tandem System Availability Between 1985 and 1990," *IEEE Transactions on Reliability*, vol. 39, no. 4, IEEE, New York, October 1990.

1.8 BIBLIOGRAPHY

Babel, Philip S.: "Software Development Integrity Program," briefing paper for the Aeronautical Systems Division, Air Force Systems Command. From Yates, W., and D. Shaller: "Reliability Engineering as Applied to Software," *Proceedings of the IEEE Reliability and Maintainability Symposium*, IEEE, New York, 1990.

Brauer, D. C., and G. D. Brauer: "Reliability-Centered Maintenance," *Proceedings of the IEEE Reliability and Maintainability Symposium*, IEEE, New York, 1987.

Cameron, D., and R. Walker: "Run-in Strategy for Electronic Assemblies," *Proceedings of the IEEE Reliability and Maintainability Symposium*, IEEE, New York, 1986.

Capitano, J., and J. Feinstein: "Environmental Stress Screening Demonstrates Its Value in the Field," *Proceedings of the IEEE Reliability and Maintainability Symposium*, IEEE, New York, 1986.

Clark, Richard J.: "Electronic Packaging and Interconnect Technology Working Group Report (IDA/OSD R&M Study)," IDA Record Document D-39, August 1983.

DesPlas, Edward: "Reliability in the Manufacturing Cycle," *Proceedings of the IEEE Reliability and Maintainability Symposium*, IEEE, New York, 1986.

Devaney, John: "Piece Parts ESS in Lieu of Destructive Physical Analysis," *Proceedings of the IEEE Reliability and Maintainability Symposium*, IEEE, New York, 1986.

Doyle, Edgar Jr.: "How Parts Fail," *IEEE Spectrum*, IEEE, New York, October 1981.

Ferrara, K. C., S. J. Keene, and C. Lane: "Software Reliability from a System Perspective," *Proceedings of the IEEE Reliability and Maintainability Symposium*, IEEE, New York, 1988.

Fink, D., and D. Christiansen: *Electronics Engineers' Handbook*, 3rd ed., McGraw-Hill, New York, 1989.

Fortna, H., R. Zavada, and T. Warren: "An Integrated Analytic Approach for Reliability Improvement," *Proceedings of the IEEE Reliability and Maintainability Symposium*, IEEE, New York, 1990.

Frey, R., G. Ratchford, and B. Wendling: "Vibration and Shock Testing for Computers," *Proceedings of the IEEE Reliability and Maintainability Symposium*, IEEE, New York, 1990.

Griffin, P.: "Analysis of the F/A-18 Hornet Flight Control Computer Field Mean Time Between Failure," *Proceedings of the IEEE Reliability and Maintainability Symposium*, IEEE, New York, 1985.

Hall, F., R. A. Paul, and W. E. Snow: "R&M Engineering for Off-the-Shelf Critical Software," *Proceedings of the IEEE Reliability and Maintainability Symposium*, IEEE, New York, 1988.

Hansen, M. D., and R. L. Watts: "Software System Safety and Reliability," *Proceedings of the IEEE Reliability and Maintainability Symposium*, IEEE, New York, 1988.

Hermason, Sue E., Major, USAF: Letter dated Dec. 2, 1988. From Yates, W., and D. Shaller: "Reliability Engineering as Applied to Software," *Proceedings of the IEEE Reliability and Maintainability Symposium*, IEEE, New York, 1990.

Hobbs, Gregg K.: "Development of Stress Screens," *Proceedings of the IEEE Reliability and Maintainability Symposium*, IEEE, New York, 1987.

Horn, R., and F. Hall: "Maintenance Centered Reliability," *Proceedings of the IEEE Reliability and Maintainability Symposium*, IEEE, New York, 1983.

Irland, Edwin A.: "Assuring Quality and Reliability of Complex Electronic Systems: Hardware and Software," *Proceedings of the IEEE, vol. 76, no. 1, IEEE, New York, January 1988.*

Kenett, R., and M. Pollak: "A Semi-Parametric Approach to Testing for Reliability Growth, with Application to Software Systems," *IEEE Transactions on Reliability*, IEEE, New York, August 1986.

Maynard, Eqbert (OUSDRE Working Group Chairman): VHSIC Technology Working Group Report (IDA/OSD R&M Study), Document D-42, Institute of Defense Analysis, November 1983.

Neubauer, R. E., and W. C. Laird: "Impact of New Technology on Repair," *Proceedings of the IEEE Reliability and Maintainability Symposium*, IEEE, New York, 1987.

Powell, Richard: "Temperature Cycling Vs. Steady-State Burn-In," *Circuits Manufacturing*, Benwill Publishing, September 1976.

Robinson, D., and S. Sauve: "Analysis of Failed Parts on Naval Avionic Systems," Report no. D180-22840-1, Boeing Company, October 1977.

Smith, A., R. Vasudevan, R. Matteson, and J. Gaertner: "Enhancing Plant Preventative Maintenance via RCM," *Proceedings of the IEEE Reliability and Maintainability Symposium,* IEEE, New York, 1986.

Spradlin, B.C.: "Reliability Growth Measurement Applied to ESS," *Proceedings of the IEEE Reliability and Maintainability Symposium,* IEEE, New York, 1986.

Tustin, Wayne: "Recipe for Reliability: Shake and Bake," *IEEE Spectrum,* IEEE, New York, December 1986.

Wei, Benjamin C., "A Unified Approach to Failure Mode, Effects and Criticality Analysis," *Proceedings of the Reliability and Maintainability Symposium,* IEEE, New York, 1991.

Wilson, M. F., and M. L. Woodruff: "Economic Benefits of Parts Quality Knowledge," *Proceedings of the IEEE Reliability and Maintainability Symposium,* IEEE, New York, 1985.

Wong, K. L., I. Quart, L. Kallis, and A. H. Burkhard: "Culprits Causing Avionics Equipment Failures," *Proceedings of the IEEE Reliability and Maintainability Symposium,* IEEE, New York, 1987.

Wong, Kam L.: "Demonstrating Reliability and Reliability Growth with Environmental Stress Screening Data," *Proceedings of the IEEE Reliability and Maintainability Symposium,* IEEE, New York, 1990.

Worm, Charles M.: "The Real World: A Maintainer's View," *Proceedings of the IEEE Reliability and Maintainability Symposium,* IEEE, New York, 1987.

Yates, W., and D. Shaller: "Reliability Engineering as Applied to Software," *Proceedings of the IEEE Reliability and Maintainability Symposium,* IEEE, New York, 1990.

Technical staff, Military/Aerospace Products Division: *The Reliability Handbook,* National Semiconductor, Santa Clara, Calif., 1987.

2

SEMICONDUCTOR FAILURE MODES

2.1 INTRODUCTION

Engineers have been working to improve the performance and reliability of semiconductor devices for decades. The first formal meeting of the IEEE-sponsored Reliability Physics Symposium, in 1962, marked the beginning of a long process that has taken the semiconductor industry from single-device packages to packages containing 100,000, or even 1,000,000, devices on a single chip. Within the foreseeable future, device density is expected to increase by a factor of 10 to 100. A single chip will constitute an entire system. Such advancements place exacting requirements on reliability engineering.

The reliability of semiconductors today is determined by the materials, processes, and quality control of chip manufacturers. Performance in the field is determined by hardware design and the operating environment.

2.1.1 Terminology

Following are definitions of key terms that will be used throughout this examination of the mechanics of semiconductor failure:

- Acceptance quality level (AQL): Sampling based upon total lot size, as opposed to lot tolerance percent defective (LTPD) sampling, which is independent of lot size.
- Active device: A mechanism that converts input signal energy into output signal energy through interaction with energy from an auxiliary source.
- Bimetallic contamination: Corrosion that results from the interaction of gold with aluminum containing more than 2 percent silicon. At temperatures greater than 167°C, the silicon will act as a catalyst to create an aluminum-

gold alloy. The resulting metallic migration causes gaps in the gold-aluminum interface. Bimetallic contamination (also referred to as "purple plague") can decrease bond-wire adhesion and lower current-carrying capacity at the interface.

- Coefficient of thermal expansion: The measurement of the rate at which a given material will expand or contract with a change in temperature. Where two materials with different rates of thermal expansion are joined, expansion or contraction will strain the bond interface.
- Contact step: The drop from the surface of the passivation of the semiconductor die to the surface of the contact.
- Contact window: The opening in the passivation through which the metallization makes contact with the circuit elements.
- Crazing: The propagation of small cracks in the surface of the glassivation.
- Current-carrying edge: That portion of the metal over a contact window that is closest to the incoming metallization.
- Die: The active silicon-based element that is the heart of a semiconductor product. (The plural form of die is *dice*.)
- Diffused area: A portion of the die in which impurities diffused into the surface of the silicon have changed the electrical characteristics of the die.
- Dual in-line package (DIP): A rectangular integrated circuit package with leads protruding from two opposite sides. The DIP is, by far, the most popular package in use today. Pinouts range from 8 to 40 or more.
- Eutectic: An alloy of two or more metals providing the lowest possible melting point, usually lower than the melting points of any of the constituent metals.
- FIT: Failures in time at 10^9 hours, a standard measure of reliability in semiconductor devices.
- Glassivation: The protective coating that is placed over the entire surface of the completed die.
- Hybrid microcircuit: A collection of components from one or more technologies combined into a single functional device.
- Junction: The boundary between a P region and an N region in a silicon substrate.
- Land: A portion of the package lead that extends into the package cavity.
- MOSFET: Metal-oxide semiconductor field-effect transistor.
- Pad: An expanded metal area on the die that is not covered with glassivation.
- Parasitic elements: Interactions among elements of a semiconductor to produce additional, usually undesirable, effects. The nature of semiconductor design makes it impossible to eliminate all parasitics.
- Passivation: The surface coat of silicon dioxide that is deposited over the surface of the die at various diffusion steps.

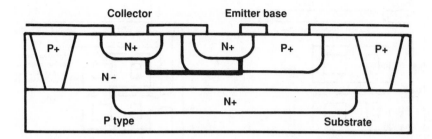

Figure 2.1 The construction of a semiconductor device. (Data from: Technical staff, Military/Aerospace Products Division, *The Reliability Handbook*, National Semiconductor, Santa Clara, Calif., 1987)

- Post: The portion of the lead inside the package cavity. Although only header packages have true posts, the term often is used to describe all internal lead surfaces.
- ULSI: Ultralarge-scale integration, the next step beyond VLSI (very large-scale integration).
- Undercutting: The inward sloping of the metal or silicon on a die, such that the surface is wider than at the base.
- VHSIC: Very high-speed integrated circuit, a relatively new class of devices designed for high-frequency operation.

2.2 SEMICONDUCTOR DEVICE ASSEMBLY

The intrinsic reliability of a semiconductor begins with the die itself. As illustrated in Figure 2.1, semiconductors are manufactured by building layer upon layer to produce the desired product. Semiconductor production is an exacting science. A reliable product cannot be assembled from an unreliable die. Reliability is governed by:

- The purity of the raw materials.
- The cleanliness of the fabrication facility (fab).
- The accuracy of process control steps.
- The design of the device itself.

Impurities, whether introduced through raw materials or the fab environment, are the enemies of efficient semiconductor production. Even a nonconductive particle can cause a minute defect during the masking process, resulting in failure that may occur immediately or during a subsequent operation. Figure 2.2 illustrates some of

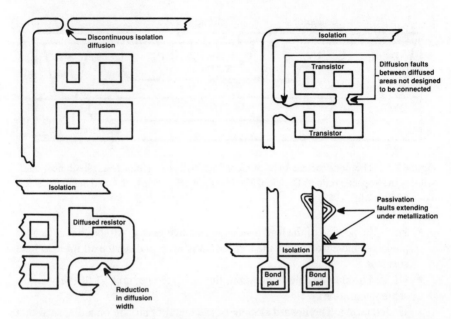

Figure 2.2   Types of manufacturing defects that may result from contamination of a semiconductor die. (Data from: Technical staff, Military/Aerospace Products Division, *The Reliability Handbook*, National Semiconductor, Santa Clara, Calif., 1987)

the defects that contamination may cause in a device. Any defects in the metallization may lead to eventual device failures. Methods for evaluating the metallization include:

- *Scanning electron microscope (SEM) inspection.* A sample of the die, still in wafer form, is checked for flaws. The SEM allows magnification of 15,000 times and greater, enabling engineers to see defects that would not be detected with an optical microscope.
- *Internal visual inspection.* An SEM can be used only to evaluate the metallization quality within a wafer run on a sample basis. A visual inspection with a microscope of 100X to 200X permits each die to be checked for scratches in the metallization surface. Scratches that decrease the total metallization cross section can result in higher-than-acceptable current density. The metallization also is examined for evidence of corrosion during this quality control step.

Glassivation typically is used to designate any inert surface layer added to a completed wafer. Glassivation protects the surface of the die from chemical contamination, handling damage, and the possibility of short circuits caused by loose particles within the package. Because of its importance, glassivation must be evaluated carefully for thickness, coverage, and integrity. Glassivation usually is checked through visual inspection. Excessive glassivation coverage, which may partially obscure bonding surfaces, can be as much of a problem as inadequate coverage.

After a die has passed the preceding quality checks, its electrical performance must be confirmed. Electrical testing begins at the wafer stage and continues throughout the assembly process. Each die is checked electrically before scribe and beak. Dice that meet the required performance characteristics are assembled into packages. The others are discarded. After final assembly, samples are subjected to burn-in qualification tests (described in Chapter 1).

2.2.1   Packaging the Die

A semiconductor package serves three important functions: It protects the die from mechanical and environmental damage. It dissipates heat generated during operation of the device. It provides connection points for electrically accessing the die. The mechanical assembly process is critical to the ultimate reliability of the device. The die is bonded to the package of lead frame using eutectic, an epoxy compound, or some other suitable medium. The strength (or adhesion) of the die to the package is an important concern, because a poorly attached die could lift later as a result of vibration. After attachment, the die is connected electrically to the package leads by bonding aluminum or gold wires to each bonding pad on the die, then to the respective package lead posts. Attachment of the lid completes the mechanical assembly. The device may be sealed hermetically (for military-grade products) using glass, solder, or welding techniques, or it may be encapsulated (for commercial-grade devices) in plastic or epoxy.

Loose particles within the package cavity may present a serious reliability hazard, because they could short-circuit exposed metallization. After attachment of the bonding wires, the die attach medium and the package itself are examined for particles that could break loose. Particles of foreign material on the surface of the die also are checked to determine whether they are embedded in the die surface or are attached in a manner that would allow them to break loose.

For hermetically sealed packages, the quality of the seal is an important consideration. Damage to the seal may allow chemicals, moisture, and other contaminants from the external environment to enter the package cavity and cause corrosive damage or electrical short-circuiting. Figure 2.3 shows a cutaway view of a DIP IC package.

Hybrid Devices. A hybrid microcircuit typically uses a number of components from more than one technology to perform a function that could not be achieved in monolithic form as efficiently and/or cost-effectively. Hybrids can be subdivided into the following general categories:

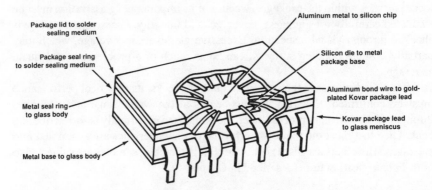

Package lid to solder sealing medium

Package seal ring to solder sealing medium

Metal seal ring to glass body

Metal base to glass body

Aluminum metal to silicon chip

Silicon die to metal package base

Aluminum bond wire to gold-plated Kovar package lead

Kovar package lead to glass meniscus

Figure 2.3   Cutaway view of a DIP integrated circuit package showing the internal-to-external interface. (Data from: Technical staff, Military/Aerospace Products Division, *The Reliability Handbook*, National Semiconductor, Santa Clara, Calif., 1987)

- Multichip device: A product that contains several chips (active or passive), and may contain a substrate. A multichip device does not include thick film elements on the substrate. Figure 2.4 illustrates a multichip hybrid.
- Simple hybrid: A product that contains one or more dice, with the dice mounted on a substrate (normally ceramic or alumina) that also has deposited

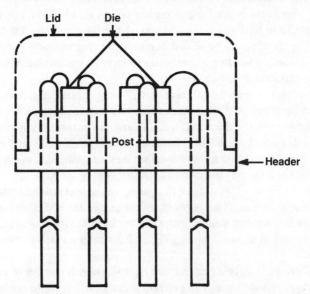

Lid   Die

Post

Header

Figure 2.4   Basic construction of a multichip hybrid device. *(Data from: Technical staff, Military/Aerospace Products Division, The Reliability Handbook,* National Semiconductor, Santa Clara, Calif., 1987)

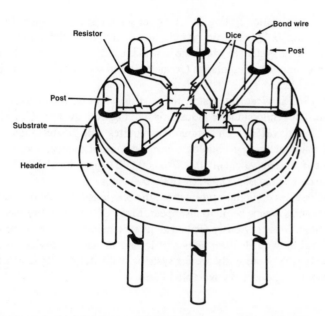

Figure 2.5 Basic construction of simple hybrid device. (Data from: Technical staff, Military/Aerospace Products Division, *The Reliability Handbook*, National Semiconductor, Santa Clara, Calif., 1987)

metallization traces and other thin or thick film components, such as resistors and capacitors. Figure 2.5 illustrates a simple hybrid device.

• Complex hybrid: A product that contains multiple dice mounted on a substrate with deposited metallization traces and other thin or thick film components. A complex hybrid typically has a seal periphery that exceeds 2 in.

Completed hybrids usually are subjected to 100 percent screening. Because of the number of dice used and the tight electrical tolerances imposed on many components, hybrid yields tend to decrease as complexity increases. Even if electrical yields of 100 percent could be guaranteed for each component used in a hybrid, the hybrids still must contend with burn-in dropouts. For example, a 10-chip hybrid in which each chip yielded 95 percent reliability through burn-in would have a cumulative yield of only 60 percent. The use of an intermediate package, referred to as a *leadless chip carrier* (LCC), for each active device in the hybrid permits the individual elements to pass burn-in qualification before being placed into the hybrid. The LCC functions like a conventional package, but is small enough that it can be installed on a hybrid substrate, which is later placed in a larger package.

When complex dice such as microprocessors are to be used as part of a hybrid, the LCC provides the hybrid manufacturer the capability to extensively test the dice before further assembly. This prevents the need for expensive rework if the dice fails during qualification testing. With proper screening before assembly, the yields for even complex hybrids can be quite good. Given a reliable hybrid, the overall reliability of the system in which the hybrid is used should improve because of the reduction in implicit components (connectors and solder joints) that circuit concentration provides.

VLSI and VHSIC Devices. Integrated circuit density has been increasing at an exponential rate. The semiconductor industry is moving from the age of VLSI into the age of ULSI and VHSIC. This transition requires still greater attention to semiconductor manufacturing technology. For the sake of comparison, a VLSI device may contain the equivalent of 10,000 to 100,000 logic gates. A ULSI device may contain the equivalent of 100,000 to 1,000,000 gates.

As advanced devices come into production, the previous clear-cut distinctions between devices and systems begin to disappear. Many of the reliability measurements and considerations previously applied at the system level will be appropriate at the device level as well. Failure rates and failure modes for devices will be categorized much as they presently are for systems, with attention to distinctions such as catastrophic failure vs. recoverable failure.

2.3 SEMICONDUCTOR RELIABILITY

Active components are the heart of any electronic product, and most—with the exception of high-power transmitter stages—employ semiconductors. Although highly reliable, discrete semiconductors and integrated circuits are vulnerable to damage from a variety of environmental and electrical forces. The circuit density of an IC can have a direct bearing on its survivability under adverse conditions in the field. As chip density increases, trace widths generally decrease. Smaller elements usually are susceptible to damage at lower stress levels than those of older design.

2.3.1 Reliability Analysis

Failure rate is the measure of semiconductor reliability. Techniques are available for predicting device failure rates and for controlling those rates in future designs. Environmental stress screening (described in Chapter 1) is used extensively during product development and manufacture to identify latent defects. The two major classifications of infant mortality are:

- *Dead on arrival (DOA)*. Devices that fail when initially tested, after shipment, or incorporation in the next assembly level.
- *Device operating failure (DOF)*. To devices that fail after some period of operating time.

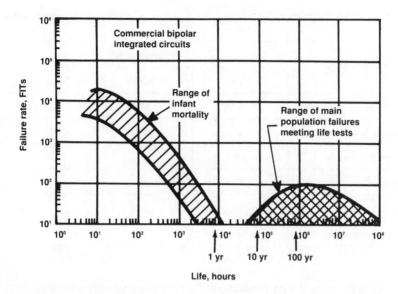

Figure 2.6   Operating failure rate for commercial bipolar ICs as a function of time. (Data from: D. Fink and D. Christiansen (eds.), Electronics Engineers' Handbook, 3rd ed., McGraw-Hill, New York, 1989)

Experience has shown that although DOA levels may range from 0.3 to 1.0 percent of shipped product, DOF levels observed in normal factory testing and equipment burn-in typically are smaller, ranging from 0.01 to 0.5 percent. The length of the infant mortality period cannot be defined precisely, but infant mortality clearly has the greatest impact in the first year of device use. Figure 2.6 plots failure rates vs. time for commercial bipolar ICs. Infant mortality failures bring added costs for both original equipment manufacturers (OEMs) and users in the field. Higher costs result from failures at incoming test, circuit-package test, system test and burn-in, and field installation, as well as in-service failures. Equipment-repair costs increase as the device moves from one stage to the next higher in the production chain.

Infant mortality defects have a variety of causes related to device design, manufacturing, handling, and application. Failure mechanisms differ in character and degree from one product type to another, and from one lot to another. Failures may result from one or more of the following:

- Manufacturing defects, including oxide pinholes, photoresist or etching defects, scratches, weak bonds, conductive debris, or partially cracked chips or ceramics.

Table 2.1   Infant Mortality Failures for a Sample of TTL, CMOS, and Memory Devices.

Failure mode	Product type		
	TTL	**CMOS**	**Memory**
Overstress	4	60	17
Oxide defects	2	1	51
Surface defects	18	0	24
Bonds, beams	37	5	7
Metallization	30	34	0
Other failure modes	9	0	1

- Operator errors during production. Wafer contamination (as described previously).
- Assembly operations, including physical stress on the die and incomplete sealing or encapsulation.

Small variations in the manufacturing process may cause device lots to exhibit infant mortality characteristics that are significantly different from the norm. For example, Table 2.1 lists device failures resulting from various mechanisms for a lot of TTL (transistor-transistor logic), CMOS (complementary metal-oxide semiconductor), and memory devices. The table demonstrates the variability of infant mortality failures. There is no fundamental reason for TTL circuits to have a greater percentage of bond failures than CMOS devices. Samples taken from a different lot at a different time might show a significantly different failure mix.

Semiconductor Burn-In. Burn-in is an effective means of screening out defects in semiconductor devices. There are two primary burn-in classifications:

- Static burn-in. A dc bias is applied to the device under test at an elevated temperature. The voltage is used to reverse-bias as many junctions as possible within the device. Static burn-in is particularly effective in identifying defects resulting from corrosion or contamination.
- Dynamic burn-in. The device under test is exercised to simulate actual system operation. Dynamic burn-in provides more complete access to the internal device elements.

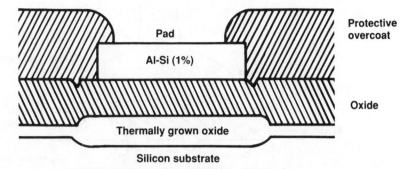

Figure 2.7 Cutaway structure of the bonding pad of a semiconductor device. (Data from: T. B. Ching and W. H. Schroen, "Bond Pad Structure Reliability," *Proceedings of the IEEE Reliability Physics Symposium,* IEEE, New York, 1988)

2.3.2 Bond Pad Structure

Wire bonding is the process by which connections are made between the die and the lead frame using bond wires. During this process, some amount of stress is imparted onto the bond pad. Excessive bonding stress may lead to cracks on the bond pads or in the silicon substrate. Figure 2.7 shows a cutaway view of a bonding pad.

Stress during the bonding process is the result of applied force and ultrasonic vibration. This delicate procedure requires a controlled process to enhance reliability. A crack in the bond is a serious failure mechanism. Microscopic cracks on the underlying layers of the bond pad, sometimes extending to the silicon substrate, give rise to functional failures. This class of defect usually is not detected visually unless the crack is continuous, causing the bond to be detached.

Plastic-encapsulated chips are susceptible to mechanical-stress-induced failures as a result of cracks in the plastic, the die, or the thin film, and/or wire problems. These failure modes result from *thermal coefficient of expansion* (TCE) mismatch among the materials of the package. Stress-related problems become more severe as the size of the die increases. Failures may occur during any of the thermal excursions to which the device is subjected during manufacture and use.

Cratering. The failure mechanism known as cratering is caused by a combination of (1) water contamination of the bonding surface and (2) heat conducted to the bonding surface during soldering of the component. Figure 2.8 shows the mechanisms involved. The problems begin with a small damage point at the bonding surface. If contaminated water subsequently penetrates to the die, it can collect between the bonding ball and the metallization. The application of heat during soldering can cause the water to vaporize, placing significant pressure on the die. Cratering of the silicon substrate may result.

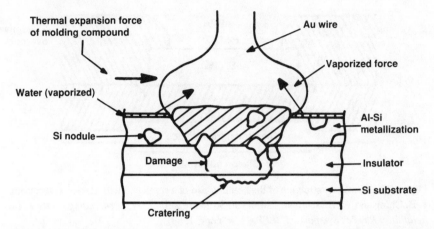

Figure 2.8   The physical mechanisms of cratering. (Data from: H. Koyama, H. Shiozaki, I. Okumura, S. Mizugashira, H. Higuchi, and T. Ajiki, "A New Bond Failure: Wire Crater in Surface Mount Device," Proceedings *of the IEEE Reliability Physics Symposium,* IEEE, New York, 1988)

2.3.3   Reliability of VLSI Devices

The circuit density of VLSI devices today has pushed semiconductor manufacturing technology close to its fundamental reliability limits. VLSI devices are difficult to manufacture, and they are sensitive to subtle defects. Still, VLSI reliability has improved steadily over the past two decades. Insofar as VLSI technology is concerned, reliability is a moving target. In the early 1970s, VLSI devices with failure rates of 1000 to 2000 FIT were considered acceptable. As system chip counts grew, customers began demanding lower failure rates. Meanwhile, chip manufacturers had gained a better understanding of reliability physics, and lower failure rates became common. Figure 2.9 plots the actual and projected long-term failure rates of one manufacturer (Intel) through the end of this century. Other device manufacturers have their own targets, but it is clear that customer demands will drive all producers to deliver devices that are more reliable.

Integrated circuits intended for microcomputer applications have been a driving force in the semiconductor industry. Reliability has become an important selling point for OEMs. Figure 2.10 plots the dramatic increase in device counts that has occurred during the past two decades. The 80286 microprocessor chip, for example, contains the equivalent of more than 1.2 million transistors.

From a reliability standpoint, achieving such high-density products does not come free. Each new generation of product demands more from the intrinsic reliability margin present in a given design. As device geometries scale down, small particles that would have presented no problems for earlier chips now can cause interlayer short circuits, metal line opens, and other defects.

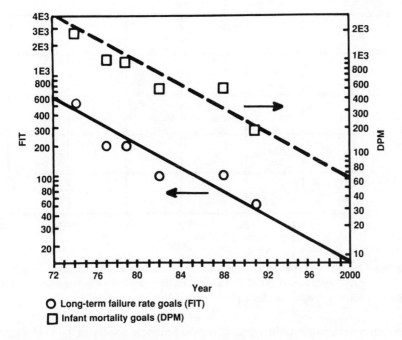

Figure 2.9 Long-term failure rate goals for VLSI devices. (Data from: D. L. Crook, "Evolution of VLSI Reliability Engineering," *Proceedings of the IEEE Reliability Physics Symposium*, IEEE, New York, 1990)

2.3.4 Case Histories

Although much effort has gone into producing semiconductors with no latent defects, problems still occur in the field. The following examples illustrate the types of failures that technicians may encounter:

- A microprocessor used in the doppler radar of a strategic bomber was found to be failing in the field at a rate greater than 50 percent. At least half of the times this radar was returned for repair, the microprocessor was found faulty and was replaced. The microprocessor was mounted on a circuit card that runs hot, and it probably saw at least 3000 temperature cycles before failure. Of the 19 chips analyzed, six had cracks, three of them internal to the device. The cracks, near the output pins, caused the microprocessors to fail, bringing down the radar system. The solution: Replace the microprocessor chip with a *hardened* device, one that is designed to operate over a wider temperature range. (See Ref. 1.)

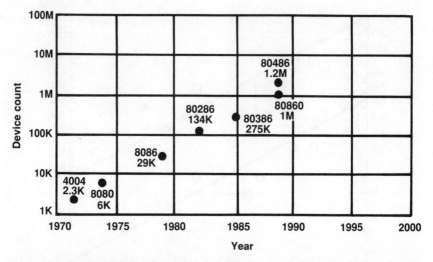

Figure 2.10   Microcomputer transistor count per chip as a function of time. (Data from: D. L. Crook, "Evolution of VLSI Reliability Engineering," *Proceedings of the IEEE Reliability Physics Symposium*, IEEE, New York, 1990)

- An operational amplifier that drives the heads-up display in the F-15 fighter aircraft was identified for failure analysis because of its high removal rate. The operational amplifier is a 3-resistor, 3-transistor, plastic-potted 8-pin cylindrical component. The failure mode was described as intermittent. Electrical testing, before and after the plastic potting was etched away by a maintenance technician, demonstrated that the failure was the result of a cracked solder joint. The root cause of the joint failure was identified as thermal cycling. At 50,000 ft, the ambient temperature easily can be - 50°C, and the F-15 probably makes that climb in less than 5 minutes. An aircraft sitting idle on the runway may experience 50°C or greater. During any one mission, the operational amplifier might experience ten 100°C thermal cycles. The solution: Replace the operational amplifier with a hardened device. (See Ref. 1.)

- Power-supply failures were experienced in the field by a manufacturer. An investigation of failed units revealed the cause to be high-voltage breakdown in a hybrid semiconductor chip. The microphotographs in Figure 2.11 (*a and b*) show the damage points. Failure analysis of several units demonstrated that the regulator pass transistor was overstressed because of excessive input/output voltage differential. The manufacturer determined that during testing of the system, the regulator load would be removed, causing the unregulated supply voltage to rise, exceeding the input/output capability of the pass transistor. Most of the failures occurred only after the supply was

Figure 2.11   Two views of a hybrid voltage regulator that failed because of a damaged pass transistor. (*a–left*) The overall circuit geometry. (*b–right*) A closeup of the damaged pass transistor area.

placed into service by the end-user. The solution: Change the testing procedures. (See Ref. 2.)

- Excessive transistor failures were experienced in the predriver stage of an RF power amplifier. A series of tests were conducted on the failed devices. Analysis showed the root cause of the problem to be overvoltage stress caused by parasitic oscillations in the stage. The solution: Place additional shielding around the transmitter assembly. (See Ref. 3.)

2.4   DEVICE RUGGEDNESS

The most well-constructed device will fail if exposed to stress exceeding its design limits. The safe operating area (SOA) of a power transistor is the single most important parameter in the design of a solid-state amplifier or controller. Fortunately, advances in diffusion technology, masking, and device geometry have enhanced the power-handling capabilities of semiconductor devices.

Two regions of operation must be avoided for a bipolar transistor:

- Dissipation region. Where the voltage-current product remains unchanged over any combination of voltage (V) and current (I). Gradually, as the collector-to-emitter voltage increases, the electric field through the base region causes hot spots to form. The carriers actually can punch a hole in the junction by melting silicon. The result is a dead (short-circuited) transistor.

- Second breakdown ($I_{s/b}$) *region.* Where power transistor dissipation varies in a nonlinear, inverse relationship with the applied collector-to-emitter voltage when the transistor is forward-biased.

To get SOA data into some type of useful format, a family of curves at various operating temperatures must be developed and plotted. This exercise gives a clear picture of what the data sheet indicates, compared with what happens in actual practice.

2.4.1   Forward Bias Safe Operating Area

The forward bias safe operating area (FBSOA) describes the ability of a transistor to handle stress when the base is forward-biased. Manufacturer FBSOA curves detail maximum limits for both steady-state dissipation and turn-on load lines. Because it is possible to have a positive base-emitter voltage and negative base current during the device storage time, forward bias is defined in terms of base current.

Bipolar transistors are particularly sensitive to voltage stress, more so than to stress induced by high currents. This situation is particularly true of switching transistors, and it shows up on the FBSOA curve. Figure 2.12 shows a typical curve

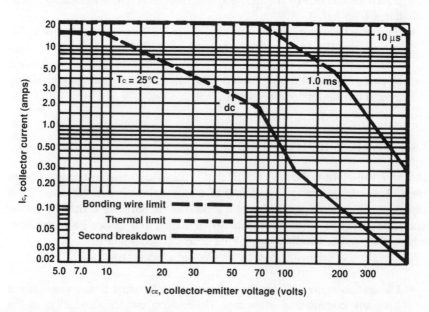

Figure 2.12   Forward bias safe operating area (SOA) curve for a bipolar transistor (MJH16010A). (Courtesy of Motorola)

for a common power transistor. In the case of the dc trace, the following observations can be made:

- The power limit established by the *bonding-wire limit* portion of the curve permits 135 W maximum dissipation (15 A x 9 V).
- The power limit established by the *thermal limit* portion of the curve permits (at the maximum voltage point) 135 W maximum dissipation (2 A x 67.5 V).
- There is no change in maximum power dissipation.
- The power limit established by the *second breakdown* portion of the curve decreases dramatically from the previous two conditions.
- At 100 V, the maximum current is 0.42 A, for a maximum power dissipation of 42 W.

2.4.2 Reverse Bias Safe Operating Area

The reverse bias safe operating area (RBSOA) describes the ability of a transistor to handle stress with its base reverse-biased. As with FBSOA, RBSOA is defined in terms of current. In many respects, RBSOA and FBSOA are analogous. First among these is voltage sensitivity. Bipolar transistors exhibit the same sensitivity to voltage stress in the reverse bias mode as in the forward bias mode. A typical RBSOA curve is shown in Figure 2.13. Note that maximum allowable peak instantaneous power decreases significantly as voltage is increased.

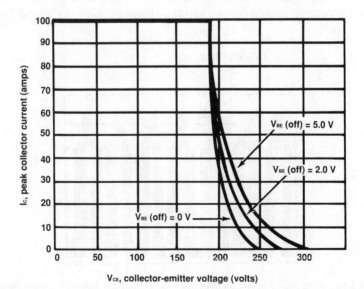

Figure 2.13 Reverse bias SOA curve for a bipolar transistor (MJH10610A). (Courtesy of Motorola)

2.4.3   Power-Handling Capability

The primary factor in determining the amount of power a given device can handle is the size of the active junction(s) on the chip. The same device power output may be achieved through the use of several smaller chips in parallel. This approach, however, may result in unequal currents and uneven distribution of heat. At high power levels, heat management becomes a significant factor in chip design.

Specialized layout geometries have been developed to ensure even current distribution throughout the device. One approach involves the use of a matrix of emitter resistances constructed so that the overall distribution of power among the parallel emitter elements results in even thermal dissipation. Figure 2.14 illustrates this *interdigited* geometry technique.

With improvements in semiconductor fabrication processes, output-device SOA is primarily a function of the size of the silicon slab inside the package. Package type, of course, determines the ultimate dissipation because of thermal saturation with temperature rise. Properly mounted, a good TO-3 or 2-screw-mounted plastic package will dissipate approximately 350 to 375 W. Figure 2.15 demonstrates the relationships between case size and power dissipation for a TO-3 package.

2.4.4   Semiconductor Derating

Good design calls for a measure of caution in the selection and application of active devices. Unexpected operating conditions, or variations in the manufacturing process, may result in field failures unless a margin of safety is allowed. Derating

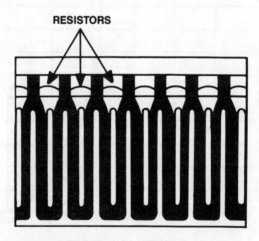

Figure 2.14   Interdigited geometry of emitter resistors used to balance currents throughout a power device chip.

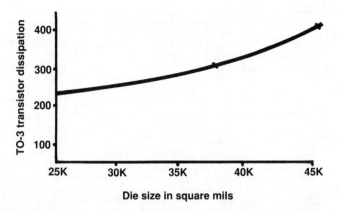

Figure 2.15   Relationship between case (die) size and transistor dissipation. (Source: *Broadcast Engineering* magazine, Intertec Publishing, Overland Park, Kan.)

is a common method of achieving such a margin. The primary derating considerations are:

- Power derating. Designed to hold the worst-case junction temperature to a value below the normal permissible rating.
- Junction-temperature derating. An allowance for the worst-case ambient temperature or case temperature that the device is likely to experience in service.
- Voltage derating. An allowance intended to compensate for temperature-dependent voltage sensitivity and other threats to device reliability as a result of instantaneous peak-voltage excursions caused by transient disturbances.

2.4.5   MOSFET Devices

Power MOSFETs have found numerous applications because of their unique performance attributes. A variety of specifications may be used to indicate the maximum operating voltages a specific device can withstand. The most common specifications include:

- Gate-to-source breakdown voltage.
- Drain-to-gate breakdown voltage.
- Drain-to-source breakdown voltage.

These limits mark the maximum voltage excursions possible with a given device before failure. Excessive voltages cause carriers within the depletion region of the reverse-biased PN junction to acquire sufficient kinetic energy to result in ioniza-

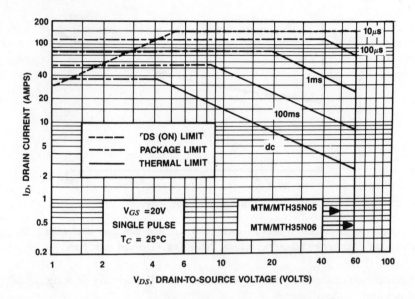

Figure 2.16 SOA curve for a power FET device. (Courtesy of Motorola)

tion. Voltage breakdown also may occur when a *critical electric field* is reached. The magnitude of this voltage is determined primarily by the characteristics of the die itself.

Safe Operating Area. The safe dc operating area of a MOSFET is determined by the rated power dissipation of the device over the entire drain-to-source voltage range (up to the rated maximum voltage). The maximum drain-source voltage is a critical parameter. If it is exceeded even momentarily, the device may be damaged permanently.

Figure 2.16 shows a representative SOA curve for a MOSFET. Notice that limits are plotted for several parameters, including drain-source voltage, thermal dissipation (a time-dependent function), package capability, and drain-source on-resistance. The capability of the package to withstand high voltages is determined by the construction of the die itself, including bonding-wire diameter, size of the bonding pad, and internal thermal resistances. The drain-source on-resistance limit is simply a manifestation of Ohm's law; with a given on-resistance, current is limited by the applied voltage.

To a large extent, the thermal limitations described in the SOA chart determine the boundaries for MOSFET use in linear applications. The maximum permissible junction temperature also affects the pulsed-current rating when the device is used as a switch. MOSFETs are, in fact, more like rectifiers than bipolar transistors, with respect to current ratings; their peak-current ratings are not gain-limited, but thermally limited.

In switching applications, total power dissipation is comprised of both switching losses and on-state losses. At low frequencies, switching losses are small. As the operating frequency increases, however, switching losses become a significant factor in circuit design.

2.5   FAILURE MODES

A semiconductor device may fail in a catastrophic, intermittent or degraded mode. Such failures usually are open circuits, short circuits, or parameters out of specifications. For integrated circuits and discrete semiconductors, the three most destructive stresses are excessive temperature, voltage, and vibration.

Semiconductor failure modes can be broken down into two basic categories: mechanical (including temperature and vibration) and electrical (including electrostatic discharge and transient overvoltage). Semiconductor manufacturers are able to increase device reliability by analyzing why good parts go bad. Figure 2.17 shows an example of failure caused by mechanical stress. The two diodes failed as

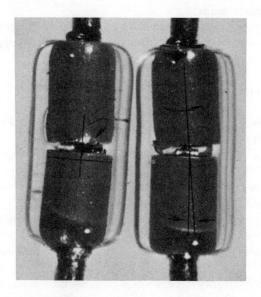

Figure 2.17   These diodes, used in a flight control computer, failed as a result of temperature cycling. The arrows show the mechanical forces at work that ultimately led to the failures. The root cause of the failures was off-center die placement and axial misalignment during assembly. (Source: Edgar Doyle Jr., "How Parts Fail," *IEEE Spectrum,* IEEE, New York, October 1981)

a result of temperature cycling that led to a cracked die and fractured metallurgical die bond. The underlying reasons for the failure were off-center die placement and axial misalignment during diode assembly. The arrows show the mechanical stresses that were present on the diode structure.

Even though misalignment during assembly was the root cause of failure, thermal cycling triggered the failures. There are, in fact, a frightening number of mechanical construction anomalies that can result in degradation or catastrophic failure of a semiconductor device. Some of the more significant threats include:

- Encapsulation failures caused by humidity and impurity penetration.
- Imperfections in termination materials.
- Stress cracks in the encapsulation material.
- Differential thermal expansion coefficients of the encapsulant, device leads, or chip.
- Wire-bond failures caused by misplaced bonds, crossed wires, and oversize bonds.
- Imperfect chip attachment to the device substrate, resulting in incomplete thermal contact, stress cracks in the chip or substrate, and solder or epoxy material short circuits.
- Aluminum conductor faults caused by metallization failures at contact windows, electromigration, corrosion, and geometrical misalignment of leads and/or the chip itself.

The principal failure modes for semiconductor devices include the following:

- Internal short circuit between metallized leads or across a junction, usually resulting in system failure.
- Open circuit in the metallization or wire bond, usually resulting in system failure.
- Variation in gain or other electrical parameters, resulting in marginal performance of the system or temperature sensitivity.
- Leakage currents across PN junctions, causing effects ranging from system malfunctions to out-of-tolerance conditions.
- Shift in turn-on voltage, resulting in random logic malfunctions in digital systems.
- Loss of seal integrity through the ingress of ambient air, moisture, and/or contaminants. The effects range from system performance degradation to complete failure.

Aluminum interconnects in a semiconductor device are the nerves that make it possible to integrate complex circuits onto blocks of silicon. The integrity of these interconnects is of critical importance to the reliable operation of a device. Figure

Figure 2.18 Die photos of two 16 K EEPROM devices. (*a*) A good part (note the connection dot shown at the arrow). (*b*) A bad part (the aluminum interconnect is missing). (Source: R. King and J. Hiatt, "A Practical VLSI Characterization and Failure Analysis System for the IC User," *Proceedings of the IEEE Reliability Physics Conference*, IEEE, New York, April 1986)

2.18 (*a*) and (*b*) show a failure mode discovered in a common 16-K EEPROM (electrically erasable programmable read-only memory) device.

2.5.1 Discrete Transistor Failure Modes

It is estimated that as much as 95 percent of all transistor failures in the field are, directly or indirectly, the result of excessive dissipation or applied voltages in excess of the maximum design limits of the device. At least four types of voltage breakdown must be considered in a reliability analysis of discrete power transistors. Although each type is not strictly independent, each can be treated separately, with the understanding that each is related to the others.

Avalanche Breakdown. Avalanche is a voltage breakdown that occurs in the collector-base junction, similar to the *Townsend effect* in gas tubes. This effect is caused by the high dielectric field strength that occurs across the collector-base junction as the collector voltage is increased. This high-intensity field accelerates the free charge carriers so that they collide with other atoms, knocking loose additional free charge carriers that, in turn, are accelerated and have further collisions.

This multiplication process occurs at an increasing rate as the collector voltage increases until at some voltage, V_a (avalanche voltage), the current suddenly tries to go to infinity. If enough heat is generated in this process, the junction will be damaged or destroyed. A damaged junction will result in higher-than-normal leakage currents, increasing the steady-state heat generation of the device, which ultimately may destroy the semiconductor junction.

Alpha multiplication. Alpha multiplication is produced by the same physical phenomenon that produces avalanche breakdown, but differs in circuit configuration. This effect occurs at a lower potential than the avalanche voltage and generally is responsible for collector-emitter breakdown when base current is equal to zero.

Punch-through. Punch-through is a voltage breakdown occurring at the collector-base junction because of high collector voltage. As collector voltage increases, the *space charge region* (collector junction width) gradually expands until it completely penetrates the base region, touching the emitter. At this point, the emitter and collector are effectively short-circuited together.

This type of breakdown occurs in some PNP junction transistors but, generally, alpha multiplication breakdown occurs at a lower voltage than punch-through. Because this breakdown occurs between collector and emitter, punch-through is more serious in the common-emitter or common-collector configuration.

Thermal runaway. Thermal runaway is a regenerative process by which an increase in temperature causes a higher leakage current, which results in increased collector current. This, in turn, leads to an increase in power dissipation, which raises the junction temperature, bringing about a further increase in leakage current.

If the leakage current is sufficiently high (resulting from high temperature or high voltage), and the current is not adequately stabilized to counteract increased collector current because of increased leakage current, this process can regenerate to a point that the temperature of the transistor rapidly rises, destroying the device. This type of effect is more prominent in power transistors, where the junction normally operates at high temperatures and where high leakage currents are present because of the large junction area. Thermal runaway is related to the avalanche effect, and is dependent upon circuit stability, ambient temperature, and transistor power dissipation.

Breakdown Effects. The effects of the breakdown modes outlined manifest themselves in various ways on the transistor: Avalanche breakdown usually results in destruction of the collector-base junction because of excessive currents. This, in turn, results in an open between the collector and base. Breakdown caused by alpha multiplication and thermal runaway most often results in destruction of the transistor because of excessive heat dissipation that shows up electrically as a short circuit between the collector and emitter. This condition, which is most common in transistors that have suffered catastrophic failure, is not always easily detected. In many cases, an ohmmeter check may indicate a good transistor. Only after operating voltages are applied will the failure mode be exhibited. Punch-through breakdown generally does not permanently damage the transistor; it can be a self-healing type of breakdown. After the overvoltage is removed, the transistor usually will operate satisfactorily.

2.5.2 Thermal-Stress-Induced Failures

Thermal fatigue represents a threat to any semiconductor component, especially power devices. This phenomenon results from the thermal mismatch between a silicon chip and the device header under temperature-related stresses. Typical failure modes include voids and cracks in solder material within the device, resulting in increased thermal resistance to the outside world and the formation of "hot spots" inside the device. Catastrophic failure will occur if sufficient stress is put on the die. Although generally considered a problem pertaining to power devices, thermal fatigue also affects VLSI ICs because of the increased size of the die. Figure 2.19 illustrates the deformation process that results from excessive heat on a semiconductor device.

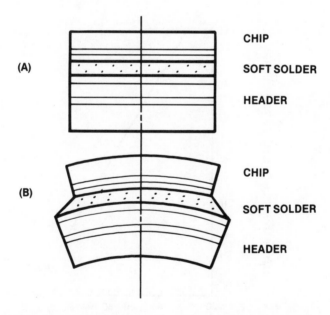

CHIP

SOFT SOLDER

HEADER

CHIP

SOFT SOLDER

HEADER

Figure 2.19 The mechanics of thermal stress on semiconductor devices. (*a*) A normal chip/solder/header composite structure. (*b*) The assembly subjected to a change in temperature. (Data from: G. Guang-bo, C. An, and G. Xiang, "A Layer Damage Model for Calculating Thermal Fatigue Lifetime of Power Devices," *Proceedings of the IEEE Reliability Physics Conference*, IEEE, New York, April 1986)

As a case in point, consider the 2N3055 power transistor. The component is an NPN device rated for 115 W dissipation at 25°C ambient temperature. The component can handle up to 100 V on the collector at 15 A. Although the 2N3055 is designed for demanding applications, the effects of thermal cycling take their toll. Figure 2.20 charts the number of predicted thermal cycles for the 2N3055 vs. temperature change. Note that the lifetime of the device increases from 9000 cycles at 120°C to 30,000 cycles at 50°C.

2.5.3 Package Failure Mechanisms

Most failures in electronic packages are caused by mechanical failure mechanisms, such as fracture, fatigue, and corrosion. These failures are primarily wearout phenomena rather than statistically random phenomena, hence they cannot be characterized as part of a constant failure-rate process during any significant segment of device life.

Fatigue failures in microelectronic packages usually are the result of thermo-mechanical loading from thermal and power excitation, and mechanical loading

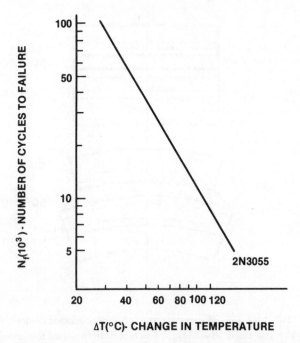

Figure 2.20 The effect of thermal cycling on the lifetime of a 2N3055 power transistor. (Data from: G. Guang-bo, C. An, and G. Xiang, "A Layer Damage Model for Calculating Thermal Fatigue Lifetime of Power Devices," *Proceedings of the IEEE Reliability Physics Conference*, IEEE, New York, April 1986)

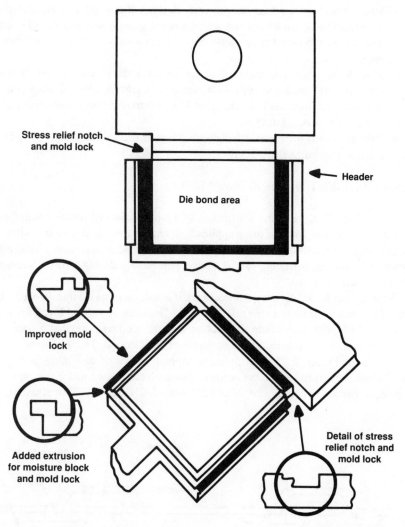

Figure 2.21   Improved mechanical packaging for a TO-220. (Courtesy of Motorola)

from vibration. Failures often occur not in the bulk of any material, but at the interface of different materials. Such problems commonly are observed at the interfaces of (1) the die and die attach, (2) substrate or case and the attach, (3) wire and bond pad, or (4) substrate and bond pad.

Package Improvements. As chip design and manufacturing processes improve, the resulting performance benefits tend to be well-publicized. At the same time, however, package technology is evolving. Figure 2.21 illustrates improvements made to the TO-220 package by one manufacturer. Changes include:

- *Stress relief.* In contrast to previous designs that used planar heat-sink surfaces, the new package has a stress-relief groove between the die bond area and the tab. Notches on both sides of this groove provide additional stress relief.
- *Mold locks.* Improved mechanical design gives the molding compound an optimal metal surface to grip as it curves. This design also extends the distance that any contaminants must travel before penetrating into the die bond area of a plastic package device.
- *Heat sink.* Advanced materials provide higher thermal conductivity, which helps to maximize thermal efficiency.

2.5.4   MOSFET Device Failure Modes

Power MOSFETs have found application in a wide variety of power-control and conversion systems. Most of these applications require that the device be switched on and off at a high frequency. The thermal and electrical stresses that a MOSFET experiences during switching can be severe, particularly during turn-off when an inductive load is present.

When power MOSFETs were introduced, it usually was stated that, because the MOSFET was a majority carrier device, it was immune to second breakdown as observed in bipolar transistors. It must be understood, however, that a parasitic bipolar transistor is inherent in the structure of a MOSFET. This phenomenon is illustrated in Figure 2.22. The parasitic bipolar transistor can allow a failure mechanism similar to second breakdown. Research has shown that if the parasitic transistor becomes active, the MOSFET may fail (Ref. 4). This situation is

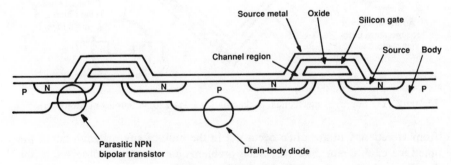

Figure 2.22   Cross section of a power MOSFET device showing the parasitic bipolar transistor and diode inherent in the structure. (Data from: David L. Blackburn, "Turn-Off Failure of Power MOSFETs," *IEEE Transactions on Power Electronics,* vol. PE-2, no. 2, IEEE, New York, April 1987)

particularly troublesome if the MOSFET drain-source breakdown voltage is approximately twice the collector-emitter sustaining voltage of the parasitic bipolar transistor. This failure mechanism results, apparently, when the drain voltage snaps back to the sustaining voltage of the parasitic device. This *negative resistance characteristic* can cause the total device current to constrict to a small number of cells in the MOSFET structure, leading to device failure.

The precipitous voltage drop synonymous with second breakdown is a result of avalanche injection and any mechanism, electrical or thermal, that may cause the current density to become large enough for avalanche injection to occur.

2.5.5   Surface-Mounted Device Failure Modes

Surface-mounted devices are used widely today in products ranging from consumer television sets to airborne radar. SMD components are mounted directly onto a printed wiring board and soldered to its surface. This process differs from through-hole mounting, where the device lead is inserted through a plated hole in the PWB and soldered into place. Some SMDs have leads, and others are considered leadless, with the edge of the package attached directly to the board by solder. Three popular SMD designs are shown in Figure 2.23.

When an SMD component is exposed to thermal cycling, the package and the board materials expand at different rates, causing strain on the solder connection. Cracking of the solder joint is the major failure mode for an SMD. The main design parameters that affect the integrity of the connections include the lead design, package size, board material, and the ability to dissipate heat.

The main failure mechanisms for the solder connection are thermal fatigue and *creep*. Creep is defined as the slow and progressive deformation of a material with time under a constant stress. Combined, these phenomena accelerate the failure of the solder.

Finite element analysis (FEA) has been applied to study the reliability of SMD components (Ref. 5). FEA is a computer simulation technique that may be used to predict the thermal and mechanical response of structures exposed to various environmental conditions. To assess reliability, a comparison is made between the stresses in the materials and the corresponding material strengths. In Ref. 5 FEA was performed on three different surface-mount designs: the leadless chip carrier, the S-lead chip carrier, and the gull-wing chip carrier. For each design, two computer models were created. The first represented the package mounted onto a board, and the second represented a single lead (see Figure 2.24). The package/board model was cycled thermally, and the results were incorporated into the lead model to obtain a detailed stress-distribution plot for the solder connection points. The results indicated that the gull-wing and S-lead chip carriers would be reliable when placed in a given temperature environment, but that the leadless chip carrier would have reliability problems after a short period of time.

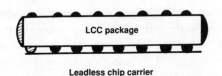

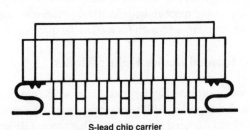

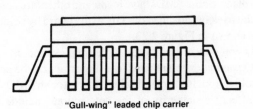

Figure 2.23   Three common surface-mounted chip designs. (Data from: Gretchen A. Bivens, "Predicting Time-to-Failure Using Finite Element Analysis," *Proceedings of the IEEE Reliability and Maintainability Symposium,* IEEE, New York, 1990)

2.6   HEAT SINK CONSIDERATIONS

Heat generated in a power transistor (primarily at the collector junction) must be removed at a sufficient rate to keep the junction temperature within a specific upper limit. This is accomplished primarily by conduction from the junction through the transistor material to a metal mounting base, which is designed to provide good thermal contact, to an external heat dissipator or heat sink.

Because heat transfer is associated with a temperature difference, a differential will exist between the collector junction and the transistor mounting surface. A temperature differential also will exist between the device mounting surface and the heat sink. Ideally, these differentials will be small, but they will exist. It follows, therefore, that an increase in dissipated power at the collector junction will result in a corresponding increase in junction temperature. In general, assessing the heat

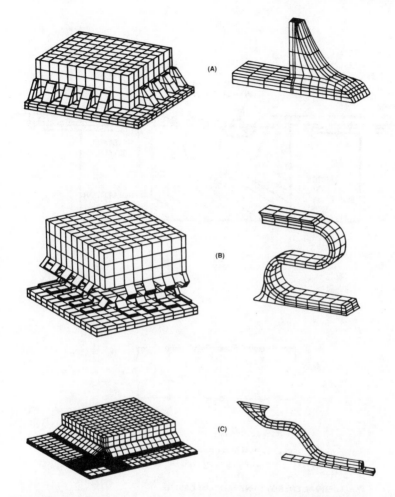

Figure 2.24 Finite element analysis plots for three types of SMD chips. (*a*) Leadless chip carrier. (*b*) S-lead chip carrier. (*c*) Gull-wing leaded chip carrier. (Data from: Gretchen A. Bivens, "Predicting Time-to-Failure Using Finite Element Analysis," *Proceedings of the IEEE Reliability and Maintainability Symposium,* IEEE, New York, 1990)

sink requirements of a device or system (and the potential for problems) is a difficult proposition.

Figure 2.25 shows some of the primary elements involved in thermal transmission of energy from the silicon junction to the external heat sink. An electrical analog of the process is helpful for illustration. The model shown in (*b*) includes two primary elements: *thermal capacitance* and *thermal resistance*. The energy storage property of a given mass, expressed as *C*, is the basis for the transient

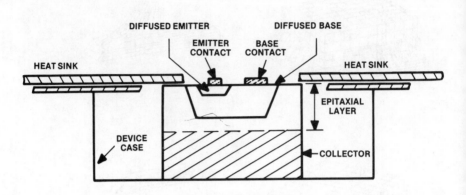

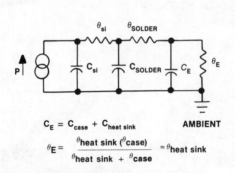

$$C_E = C_{case} + C_{heat\ sink}$$

$$\theta_E = \frac{\theta_{heat\ sink}\ (\theta_{case})}{\theta_{heat\ sink}\ +\ \theta_{case}} \approx \theta_{heat\ sink}$$

P = THERMAL ENERGY GENERATED BY DEVICE

θ = THERMAL RESISTANCE (MEASURED IN °C/WATT)

C = THERMAL CAPACITANCE (MEASURED IN WATT-SECONDS/°C)

Figure 2.25 Simplified model of thermal transmission from the junction of a power transistor (TO-3 case) to a heat sink. (*a*) The structure of a double-diffused epitaxial planar device. (*b*) Simplified electrical equivalent of the heat-transfer mechanism. (Data from: RCA Power Circuits: DC to Microwave, *RCA Electronic Components,* Harrison, N.J., 1969)

thermal properties of transistors. The thermal transmission loss from one surface or material to another, expressed as θ, causes a temperature differential among the various components of the semiconductor model shown in the figure.

Although this model may be used to predict the rise in junction temperature, the result of a given increase in power dissipation, it is an extreme oversimplification of the mechanics involved. The elements considered in our example include the silicon transistor die (Si); the solder used inside the transistor to bond the emitter, base, and collector to the outside-world terminals; and the combined effects of the heat sink and transistor case. This model assumes the transistor is mounted directly onto a heat sink, not through a mica (or other type of) insulator.

The primary purpose of a heat sink is to increase the effective heat-dissipation area of the transistor. If the full power-handling capability of a transistor is to be achieved, there must be zero temperature differential between the case and the ambient air. This condition exists only when the thermal resistance of the heat sink is zero, which would require an infinitely large heat sink. Although such a device never can be realized, the closer the approximation of actual conditions to ideal conditions, the greater the maximum possible operating power.

In typical power-transistor applications, the case must be insulated electrically from the heat sink (except for circuits using a grounded-collector configuration). The thermal resistance from case to heat sink, therefore, includes two components: surface irregularities of the insulating material, transistor case, and heat sink; and the insulator itself. Thermal resistance resulting from surface irregularities may be minimized through the use of silicon grease compounds. The thermal resistance of the insulator itself, however, may represent a significant problem. Unfortunately, materials that are good electrical insulators usually are also good thermal insulators. The best materials for such applications are mica, beryllium oxide, and anodized aluminum.

2.6.1 Mounting Power Semiconductors

Semiconductor current and power ratings are linked inseparably to their thermal environment. Except for lead-mounted parts used at low currents, a heat sink (or heat exchanger) is required to prevent the junction temperature from exceeding its rated limits. Experience has shown that most early-life field failures of power semiconductors can be traced to faulty mounting procedures. High junction temperatures can result in reduced component life. Failures also may result from improperly mounting a device to a warped surface; mechanical damage can cause cracks in the case, resulting in moisture contamination, or a crack in the semiconductor die itself.

Figure 2.26 shows an extreme case of improper mounting for a TO-220 package. Excessive force on the fastening screw has warped the device heat sink, and possibly resulted in a crack in the package. The die will experience excessive heating because of poor heat transfer. Note that only a small portion of the device heat sink is in direct contact (through the mica washer) with the equipment heat sink.

Proper mounting of a power device to a heat sink requires attention to the following steps:

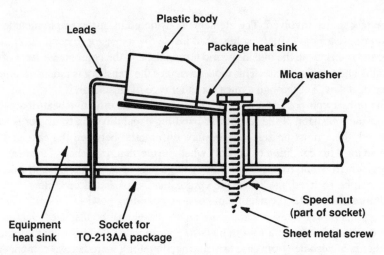

Figure 2.26 Extreme case of improper mounting of a TO-220 package semiconductor device. (Courtesy of Motorola)

- Prepare the mounting surface. The surface must be reasonably flat to facilitate efficient heat transfer.
- Apply thermal grease. When a significant amount of power must be radiated, something must be done to fill the air voids between the mating surfaces in the thermal path. The use of thermal joint compounds (thermal grease) is common for this purpose. To prevent accumulation of airborne particulate matter, wipe away any excess compound using a cloth moistened with acetone or alcohol.
- Install the insulator, if it is to be used. Devices using the case as a terminal usually must be insulated from the equipment heat sink. Insulating materials include mica, a variety of proprietary films, and beryllium oxide. From the standpoint of heat transfer, beryllium is the best. It is expensive, however, and must be handled with care; beryllium dust is highly toxic.
- Secure the device to the heat sink. Mounting holes should be only large enough to allow clearance of the fastener. Punched holes should be avoided, if possible, because of the potential for creating a depression around the hole. Such a crater in the heat sink may distort the semiconductor package and/or result in an inefficient transfer of heat.
- Connect the device terminals to the circuit.

Fastening Techniques. Each type of power semiconductor package requires the use of unique fastening hardware. Following are guidelines pertaining to various devices:

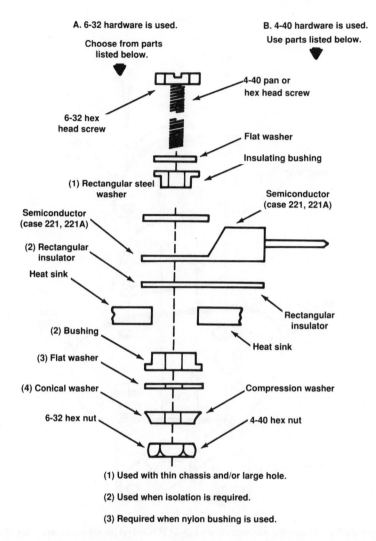

A. 6-32 hardware is used.
Choose from parts
listed below.

B. 4-40 hardware is used.
Use parts listed below.

4-40 pan or
hex head screw

6-32 hex
head screw

Flat washer

Insulating bushing

(1) Rectangular steel
washer

Semiconductor
(case 221, 221A)

Semiconductor
(case 221, 221A)

(2) Rectangular
insulator

Heat sink

Rectangular
insulator

(2) Bushing

Heat sink

(3) Flat washer

(4) Conical washer

Compression washer

6-32 hex nut

4-40 hex nut

(1) Used with thin chassis and/or large hole.

(2) Used when isolation is required.

(3) Required when nylon bushing is used.

Figure 2.27 Mounting hardware for a TO-220 package. (*a*) Preferred arrangement for isolated or nonisolated mounting (screw is at semiconductor case potential). (*b*) Alternate arrangement for isolated mounting (screw is at heat sink potential). (Courtesy of Motorola)

- Tab-mount device. This class of device, characterized by the popular TO-220 package, often is mounted with improper hardware. The correct method of mounting a TO-220 is shown in Figure 2.27. The rectangular washer shown in (*a*) is used to minimize distortion of the mounting flange. Be certain that

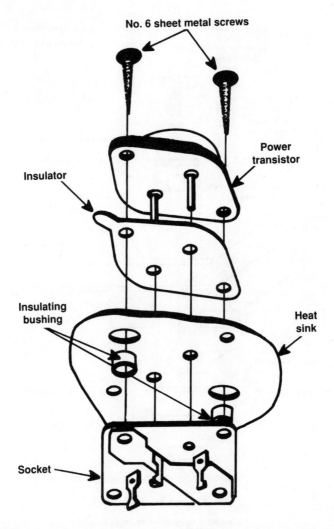

No. 6 sheet metal screws

Power transistor

Insulator

Insulating bushing

Heat sink

Socket

Figure 2.28 Mounting hardware for a TO-3 (TO-204A) power semiconductor. (Courtesy of Motorola)

the tool used to drive the mounting screw does not apply pressure to the body of the device; the result could be a crack in the package.

• Flange-mount device. A large variety of devices fall into the flange-mount category, including the popular TO-3 (also known as the TO-204A). The rugged base and the distance between the die and the mounting holes combine to make mounting practices less critical for most flange-mounted power semiconductors. Figure 2.28 shows the mounting hardware used for a TO-3

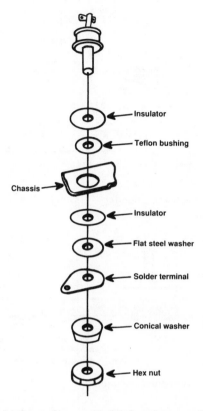

Figure 2.29 Mounting hardware for a nonisolated stud-mounted device. (Courtesy of Motorola)

package. RF power devices are more sensitive to heat sink surface flatness than other packages. Specific mounting requirements for RF devices must be followed carefully to ensure reliable operation.

- Stud-mount device. Errors with noninsulated stud-mounted devices generally are caused by applying excessive torque or tapping the stud into a threaded heat sink hole. Either practice may result in warping of the hex base, which may crack the semiconductor die. The proper hardware components for mounting a noninsulated device to a heat sink or chassis are shown in Figure 2.29.
- Press-fit device. Mounting a press-fit device into a heat sink is straightforward. Figure 2.30 illustrates the approach for mounting on a heat sink, and for mounting on a chassis. Apply force evenly on the shoulder ring to avoid tilting the case in the hole during assembly. Use thermal joint compound. Pressing force will vary from 250 to 1000 pounds, depending on the heat sink material.

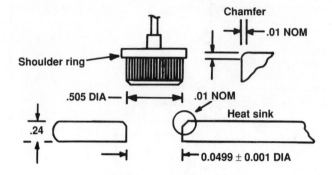

Heat sink mounting

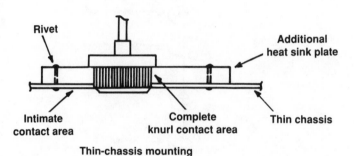

Thin-chassis mounting

Figure 2.30   Mounting a press-fit device. (a) Heat sink mounting. (b) Thin-chassis mounting. (Courtesy of Motorola)

2.6.2   Air-Handling Systems

There are two basic methods of cooling electronic equipment:

- Active cooling, where the coolant uses machinery (fans, blowers or pumps) to remove heat.
- Passive cooling, where no machinery is used to effect the heat removal process.

Passive cooling can involve conduction, radiation, and natural convection. One major disadvantage of active cooling is that once incorporated, the equipment may be dependent upon the cooling system for survival, thus introducing an additional

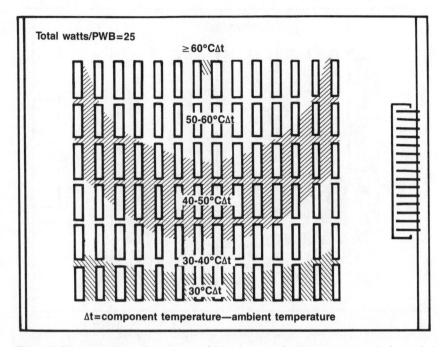

Figure 2.31 Operating temperatures of board-mounted components, passively cooled, with the power dissipation set uniformly at 0.5 W/sq. inch. (Data from: Charles T. Leonard and Michael G. Pecht, "Improved Techniques for Cost Effective Electronics," *Proceedings of the Reliability and Maintainability Society,* IEEE, New York, 1991.)

source of potential failure. Passive cooling, therefore, is generally preferred unless large amounts of heat must be dissipated.

Research into passive cooling methods for electronic systems shows that — on average — circuit boards should be mounted vertically with the long dimension horizontal. Unrestricted free air passageways must be provided between circuit boards, including adequate ventilation holes at the top and bottom of the card cage. High wattage components should be mounted along the bottom edge, and low wattage devices should be mounted at the top of the board. Assuming uniform dissipation over the assembly, the hottest point is the top-center region, as shown in Figure 2.31. Note that the hottest components exceed a 60°C temperature rise, with the power to the board at 25 W. If the same board is redesigned for optimum power dissipation, with the hottest components at the bottom and sides of the board, more efficient cooling is possible. As shown in Figure 2.32, optimizing for power dissipation permits greater total power input to the board, with cooler operation overall. All components in the redesigned board operate at a 50°C temperature rise.

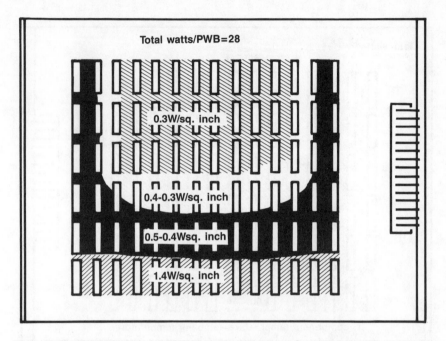

Figure 2.32 Printed wiring board component arrangement for constant temperature operation as a passively cooled assembly. Overall power dissipation is adjusted by changing component placement. (Data from: Charles T. Leonard and Michael G. Pecht, "Improved Techniques for Cost Effective Electronics," *Proceedings of the Reliability and Maintainability Society,* IEEE, New York, 1991.)

Additional reliability is achieved because uniform temperatures across the assembly reduce thermal gradient stresses, thus improving the fatigue life of the connections.

The construction of an equipment housing may also have a significant effect on the internal operating temperature of an electronic system. For example, excessive failures were noted on a ship-based missile defense control system. Excessive heating of critical components was suspected. Temperature measurements at various points in the equipment rack showed poor air circulation, as illustrated in Figure 2.33. Because of the chassis layout, virtually no air circulated through the card cage area. Further, heat from the power supply mounted below the assembly added to the heat build-up. A redesign of the ventilation system solved the reliability problem.

The fans and filters used in an electronic system are an important to long term reliability in the field. Depending on the pores per inch rating (PPI) of the material used, fan filters can cause a significant difference in ambient temperature versus

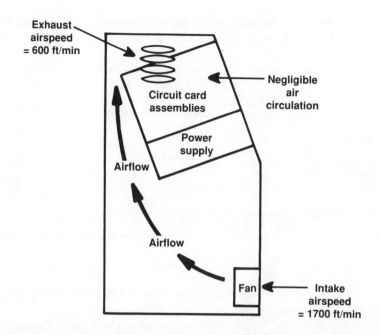

Figure 2.33 Control-panel airflow for a ship-based missile-control system. (Data from: H. Fortna, R. Zavada, and T. Warren, "An Integrated Analytic Approach for Reliability Improvement," *Proceedings of the IEEE Reliability and Maintainability Symposium*, IEEE, New York, 1990)

internal equipment air temperature, even when clean. Table 2.2 documents typical measurements taken on a piece of electronic equipment. Clearly, the *porosity* (and thickness of the filter material selected) influences the best-case internal temperature of an electronic system. The cubic feet per minute (CFM) rating of the fan or blower used must provide for operation under a wide variety of conditions and temperatures.

2.7 REFERENCES

[1] Thomas J. Green, "A Review of IC Fabrication Design and Assembly Defects Manifested as Field Failures in Air Force Avionic Equipment," *Proceedings of the IEEE Reliability Physics Symposium*, IEEE, New York, 1988.

[2] Alvin Sydnor, "Voltage Breakdown in Transistors," *Electronic Servicing & Technology* magazine, Intertec Publishing, Overland Park, Kan., July 1986.

[3] Paul Dick, "System Effectiveness: A Key to Assurance," *Proceedings of the IEEE Reliability and Maintainability Symposium*, IEEE, New York, 1988.

Table 2.2 Temperature Measurements (in Degrees Fahrenheit) on an Electronic System, Showing Cooling Performance without Fan Filters and with Filters of Various PPI Ratings.

		FOAM FILTERS		
	WITHOUT FILTER	4OPPI-¼THK	2OPPI-¼THK	1OPPI-¼THK
ROOM AMBIENT	71°	73°	75°	77°
EQUIPMENT AIR TEMPERATURE	81°	97°	94°	93°
TEMPERATURE RISE	10°	24°	19°	16°

NOTE: The more pores per inch, the greater the pressure drop. Density of material is not related to pore size.

[4] David L. Blackburn, "Turn-Off Failure of Power MOSFETs," *IEEE Transactions on Power Electronics*, Vol. PE-2, No. 2, IEEE, New York, April 1987.

[5] Gretchen A. Bivens, "Predicting Time-to-Failure Using Finite Element Analysis," *Proceedings of the IEEE Reliability and Maintainability Symposium*, IEEE, New York, 1990.

2.8 BIBLIOGRAPHY

Benson, K. B., and J. Whitaker: *Television and Audio Handbook for Engineers and Technicians,* McGraw-Hill, New York, 1989.

Bivens, Gretchen A.: "Predicting Time-to-Failure Using Finite Element Analysis," *Proceedings of the IEEE Reliability and Maintainability Symposium,* IEEE, New York, 1990.

Blackburn, David L.: "Turn-Off Failure of Power MOSFETs," *IEEE Transactions on Power Electronics,* Vol. PE-2, No. 2, IEEE, New York, April 1987.

Ching, T. B., and W. H. Schroen: "Bond Pad Structure Reliability," *Proceedings of the IEEE Reliability Physics Symposium,* IEEE, New York, 1988.

Crook, D. L.: "Evolution of VLSI Reliability Engineering, *Proceedings of the IEEE Reliability Physics Symposium,* IEEE, New York, 1990.

Dasgupta, A., D. Barker, and M. Pecht: "Reliability Prediction of Electronic Packages," *Proceedings of the IEEE Reliability and Maintainability Symposium,* IEEE, New York, 1990.

Dick, Paul: "System Effectiveness: A Key to Assurance," *Proceedings of the IEEE Reliability and Maintainability Symposium,* IEEE, New York, 1988.

Fink, D., and D. Christiansen (eds.): *Electronics Engineers' Handbook*, 3rd ed., McGraw-Hill, New York, 1989.

Green, Thomas J.: "A Review of IC Fabrication Design and Assembly Defects Manifested as Field Failures in Air Force Avionic Equipment," *Proceedings of the IEEE Reliability Physics Symposium*, IEEE, New York, 1988.

Guang-bo, G., C. An, and G. Xiang: "A Layer Damage Model for Calculating Thermal Fatigue Lifetime of Power Devices," *Proceedings of the IEEE Reliability Physics Conference*, IEEE, New York, April 1986.

Jordan, Edward C. (ed.): *Reference Data for Engineers: Radio, Electronics, Computer, and Communications*, 7th ed., Howard W. Sams, Indianapolis, Ind., 1985.

King, R., and J. Hiatt: "A Practical VLSI Characterization and Failure Analysis System for the IC User," *Proceedings of the IEEE Reliability Physics Conference*, IEEE, New York, April 1986.

Koch, T., W. Richling, J. Witlock, and D. Hall: "A Bond Failure Mechanism," *Proceedings of the IEEE Reliability Physics Conference*, IEEE, New York, April 1986.

Koyama, H., Y. Mashiko, and T. Nishioka: "Suppression of Stress Induced Aluminum Void Formation," *Proceedings of the IEEE Reliability Physics Conference*, IEEE, New York, April 1986.

_____, H. Shiozaki, I. Okumura, S. Mizugashira, H. Higuchi, and T. Ajiki: "A New Bond Failure: Wire Crater in Surface Mount Device," *Proceedings of the IEEE Reliability Physics Symposium*, IEEE, New York, 1988.

Lau, J., G. Harkins, D. Rick, and J. Kral: "Thermal Fatigue Reliability of SMD Packages and Interconnections," *Proceedings of the IEEE Reliability Physics Symposium*, IEEE, New York, 1987.

Leonard, Charles T., and Michael G. Pecht, "Improved Techniques for Cost Effective Electronics," *Proceedings of the Reliability and Maintainability Society*, IEEE, New York, 1991.

Meeldijk, Victor: "Why Do Components Fail?" *Electronic Servicing & Technology* magazine, Intertec Publishing, Overland Park, Kan., November 1986.

Pantic, Dragan: "Benefits of Integrated Circuit Burn-In to Obtain High Reliability Parts," *IEEE Transactions on Reliability*, vol. R-35, no. 1, IEEE, New York, 1986.

Roehr, Bill: "Mounting Considerations for Power Semiconductors," *TMOS Power MOSFET Data Handbook*, Motorola Semiconductor, Phoenix, Ariz., 1989.

Shirley, C. G., and R. C. Blish: "Thin-Film Cracking and Wire Ball Shear in Plastic DIPs Due to Temperature Cycle and Thermal Shock," *Proceedings of the IEEE Reliability Physics Symposium*, IEEE, New York, 1987.

Smith, William B.: "Integrated Product and Process Design to Achieve High Reliability in Both Early and Useful Life of the Product," *Proceedings of the IEEE Reliability and Maintainability Symposium*, IEEE, New York, 1987.

Sydnor, Alvin: "Voltage Breakdown in Transistors," *Electronic Servicing & Technology* magazine, Intertec Publishing, Overland Park, Kan., July 1986.

Whitaker, Jerry: *Radio Frequency Transmission Systems: Design and Operation*, McGraw-Hill, New York, 1990.

Zins, E., and G. Smith: "R&M Attributes of VHSIC/VLSI Technology," *Proceedings of the IEEE Reliability and Maintainability Symposium*, IEEE, New York, 1987.

Technical staff: *Bipolar Power Transistor Reliability Report*, Motorola Semiconductor, Phoenix, Ariz., 1988.

Technical staff, Military/Aerospace Products Division: *The Reliability Handbook*, National Semiconductor, Santa Clara, Calif., 1987.

Technical staff: *RCA Power Circuits: DC to Microwave*, RCA Electronic Components, Harrison, N.J., 1969.

3

EFFECTS OF OVERVOLTAGES ON SEMICONDUCTORS

3.1 INTRODUCTION

Semiconductor failures caused by high-voltage stresses are becoming a serious concern for engineers, operators, and technical managers as new, high-density integrated circuits are placed into service. Internal IC connection lines have been reduced from 5 microns, common a few years ago, to less than 1 micron today. Spacing between leads has been reduced by a factor of 4 or more. In the past, the overvoltage peril was primarily to semiconductor substrates. Now, however, the metallization itself—the points to which leads connect—is subject to damage. Failures are the result of three primary overvoltage sources:

1. *External man-made:* Overvoltages coupled into electronic hardware from utility company ac power feeds, or other ac or dc power sources.
2. *External natural:* Overvoltages coupled into electronic hardware as a result of natural sources, such as lightning and other atmospheric disturbances.
3. *Electrostatic discharge (ESD):* Overvoltages coupled into electronic hardware as a result of static generation and subsequent discharge. Depending upon the local weather conditions, maintenance personnel can develop damaging ESD potentials simply by walking across a carpeted floor, then touching a component.

3.2 TRANSIENT DISTURBANCES

Every piece of electronic equipment requires a steady supply of clean power in order to function properly. Recent advances in technology have made the issue of

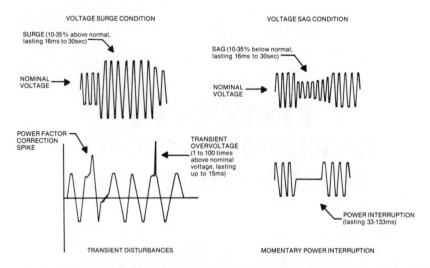

Figure 3.1 The four basic classifications of short-term power-line disturbances. (Source: *Broadcast Engineering* magazine, Intertec Publishing, Overland Park, Kan.)

ac power quality even more important, as microcomputers are integrated into a wide variety of electronic products. The high-speed logic systems prevalent today can garble or lose data because of power-supply disturbances or interruptions.

The ac power line into a facility is the lifeblood of any operation. It is also, however, a frequent source of equipment malfunctions and component failures. The utility company ac feed contains not only the 60 Hz power needed to run the facility, but also a variety of voltage sags, surges, and transients. These abnormalities cause different problems for different types of equipment. Short-term ac-voltage disturbances can be divided into four basic categories, as shown in Figure 3.1. The generally accepted definitions for these disturbances are:

- Voltage surge: An increase of 10 to 35 percent above the normal voltage for a period of 16 ms to 30 s.
- Voltage sag: A decrease of 10 to 35 percent below the normal voltage for a period of 16 ms to 30 s.
- Transient disturbance: A voltage pulse of high energy and short duration impressed upon the input ac waveform. The overvoltage pulse may be 1 to 100 times the normal potential and may last up to 15 ms. Rise times can measure in the nanosecond range.
- Momentary power interruption: A decrease to zero voltage of the ac-power-line potential, lasting from 33 to 133 ms. (Longer-duration interruptions are considered power outages.)

Voltage surges and sags occasionally result in operational problems for equipment on-line, but automatic protection or correction circuits generally take appropriate actions to prevent equipment damage. Such disturbances can, however, garble computer-system data if the disturbance *transition time* (the rise or fall time of the disturbance) is sufficiently fast. System hardware also may be stressed if there is only a marginal power-supply reserve, or if the disturbances are frequent. The possibility of complete system failure because of voltage sag or surge conditions, however, is relatively small. The greatest threat to the proper operation of electronic equipment rests with transient overvoltages. Transients are difficult to identify and difficult to eliminate. Many devices commonly used to correct for sag and surge conditions, such as series-regulated power supplies and ferroresonant transformers, are of limited value in protecting a load from high-energy, fast-rise-time disturbances on the ac line.

In the computer industry, the majority of unexplained problems resulting in disallowed states of operation actually are caused by transients. With the increased use of microcomputers in industry, this warning cannot be ignored. Because of the high potential that transient disturbances typically exhibit, they not only cause data and program errors, but also can damage or destroy electrical components. This threat to electronic equipment involves sensitive integrated circuits and many other common devices, including capacitors, transformers, rectifiers, and power semiconductors. What's more, the effects of transients often are cumulative, resulting in gradual deterioration and, ultimately, catastrophic failure. Figure 3.2 illustrates the susceptibility of semiconductors to failure as a result of transient overvoltages.

3.2.1   Lightning Effects

A lightning storm is dramatic. It also may be damaging to electrical equipment. A typical lightning strike consists of a stepped leader that progresses toward the ground at a velocity that may exceed 50 m/s. When sufficient potential difference exists between the cloud and the ground, arcs move from the ground to the leader column, completing the ionized column from upward cloud to ground. A fast, bright return stroke will then move along the leader column at about one-third the speed of light. Peak currents from a typical lightning strike can exceed 100 kA, with a total charge as high as 100 coulomb. (A coulomb is the unit of electrical charge that is transferred in one second by an electric current of one ampere. It is approximately equal to 6.24×10^{18} electrons.)

A lightning strike may be coupled into a piece of equipment through induction or direct-charge injection. A lightning strike a mile away from a utility company power line can create an electromagnetic field with a strength of as much as 70 V/m. Given a sufficiently long line, substantial voltages can be coupled to the primary power system without a direct hit.

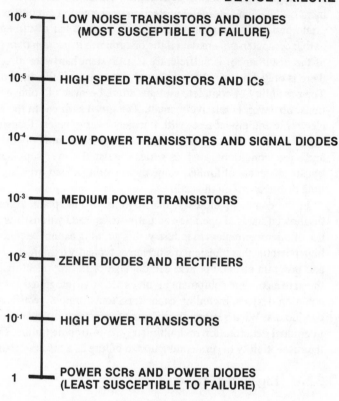

JOULES   **DEVICE FAILURE**

10^{-6} — LOW NOISE TRANSISTORS AND DIODES (MOST SUSCEPTIBLE TO FAILURE)

10^{-5} — HIGH SPEED TRANSISTORS AND ICs

10^{-4} — LOW POWER TRANSISTORS AND SIGNAL DIODES

10^{-3} — MEDIUM POWER TRANSISTORS

10^{-2} — ZENER DIODES AND RECTIFIERS

10^{-1} — HIGH POWER TRANSISTORS

1 — POWER SCRs AND POWER DIODES (LEAST SUSCEPTIBLE TO FAILURE)

Figure 3.2 An estimate of the susceptibility of semiconductor devices to failure because of transient energy. This estimate assumes a transient duration of several microseconds. (Source: *Broadcast Engineering* magazine, Intertec Publishing, Overland Park, Kan.)

Communications equipment is particularly vulnerable to lightning-related failures because an antenna of some kind is invariably connected to the gear. Lightning is a point charge-injection process, with pulses moving away from the point of injection. The amount of total energy (voltage and current) and the rise and decay times seen at the load as a result of a lightning strike is a function of (1), the distance between the strike and the load and (2), the physical characteristics of the power or antenna system (wire size, number and sharpness of bends, types of transformers, types of insulators, and lighting suppressors).

3.2.2 Failure Modes

Semiconductor devices can be destroyed or damaged by transient disturbances in one of several ways. The primary failure mechanisms include:

- Avalanche-related failure.
- Thermal runaway.
- Thermal second breakdown.
- Metallization failure.
- Polarity reversals.

When a semiconductor junction fails because of overstress, a low-resistance path is formed that shunts the junction. This path is not a true short circuit, but it is a close approximation. The shunting resistance may be less than 10 Ω in a junction that has been heavily overstressed. By comparison, the shunting resistance of a junction that has been only mildly overstressed may be as high as 10 MΩ. The formation of low-resistance shunting paths is the result of a junction's electrothermal response to overstress.

Avalanche-Related Failure. A high reverse voltage applied to a nonconducting PN junction can cause *avalanche* currents to flow. Avalanche is the process resulting from high fields in a semiconductor device in which an electron, accelerated by the field, strikes an atom and releases more electrons, which continue the sequence. If enough heat is generated in this cycle, the junction may be damaged or destroyed.

If such a process occurs between the base and emitter junction of a transistor, the effect may be either minor or catastrophic. With a minor failure, the gain of the transistor can be reduced through the creation of *trapping centers*, which restrict the free flow of carriers. With a catastrophic failure, the transistor will cease to function.

The most common cause of failure in a power MOSFET is an overvoltage that exceeds the maximum rated drain-source voltage of the device. Load transients caused by switching high (inductive) currents or by lightning discharges may contain enough energy to destroy a MOSFET if it begins to avalanche.

Thermal Runaway. A thermal runaway condition can be triggered by a sudden increase in gain resulting from the heating effect of a transient on a transistor. The transient can bring the device (operating in the active region) out of its safe operating area and into an unpredictable operating mode.

Thermal Second Breakdown. Junction burnout is a significant failure mechanism for bipolar devices, particularly JFET and Schottky devices. The junction between a *P*-type diffusion and an *N*-type diffusion normally has a positive temperature coefficient at low temperatures. Increased temperature will result in increased resistance. When a reverse-biased pulse is applied, the junction dissipates heat in a narrow *depletion region*, and the temperature in that area increases rapidly. If enough energy is applied in this process, the junction will reach a point at which the temperature coefficient of the silicon will turn negative. In other words, increased temperature will result in decreased resistance. A thermal runaway condition may then ensue, resulting in localized melting of the junction. If sustaining energy is available after the initial melt, the hot spot can grow into a *filament short circuit.* The longer the energy pulse, the wider the resulting filament short circuit. *Current filamentation* is a concentration of current flow in one or more narrow regions, which leads to localized heating.

After the transient has passed, the silicon will resolidify. The effect on the device may be catastrophic, or the performance of the component simply may be degraded. With a relatively short pulse, a hot spot may form, but not grow completely across the junction. As a result, the damage may not appear immediately as a short circuit, but may manifest itself at a later time as a result of *electromigration* or another failure mechanism.

Metallization Failure. The smaller device geometry required by high-density integrated circuits has increased the possibility of metallization failure resulting from transient overvoltages. Metallization melt is a power-dependent failure mechanism. It is more likely to occur during a short-duration, high-current pulse. Heat generated by a long pulse tends to be dissipated in the surrounding chip die.

Metallization failure also may occur as a side effect of junction melt. The junction usually breaks down first, opening the way for high currents to flow. The metallization then heats until it reaches the melting point. Metallization failure results in an open circuit. A junction short, therefore, can lead to an open-circuit failure.

Polarity Reversal. Transient disturbances typically build rapidly to a peak voltage, then decay slowly. If enough inductance and/or capacitance is present in the circuit, the tail will oscillate as it decays. This concept is illustrated in Figure 3.3. The oscillating tail can subject semiconductor devices to severe voltage polarity reversals, forcing the components into or out of a conducting state. This action may damage the semiconductor junction or result in catastrophic failure.

3.2.3 Power-Supply Failure Modes

The first line of defense in the protection of electronic hardware from damaging transient overvoltages is the ac-to-dc power supply. The power-supply components most vulnerable to failure from an ac line spike are rectifier diodes and filter capacitors. Diodes occasionally will fail from one large transient, but many more fail because of smaller, more frequent spikes that punch through the device junction. Such occurrences explain why otherwise reliable systems fail "without apparent reason."

Silicon controlled rectifiers (SCRs), like diodes, are subject to damage from transients because the peak inverse voltage or instantaneous forward voltage (or current) rating of the device may be exceeded. SCRs face an added problem from transient occurrences because of the possibility of device misfiring. An SCR can break over into a conduction state regardless of gate drive if:

- A too-high positive voltage is applied between the anode and cathode.
- A positive anode-to-cathode voltage is applied too quickly, exceeding the *dv/dt* (delta voltage/delta time) rating.

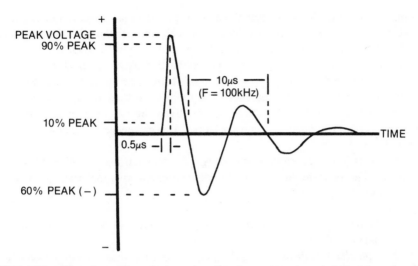

Figure 3.3 Waveshape of a typical transient disturbance. Note how the tail of the transient oscillates as it decays.

If the leading edge is sufficiently steep, even a small voltage pulse can turn on an SCR. This represents a threat not only to the device, but to the load that it controls.

Capacitors may be damaged by a transient if the working voltage of the device is exceeded during the occurrence. An overvoltage can punch a hole in the dielectric, leaving the capacitor useless at its normal operating value.

Transient overvoltages also may result in insulation breakdown in transformers, cables, and connectors. The breakdown of a solid insulating material usually results in localized carbonization. This may be catastrophic, or it may simply result in decreased dielectric strength of the material at the arc-over point. The occurrence of additional transients often causes a breakthrough at the weakened point in the insulating material, eventually resulting in catastrophic failure of the insulation. Printed wiring board arcing may result in similar failure modes. A breakdown induced by high voltages along the surface of a PWB can create a conductive path of carbonized insulation and vaporized metal from the wiring traces or component leads.

The greatest damage to equipment from insulation breakdown generally occurs after the spike has passed. The follow-on steady-state current that can flow through fault paths created by a transient often cause the majority of component damage and system failure.

3.3   ELECTROSTATIC DISCHARGE

A static charge is the result of an excess or a deficiency of electrons on a given surface. The relative level of electron imbalance determines the static charge.

Simply stated, a charge is generated by physical contact between, and then separation of, two materials. One surface loses electrons to the other. The types of materials involved and the speed and duration of motion between the materials determine the charge level. Common nonconductive plastic packaging materials, such as polyethylene, polystyrene, or mylar films, are prone to such electron imbalance. The normal movement of people in a work environment easily can generate static levels of 25 kV or higher. Even the handling of a seemingly innocuous plastic-foam coffee cup at a workstation can generate a damaging static charge.

Electrostatic energy is a stationary charge phenomenon that can build up in either a nonconductive material or in an ungrounded conductive material. The charge may occur in one of two ways:

1. *Polarization:* Charge buildup when a conductive material is exposed to a magnetic field.
2. *Triboelectric effects:* Charge buildup that occurs when two surfaces contact and then separate, leaving one positively charged and one negatively charged. Friction between two materials increases the triboelectric charge by increasing the surface area that experiences contact. For example, a person accumulates charge by walking across a nylon carpet; discharge occurs when the person touches a conductive surface. Electrostatic discharge resulting from triboelectric effects is the primary area of concern to maintenance engineers.

It is estimated that the value of semiconductor devices lost to ESD could run as high as $1 billion per year. Most ESD damage is preventable; it is simply the result of improper device-handling.

3.3.1   Triboelectric Effects

Different materials have differing potentials for charge. Nylon, human and animal hair, wool, and asbestos have high positive triboelectric potential. Silicon, polyurethane, rayon, and polyester have negative triboelectric potentials. Cotton, steel, paper, and wood all tend to be relatively neutral materials. The intensity of the triboelectric charge is inversely proportional to the relative humidity. As humidity increases, ESD problems decrease. For example, a person walking across a carpet can generate a 1.5 kV charge at 90 percent RH, but will generate as much as 35 kV at 10 percent RH.

When a charged object comes in contact with another object, the electrostatic charge will attempt to find a path to ground, discharging into the contacted object. The current level is very low (typically less than 0.1 nA), but the voltage can be high (25 to 50 kV).

ESD also can collect on metallic furnishings, such as chairs and equipment racks. Sharp corners and edges, however, encourage a corona that tends to bleed the charge off such objects. The maximum voltage normally expected for furniture-related ESD is about 6 to 8 kV. Because metallic furniture is much more conductive than humans, however, furniture-related ESD generally will result in higher peak discharge currents.

Low-power semiconductors are particularly vulnerable to damage from ESD discharges. MOS devices tend to be more vulnerable than other components. The gate of a MOS transistor is especially sensitive to electrical overstress. Application of excessive voltage may exceed the dielectric standoff voltage of the chip structure and punch through the oxide, forming a permanent path from the gate to the semiconductor below. A pulse of 25 kV usually is sufficient to rupture the gate oxide. The scaling of device geometry that occurs with LSI and VLSI components complicates this problem. The degree of damage caused by electrostatic discharge is a function of the following parameters:

- Size of the charge, determined by the capacitance of the charged object.
- Rate at which the charge is dissipated, determined by the resistance into which it is discharged. This relationship is illustrated in Figure 3.4 (a double exponential decay pulse).

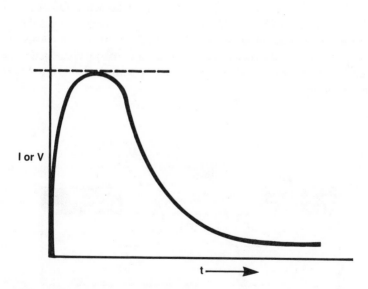

Figure 3.4 Discharge waveform for an ESD event. (Data from: Technical staff, Military/Aerospace Products Division, *The Reliability Handbook,* National Semiconductor, Santa Clara, Calif., 1987)

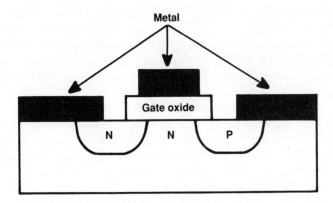

Figure 3.5   Construction of a metal gate NMOS transistor. (Data from: Technical staff, Military/Aerospace Products Division, *The Reliability Handbook*, National Semiconductor, Santa Clara, Calif., 1987)

3.3.2   ESD Failure Mechanisms

Destructive voltages and/or currents from an ESD event may result in device failure because of thermal fatigue and/or dielectric layer breakdown. MOS transistors are normally constructed with an oxide between the gate conductor and the source-drain channel region, as illustrated in Figure 3.5 for a metal gate NMOS device, and in Figure 3.6 for a silicon gate device. Bipolar transistor construction, shown in Figure 3.7, is less susceptible to ESD damage because the oxide is used only for surface insulation.

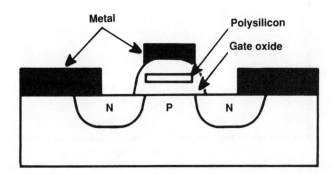

Figure 3.6   Construction of a silicon gate NMOS transistor. (Data from: Technical staff, Military/Aerospace Products Division, *The Reliability Handbook*, National Semiconductor, Santa Clara, Calif., 1987)

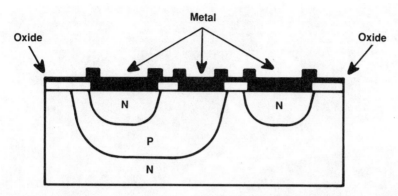

Figure 3.7 Construction of a bipolar transistor. (Data from: Technical staff, Military/Aerospace Products Division, *The Reliability Handbook*, National Semiconductor, Santa Clara, Calif., 1987)

Oxide thickness is the primary factor in MOS ruggedness. A thin oxide is more susceptible to electrostatic punch-through, which results in a permanent low-resistance short circuit through the oxide. Where pinholes or other weaknesses exist in the oxide, damage is possible at a lower charge level. Semiconductor manufacturers have reduced oxide thickness as they have reduced device size. This trend has resulted in a significant increase in sensitivity to ESD damage.

Detecting an ESD failure in a complex device may present a significant challenge for quality control engineers. For example, erasable programmable read-only memory (EPROM) chips use oxide layers as thin as 100 angstroms, making them susceptible to single-cell defects that can remain undetected until the damaged cell itself is addressed. An electrostatic charge, which may not be large enough to cause oxide breakdown, still can cause lattice damage in the oxide, lowering the ability to withstand subsequent ESD exposure. A weakened lattice will have a lower breakdown threshold voltage.

Latent Failures. Immediate failure resulting from ESD exposure is easy to determine: the device no longer works. A failed component may be removed from the subassembly in which it is installed, representing no further reliability risk to the system. Not all devices exposed to ESD, however, fail immediately. Unfortunately, there is little data dealing with the long-term reliability of devices that have survived ESD exposure. Some experts think that two to five devices are degraded for every one that fails. Available data indicates that latent failures can occur in both bipolar and MOS chips, and that there is no direct relationship between the susceptibility of a device to catastrophic failure and its susceptibility to latent failure. Damage can manifest itself in one of two primary mechanisms:

Figure 3.8   A scanning electron microscope photo illustrating ESD damage to the metallization of a MOSFET device.

- Shortened lifetime (a possible cause of many infant mortality failures seen during burn-in).
- Electrical performance shifts, many of which may cause the device to fail electrical limit tests.

Case in Point. Figure 3.8 shows an electron microscope photo of a chip that failed because of an overvoltage condition. An ESD to this MOSFET damaged one of the metallization connection points of the device, resulting in catastrophic failure. Note the spot where the damage occurred. The objects in the photo that look like bent nails actually are gold lead wires with a diameter of 1 mil. By contrast, the diameter of a typical human hair is about 3 mil. The photo was shot at 200X magnification. Figure 3.9 presents another view of the MOSFET damage point, but at 5000X. The character of the damage can be observed. Some of the aluminum metallization has melted and can be seen along the bottom edge of the hole.

Figure 3.9   The device shown in Figure 3.8, at 5000X magnification. The character of the damage can be observed.

3.3.3   ESD Protection

With the push for faster and more complex ICs, it is unlikely that semiconductor manufacturers will return to thicker oxide layers or larger junctions. ESD protection will come, instead, from circuitry built into individual chips to shunt transient energy to ground.

Most MOS circuits incorporate protective networks. These circuits can be made to be quite efficient, but there is a tradeoff between the amount of protection provided and the speed and packing density of the device. Protective elements, usually diodes, must be physically large if they are to clamp adequately. Such elements take up a significant amount of chip space. The *RC* time constants of protective circuits also can place limits on switching speeds.

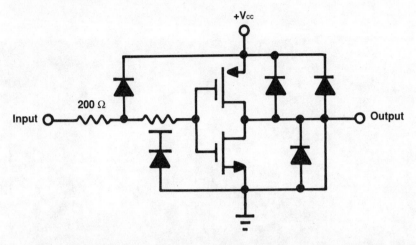

Figure 3.10 CMOS transistor with built-in ESD protection circuitry. (Data from: Technical staff, Military/Aerospace Products Division, *The Reliability Handbook,* National Semiconductor, Santa Clara, Calif., 1987)

Protective networks for NMOS devices typically use MOS transistors, rather than diodes, as shunting elements. Although diodes are more effective, fewer diffusions are available in the NMOS process. Consequently, not as many forward-biased diodes can be constructed. Off-chip protective measures, including electromagnetic shielding, filters, and discrete diode clamping, seldom are used because they are bulky and expensive.

Figure 3.10 shows the protective circuitry used in a 54HC high-speed CMOS device. Polysilicon resistors are placed in series with each input pin, and relatively large-geometry diodes are added as clamps on the IC side of the resistors. Clamping diodes also are used at the output. The diodes restrict the magnitude of the voltages that can reach the internal circuitry. Protective features such as these have allowed CMOS devices to withstand ESD test voltages in excess of 2 kV.

3.4 PREVENTING ESD FAILURES

Imagine this scenario: A maintenance technician, wearing synthetic clothing, walks across a carpeted floor and sits down at an uncovered work bench littered with used plastic-foam coffee cups, some empty cardboard boxes, and an equipment instruction manual. Then our technician touches the CPU board of a personal computer being serviced. In the old days of electronic servicing, this chain of events would have posed no particular problem. Today, however, such a routine invites disaster.

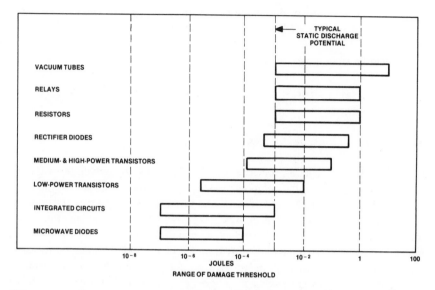

Figure 3.11 Estimation of the vulnerability of various components to ESD-induced damage. (Source: *Broadcast Engineering* magazine, Intertec Publishing, Overland Park, Kan.)

ESD poses an invisible, but very real, threat to electronic equipment of all types. In servicing a minor problem, the technician may inadvertently create new, major problems: fried ICs, short-circuited diodes, and other, more sinister component-degradation troubles that can plague the equipment long after servicing is complete. New technology and radically downsized microprocessor chips require new servicing methods. Static control has become a major concern for equipment manufacturers and facility technical managers. Figure 3.11 graphically illustrates the susceptibility of various components to ESD damage.

3.4.1 Technician Training

Although the need for static-control measures has been known to the electronics industry for more than a decade, many people in the sequence of equipment manufacturing, assembly, transport, installation, and service have a limited understanding of the problem and its solutions. Add the influx of new and untrained people joining the work force each year, and it is clear that training cannot be considered simply a onetime effort. It must be an ongoing process.

A high level of awareness by maintenance engineers is unquestionably one of the most critical elements in the static-control process. Unless technicians have a thorough understanding of the problem, it is unlikely that they can avoid making costly static-control errors. There are two basic rules for static-control:

1. Handle all static-sensitive components only at a static-safeguarded workstation.
2. Transport all static-sensitive components only in static-shielded containers or packages.

3.4.2   Static-Control Process

Electronic devices require protection from static discharge at each phase of the product's lifetime, from the component-assembly level at the factory to the point at which the product is taken out of service and thrown away. There can be no weak links or lapses in the process. A chip begins life as a silicon substrate onto which multiple layers of oxides and conductive materials are screened in a precise microscopic process. To guard against static problems, chip manufacturing takes place in a clean-room environment with grounded, conductive work surfaces. Individual chips then are packaged in plastic or ceramic housings, with leads bonded to pins or pads on the housing, again under conditions that protect against static buildup. Completed ICs generally are stored in conductive plastic sleeves, commonly referred to as *DIP tubes*. Individual pins also may be shorted together with conductive foam for protection when the devices are not in DIP tubes.

Next, the static-sensitive device is inserted into a socket or mounted on a circuit board and soldered into place. Whether manual or automated, this process requires continuing care to avoid static damage. Conductive floor and benchtop mats, grounded wrist straps, grounding shoe straps, conductive foam, static-dissipative bags, and conductive plastic carriers protect boards during the construction phase. Once a system has been assembled and is packaged for storage and shipment, sensitive circuits enjoy reasonably good static protection. However, certain precautions, such as the use of static-shielding cushioning materials, still may be appropriate.

Installation of new equipment demands continuing vigilance insofar as static suppression is concerned. A grounded wrist strap should be worn by the technician during setup or adjustment. Precautions during day-to-day use of the equipment by operators may require other static-control products, such as grounded floor mats or conductive, self-adhesive keyboard strips, depending on the environment and the system.

No piece of equipment will run forever without repair, and it is up to maintenance engineers to preserve the ESD chain of protection. Remove boards from the card cage only with the proper ESD protection in place, including a grounding wrist strap and floor mat. Whenever a board is transported within the facility or shop area, place it in a static-shielding bag. Boards to be sent back to the manufacturer or to a module repair company for rework should be placed in static-shielding bags and shipped in conductive transport cases.

Periodically test each product used for static control, to confirm that it is functioning properly. Wrist straps, for example, may no longer be able to provide good electrical interface between the conductive surface and the skin of the wearer. The integral current-limiting resistor or snap connector also may malfunction after repeated use. The same precautions apply to portable field-service kits, static-protection bags, and other conductive items. Check these components periodically with a megohmmeter. Be alert for loose grounding lugs, cracked conductive containers, and torn conductive bags.

3.4.3   The ESD Tool Kit

Having the right tools is a key to efficient maintenance. Tool costs usually are small, compared with time and safety considerations. Consider obtaining the following ESD products:

- Static-dissipative work surface and table mat
- Conductive wrist strap
- Metallized static-shielding bags
- Static neutralizer

Static-Dissipative Work Surface/Table Mat. Conductive and static-dissipative floor and table mats help to discharge energy from charged, conductive objects that are brought into contact with the mat. Materials exhibiting high-resistivity (*static-dissipative*) characteristics are preferred to the highly conductive, carbon-loaded film used in older mats. New static-dissipative materials allow a powered-up PWB to be placed on the mat without the danger of short-circuiting. They also offer other desirable features, such as flame retardancy and cold-weather flexibility.

Conductive Wrist Strap. Wrist straps, conductive footwear, and static-dissipative lab coats aid in discharging static energy from maintenance personnel. The wrist strap can be connected to a continuous monitor to alert the technician to wrist-strap or ground-lead failures and improper use. Wrist bands and grounding coats, like conductive mats, have improved dramatically over the past few years. The newest kits offer common-point grounding systems, which connect the technician's wrist strap and the work surface to the same ground point. New wrist bands also feature a woven fabric conductor; conductive fibers are woven into the band's fabric on the inside, nearest the skin, leaving only nonconductive fabric on the outside of the band. This prevents an electrical safety hazard.

Metallized Static-Shielding Bags. Static-shielding bags and conductive totes serve to protect devices and PWBs from ESD damage during transport. Transparent static-shielding bags are commonly used for protective packaging and transport of static-sensitive components. This same technology also is available in single-track, recloseable bags with built-in cushioning for added physical protection.

Static Neutralizer. An ionized air blower may be used to dissipate static charges from nonconductive objects. The air ionizer provides equal quantities of positive and negative ions in a heated airstream. Ions are attracted from the airstream to a charged object until the charge is neutralized.

3.4.4   Safety

The safety of maintenance and operating personnel always must be the top priority for any maintenance department. Carry this concern to ESD protective devices as well. The most obvious consideration involves wrist straps. Because the concept of a wrist strap is based on connecting the wearer to ground, the potential for contact with line voltages (or other high potentials) must be considered. To protect the wearer, current-limiting resistors are placed in the ground cord of the wrist strap. A value of 1 MΩ generally is used, chosen through calculation and experimentation. In case the wearer accidentally comes into contact with a hot 120 V ac circuit, a 1-MΩ resistor will limit the current experienced by the wearer to a maximum of 0.1 mA, which is below the threshold of sensation for most people. At the same time, a resistance of 1 MΩ added to the ground path will allow almost instantaneous static-charge drainage.

3.5   TRANSIENT-GENERATED NOISE

Problems can be caused in a facility by ESD and transient overvoltages not only by device failure, but also by logic-state upsets. Studies have shown that an upset in the logic of typical digital circuitry can occur with transient energy levels as low as 1×10^{-9} J. Such logic-state upsets can result in microcomputer latchup, lost or incorrect data, program errors, and control-system shutdown.

3.5.1   Noise Sources

Utility company transients, lightning, and ESD are the primary sources of random, unpredictable noise. These sources represent the greatest threat to equipment reliability. Other, more predictable sources, such as switch contact arcing and SCR switching, present problems for equipment users as well.

ESD Noise. Depending on the voltage level, relative humidity, the speed of approach, and the shape of the charged object, the upper frequency limit for an ESD discharge may exceed 1 GHz. At such frequencies, circuit-board traces and cables function as fairly efficient receiving antennas.

Electrical noise associated with ESD may enter electronic equipment by either conduction or radiation. In the near field of an ESD (within a few tens of centimeters) the primary type of radiated coupling may be either inductive or capacitive, depending on the impedances of the ESD source and the receiver. In the far field, electromagnetic field coupling predominates. Circuit operation is upset if the ESD-induced voltages and/or currents exceed the typical signal levels in the system.

Coupling of an ESD voltage in the near field is determined by the impedance of the circuit:

- *High-impedance circuit:* Capacitive coupling will dominate; ESD-induced voltages will be the major problem.
- *Low-impedance circuit:* Inductive coupling will dominate; ESD-induced currents will be the major problem.

Contact Arcing. Switch contact arcing and similar repetitive transient-generating operations can induce significant broadband noise into an electrical system. Noise generated in this fashion is best controlled at its source, almost always an inductive load. Switch contact arcing generates an effect known as *showering,* a low-current, high-voltage series of brief discharges of a damped oscillatory nature (frequencies of 1 MHz or more are common). Figure 3.12 illustrates the mechanism involved. This shower of noise can travel through power lines and create problems for microcomputer equipment either through direct injection into the system's power supply, or through coupling from adjacent cables or PWB traces.

SCR Switching. SCR power controllers are a potential source of noise-induced microcomputer problems. Each time an SCR is triggered into its active state in a resistive circuit, the load current goes from zero to the load-limited current value in less than a few microseconds. This step action generates a broadband spectrum of energy, with an amplitude inversely proportional to frequency. Electronic equipment using full-wave SCR control in a 60-Hz ac circuit may experience such noise bursts 120 times a second.

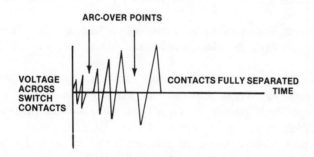

Figure 3.12   Noise-generation resulting from the separation of two contacts carrying current. (Source: *Broadcast Engineering* magazine, Intertec Publishing, Overland Park, Kan.)

In an industrial environment, where various control systems may be spaced closely, electrical noise can cause latchup problems or incorrect data in microcomputer equipment, or interaction between SCR control units in machine controllers. Power-line cables within a facility can couple noise from one area of a plant to another, further complicating the problem. To solve the SCR noise problem, look at both the source of the interference and the susceptible hardware. The use of good transient-suppression techniques in the application of SCR power controllers will eliminate noise generation in all but the most critical of systems.

As a further measure of insurance, sensitive electronic equipment should be shielded adequately against noise pickup, including metal cabinet shields, ac power-line filters, and input/output line feedthrough RF filters. Fortunately, most professional/industrial equipment is designed with RF shielding as a concern.

3.6 BIBLIOGRAPHY

Antinone, Robert J.: "How to Prevent Circuit Zapping," *IEEE Spectrum*, IEEE, New York, April 1987.

Boxleitner, Warren: "How to Defeat Electrostatic Discharge," *IEEE Spectrum*, IEEE, New York, August 1989.

Frank, Donald: "Please Keep Your EMC Out of My ESD," *Proceedings of the IEEE Reliability and Maintainability Symposium*, New York, IEEE, 1986.

Gleeson, Dixon: "Portable Field Service Kits," (parts 1, 2 and 3), *Microservice Management* magazine, Intertec Publishing, Overland Park, Kan., October, November and December 1986.

Gloer, H. Niles: "Voltage Transients and the Semiconductor," *The Electronic Field Engineer*, Vol. 2, 1979.

Jordan, Edward C.: *Reference Data for Engineers: Radio, Electronics, Computer, and Communications*, 7th ed., Howard W. Sams, Indianapolis, Ind., 1985.

Kanarek, Jess: "Protecting Against Static Electricity Damage," *Electronic Servicing & Technology* magazine, Intertec Publishing, Overland Park, Kan., March 1986.

Nenoff, Lucas: "Effect of EMP Hardening on System R&M Parameters," *Proceedings of the IEEE Reliability and Maintainability Symposium*, IEEE, New York, 1986.

Voss, Mike: "The Basics of Static Control," *Electronic Servicing & Technology* magazine, Intertec Publishing, Overland Park, Kan., July 1988.

Technical staff: *MOV Varistor Data and Applications Manual*, General Electric Company, Auburn, N.Y.

Technical staff, Military/Aerospace Products Division: *The Reliability Handbook*, National Semiconductor, Santa Clara, Calif., 1987.

4

POWER-SUPPLY
COMPONENT RELIABILITY

4.1 INTRODUCTION

The circuit elements most vulnerable to failure in any given piece of electronic hardware are those exposed to the outside world. In most systems, the greatest threat generally involves the ac-to-dc power supply. The power supply is subject to high-energy surges from lightning and other sources. For this reason, a knowledge of typical failure modes involving power-supply components is important to maintenance efforts.

4.1.1 Component Derating

Power-supply systems often are exposed to extreme voltage and environmental stresses. *Derating* of individual components is a key factor in improving supply reliability. The goal of derating, as discussed in Chapter 2, is the reduction of electrical, mechanical, thermal, and other environmental stresses on a component to decrease the degradation rate, and to prolong expected life. Through derating, the margin of safety between the operating stress level and the permissible stress level for a given part is increased. This adjustment provides added protection from system overstress, unforeseen during design.

Experience has demonstrated that types of components tend to fail in predictable ways. Table 4.1 shows the statistical distribution of failures for a transient-suppression (*electromagnetic pulse*) protection circuit. Although the data presented applies only to a specific product, some basic conclusions can be drawn:

- The typical failure mode for a capacitor is a short circuit.
- The typical failure mode for a zener diode is a short circuit.
- The typical failure mode for a connector pin is an open circuit.
- The typical failure mode for a solder joint is an open circuit.

Table 4.1 Statistical Distribution of Component Failures in an EMP Protection Circuit. Data from: Lucas Nenoff, "Effect of EMP Hardening on System R&M Parameters," *Proceedings of the 1986 Reliability and Maintainability Symposium*, IEEE, New York, 1986.

COMPONENT	MODE OF FAILURE	DISTRIBUTION
CAPACITOR (ALL TYPES)	OPEN SHORT	.01 .99
COIL	OPEN SHORT	.75 .25
DIODE (ZENER)	OPEN SHORT	.01 .99
GE-MOV	OPEN SHORT	.01 .99
TRANSZORB	OPEN SHORT	.01 .99
CONNECTOR PIN	OPEN SHORT TO GND	.99 .01
SOLDER JOINT	OPEN	1.00
LUG CONNECTION	OPEN	1.00
SURGE PROTECTOR	OPEN SHORT	.99 .01

These conclusions present no great surprises, but they point out the predictability of equipment failure modes. The first step in solving these problems is knowing what is likely to fail and what the typical failure mode will be.

4.2 RELIABILITY OF POWER RECTIFIERS

High-voltage and high-current devices present special challenges to the science of reliability engineering. Although such components are rugged because of their physical size, they usually operate in an environment that is characterized by high temperatures and exposure to transient disturbances. Power rectifiers are the foundation on which numerous products are built. An understanding of rectifier device ratings is essential to proper application and maintenance.

4.2.1 Silicon Rectifiers

Virtually all power supplies use silicon rectifiers as the primary ac-to-dc converting device. Rectifier parameters generally are expressed in terms of reverse-voltage ratings and mean-forward-current ratings in a 1/2-wave rectifier circuit operating from a 60 Hz supply and feeding a purely resistive load. The three primary reverse-voltage ratings are:

- Peak transient reverse voltage (V_{RM}): The maximum value of any nonrecurrent surge voltage. This value must never be exceeded, even for a microsecond.
- Maximum repetitive reverse voltage [$V_{RM(rep)}$]: The maximum value of reverse voltage that may be applied recurrently (in every cycle of 60 Hz power). This includes oscillatory voltages that may appear on the sinusoidal supply.
- Working peak reverse voltage [$V_{RM(wkg)}$]: The crest value of the sinusoidal voltage of the ac supply at its maximum limit. Rectifier manufacturers generally recommend a value that has a significant safety margin, relative to the peak transient reverse voltage (V_{RM}), to allow for transient overvoltages on the supply lines.

There are three forward-current ratings of similar importance in selecting rectifiers for a particular application:

- Nonrecurrent surge current [$I_{FM(surge)}$]: The maximum device transient current that must not be exceeded at any time. $I_{FM(surge)}$ is sometimes given as a single value, but often is presented in the form of a graph of permissible surge-current values vs. time. Because silicon diodes have a relatively small thermal mass, the potential for short-term current overloads must be given careful consideration.
- Repetitive peak forward current [$I_{FM(rep)}$]: The maximum value of forward current reached in each cycle of the 60 Hz waveform. This value does not include random peaks caused by transient disturbances.
- Average forward current [$I_{FM(av)}$]: The upper limit for average load current through the device. This limit is always well below the repetitive peak forward-current rating to ensure an adequate margin of safety.

Rectifier manufacturers generally supply curves of the instantaneous forward voltage vs. instantaneous forward current at one or more specific operating temperatures. These curves establish the forward-mode upper operating parameters of the device.

Heat Sink. Because silicon junctions have a low thermal mass, much lower than their associated heat sinks, the sink assembly is effective only in dissipating heat generated by steady-state operating current. The heat sink will have little effect in preventing a catastrophic failure caused by a large transient overcurrent pulse. It can be shown, in fact, that in devices with a small junction mass, the temperature of the junction will follow the cyclic variations of current at the 60 Hz power-supply frequency. This factor is, of course, taken into account by the semiconductor manufacturer in determining typical operating parameters.

4.2.2 Operating Rectifiers in Series

High-voltage power supplies (5 kV and greater) often require rectifier voltage ratings well beyond those typically available from the semiconductor industry for discrete components. To meet the requirements of the application, manufacturers commonly use silicon diodes in a series configuration to give the required working peak reverse voltage. For such a configuration to work properly, the voltage across any one diode must not exceed the rated peak transient reverse voltage (V_{RM}) at any time. The dissimilarity commonly found between the reverse leakage current characteristics of different diodes of the same type number makes this objective difficult to achieve. The problem normally is overcome by connecting shunt resistors across each rectifier in the chain, as shown in Figure 4.1. The resistors are chosen so that the current through the shunt elements (when the diodes are reverse-biased) will be several times greater than the leakage current of the diodes themselves.

The *carrier storage* effect also must be considered in the use of a series-connected rectifier stack. If precautions are not taken, different diode recovery times (caused by the carrier storage phenomenon) will effectively force the full applied reverse voltage across a small number of diodes, or even a single diode. This problem can be prevented by connecting small-value capacitors across each diode in the rectifier stack. The capacitors equalize the transient reverse voltages during the carrier storage recovery periods of the individual diodes.

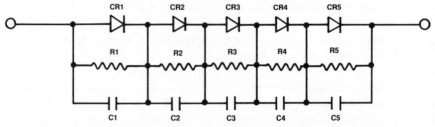

Figure 4.1 A portion of a high-voltage, series-connected rectifier stack. (Source: *Broadcast Engineering* magazine, Intertec Publishing, Overland Park, Kan.)

4.2.3   Operating Rectifiers in Parallel

Silicon rectifiers are used in a parallel configuration when a large amount of current is required from the power supply. Parallel assemblies normally are found in low-voltage, high-current supplies. *Current sharing* is the major design problem with a parallel rectifier assembly because diodes of the same type number do not necessarily exhibit the same forward characteristics.

Semiconductor manufacturers often divide production runs of rectifiers into tolerance groups, matching forward characteristics of the various devices. When parallel diodes are used, devices from the same tolerance group must be selected to avoid unequal sharing of the load current. As a margin of safety, designers allow a substantial derating factor for devices in a parallel assembly to ensure that the maximum operating limits of any one component are not exceeded.

Parallel rectifier assemblies are constructed to provide for equal heat dissipation of all devices. Ideally, the rectifiers are placed on the same heat sink to facilitate heat transfer among the individual components. The device layout should be structured so that the individual diodes are arranged symmetrically about the center of the assembly, preventing a current imbalance caused by unequal external elements.

The problems inherent in a parallel rectifier assembly can be reduced through the use of a resistance or reactance in series with each component, as shown in Figure 4.2. The buildout resistances (R1 through R4) force the diodes to share the load current equally. Such assemblies can, however, be difficult to construct and may be more expensive than simply adding diodes or going to higher-rated components.

4.2.4   Thyristor Devices

The term *thyristor* identifies a general class of solid-state silicon controlled rectifiers (SCRs). These devices are similar to normal rectifiers, but are designed to remain in a blocking state (in the forward direction) until a small signal is applied to a control electrode (the gate). After application of the control pulse, the device conducts in the forward direction and exhibits characteristics similar to a common silicon rectifier. Conduction continues after the control signal has been removed and until the current through the device drops below a predetermined threshold or until the applied voltage reverses polarity.

The voltage and current ratings for thyristors are similar to the parameters used to classify standard silicon rectifiers. Some of the primary device parameters include:

- Peak forward blocking voltage: The maximum safe value that may be applied to the thyristor while it is in a blocking state.

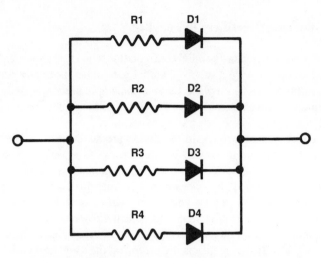

Figure 4.2   Using buildout resistances to force current-sharing in a parallel rectifier assembly. (Source: *Broadcast Engineering* magazine, Intertec Publishing, Overland Park, Kan.)

- Holding current: The minimum anode-to-cathode current that will keep the thyristor conducting after it has been switched on by the application of a gate pulse.
- Forward voltage drop: The voltage loss across the anode-to-cathode current path for a specified load current. Because the ratio of rms-to-average forward current varies with the angle of conduction, power dissipation for any average current also varies with the device angle of conduction. The interaction of forward voltage drop, phase angle, and device-case temperature generally are specified in the form of one or more graphs or charts.
- Gate trigger sensitivity: The minimum voltage and/or current that must be applied to the gate to trigger a specific type of thyristor into conduction. This value must take into consideration variations in production runs and operating temperature. The minimum trigger voltage normally is not temperature-sensitive, but the minimum trigger current can vary considerably with thyristor-case temperature.
- Turn-on time: The length of time required for a thyristor to change from a nonconducting state to a conducting state. When a gate signal is applied to the thyristor, anode-to-cathode current begins to flow after a finite delay. A second switching interval occurs between the point at which current begins to flow and the point at which full anode current (determined by the instantaneous applied voltage and the load) is reached. The sum of these two times is the turn-on time.

• Turn-off time: The length of time required for a thyristor to change from a conducting state to a nonconducting state. The turn-off time comprises two individual periods—the *storage time (similar to the storage interval of a saturated transistor) and the recovery time*. If forward voltage is reapplied before the entire turn-off time has elapsed, the thyristor will conduct again.

4.3 TRANSFORMER FAILURE MODES

Transformer failures are all too familiar to many maintenance technicians. The failure of a power transformer is almost always a catastrophic event that causes the system to fail and leaves the technician with a messy cleanup job. The two primary enemies of power transformers are heat and transient overvoltages.

Power input to a transformer is not all delivered to the secondary load. Some is expended as copper losses in the primary and secondary windings. These $I^2 R$ losses are practically independent of voltage; the controlling factor is current flow. To keep the losses as small as possible, the coils of a power transformer are wound with wire of the largest cross section that space will permit.

A practical transformer also will experience core-related losses (also known as *iron losses*). Repeated magnetizing and demagnetizing of the core (which occurs in an ac waveform) results in power loss because of the repeated realignment of the magnetic domains. This factor, known as *hysteresis loss*, is proportional to frequency and flux density. Silicon steel alloy is used for the magnetic circuit to minimize hysteresis loss. The changing magnetic flux also induces circulating currents (*eddy currents*) in the core material. Eddy-current loss is proportional to the square of the frequency, and the square of the flux density. To minimize eddy currents, the core is constructed of laminations or layers of steel that are clamped or bonded together to form a single magnetic mass.

4.3.1 Thermal Considerations

Temperature rise inside a transformer is the result of power losses in the windings and the core. The insulation within and between the windings tends to blanket these heat sources and prevent efficient dissipation of the waste energy, as illustrated in Figure 4.3. Each successive layer of windings (shown as *A, B* and *C* in the figure) acts to prevent heat transfer from the hot core to the local environment (air).

The hot spot shown in the figure can be dangerously high, even though the outside transformer case and winding are relatively cool to the touch. Temperature rise is the primary limiting factor in determining the power-handling capability of a transformer. To ensure reliable operation, a large margin of safety must be designed into a transformer. Design criteria includes winding-wire size, insulation material, and core size.

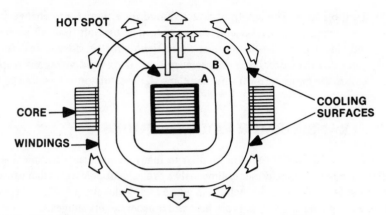

Figure 4.3 The dynamic forces of heat generation in a power transformer.

4.3.2 Voltage Considerations

Transformer failures resulting from transient overvoltages typically occur between layers of windings within a transformer. (See Figure 4.4.) At the end of each layer where the wire rises from one layer to the next, zero potential voltage exists. However, as the windings move toward the opposite end of the coil in a typical layer-wound device, there exists a potential difference of up to twice the voltage across one complete layer. The greatest potential difference is, therefore, found at the far opposite end of the layers.

This voltage distribution applies to continuous 60 Hz signals. When the transformer is switched on or when a transient overvoltage is impressed upon the device, the voltage distribution from one "hot layer" to the next can increase dramatically, raising the possibility of arc-over. This effect is caused by the inductive nature of the transformer windings and the inherent distributed capacitance of the coil. Insulation breakdown may result from one or more of the following:

- Puncture through the insulating material of the device.
- Tracking across the surface of the windings.
- Flashing through the air.

Any of these modes may result in catastrophic failure.

4.3.3 Mechanical Considerations

Current flow through the windings of a transformer applies stress to the coils. The individual turns in any one coil tend to be crushed together when current flows

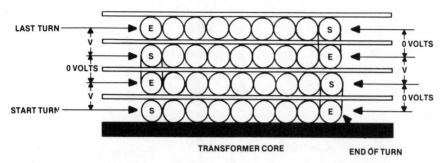

Figure 4.4 Voltage distribution between the layers of a typical layer-wound power transformer.

through them. There also may be large repulsion forces between the primary and secondary windings. These mechanical forces are proportional to the square of the instantaneous current; therefore, they are vibratory in nature under normal operating conditions. These forces, if not controlled, can result in failure of the transformer through insulation breakdown. Vibration over a sufficient period of time can wear the insulation off adjacent conductors and create a short circuit. To prevent this failure mode, power transformers routinely are coated or dipped into an insulating varnish to solidify the windings and the core into one mass.

4.4 CAPACITOR FAILURE MODES

Experience has shown that, of components prone to malfunction in electronic equipment, capacitors are second only to semiconductors and vacuum tubes. Of all the various types of capacitors used today, it is estimated that electrolytics present equipment users with the greatest potential for problems.

4.4.1 Electrolytic Capacitors

Electrolytic capacitors are popular because they offer a large amount of capacitance in a small physical size. They are used widely as filters in low-voltage power supplies and as coupling devices in audio and RF stages. An aluminum electrolytic capacitor consists of two aluminum foil plates separated by a porous strip of paper (or other material) soaked with a conductive electrolyte solution. Construction of a typical device is illustrated in Figure 4.5. The separating material between the capacitor plates does not form the dielectric, but instead serves as a spacer to prevent the plates from mechanically short-circuiting. The dielectric consists of a thin layer of aluminum oxide that is electrochemically formed on the positive foil plate. The electrolyte conducts the charge applied to the capacitor from the negative

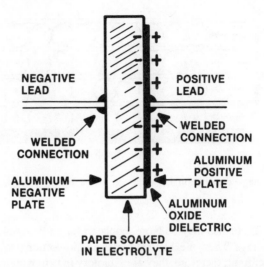

Figure 4.5   The basic design of an aluminum electrolytic capacitor. (Data from: Technical staff, *Sencore News*, issue 122, Sencore Corporation, Sioux Falls, S.D.)

plate, through the paper spacer and into direct contact with the dielectric. This sandwich arrangement of foil-spacer-foil is then rolled up and encapsulated.

Problems with electrolytic capacitors fall into two basic categories: mechanical failure and failure of electrolyte.

4.4.2   Mechanical Failure

Mechanical failures relate to poor bonding of the leads to the outside world, contamination during manufacture, and shock-induced short-circuiting of the aluminum foil plates. Typical failure modes include short circuits caused by foil impurities, manufacturing defects (such as burrs on the foil edges or tab connections), breaks or tears in the foil, and breaks or tears in the separator paper.

Short circuits are the most frequent failure mode during the useful-life period of an electrolytic capacitor. (For the purposes of this discussion, "useful life" is defined as the broad, flat portion of the bathtub curve.) Such failures are the result of random breakdown of the dielectric oxide film under normal stress. Proper capacitor design and processing will minimize such failures. Short circuits also can be caused by excessive stress, where voltage, temperature, or ripple conditions exceed specified maximum levels.

Open circuits, although infrequent during normal life, can be caused by failure of the internal connections joining the capacitor terminals to the aluminum foil. Mechanical connections can develop an oxide film at the contact interface, increasing contact resistance and eventually producing an open circuit. Defective weld connections also can cause open circuits. Excessive mechanical stress will accelerate weld-related failures.

Temperature Cycling. Like semiconductor components, capacitors are subject to failures induced by thermal cycling. Experience has shown that thermal stress is a major contributor to failure in aluminum electrolytic capacitors. Dimensional changes between plastic and metal materials may result in microscopic ruptures at termination joints, possible electrode oxidation, and unstable device termination (changing series resistance). The highest-quality capacitor will fail if its voltage and/or current ratings are exceeded. Appreciable heat rise (20°C during a 2-hour period of applied sinusoidal voltage) is considered abnormal, and may be a sign of incorrect application of the component or impending failure of the device.

Figure 4.6 illustrates the effects of high ambient temperature on capacitor life. Note that operation at 33 percent duty cycle is rated at 10 years when the ambient temperature is 35°C, but the life expectancy drops to just 4 years when the same device is operated at 55°C. A common rule of thumb states that, in the range of

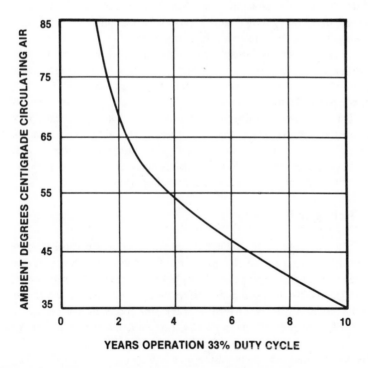

Figure 4.6   Life expectancy of an electrolytic capacitor as a function of operating temperature. (Source: *Broadcast Engineering* magazine, Intertec Publishing, Overland Park, Kan.)

Table 4.2   Safe and Unsafe Cleaning Solvents that May Be Used with Aluminum Electrolytic Capacitors.

SAFE (ACCEPTABLE) CLEANING SOLVENTS

ACETONE	METHYL ALCOHOL
TOLUOL-HEXANE-ETHYL-ACETATE	MINERAL SPIRITS
BUTYL ALCOHOL	PROPYL ALCOHOL
ETHYL ALCOHOL	XYLENE
LACQUER THINNER	

UNSAFE (UNACCEPTABLE) CLEANING SOLVENTS

CARBON TETRACHLORIDE	METHYLENE CHLORIDE
CHLOROFORM	PERCHLORTHYLENE
CHLOROTHENE	TRICHLOROETHANE
FREON	TRICHLOROETHYLENE

+75°C through the full-rated temperature, stress and failure rate double for each 10°C increase in operating temperature. Conversely, the failure rate is reduced by half for every 10°C decrease in operating temperature.

Excessive temperature during soldering also may result in device failure. Capacitors with leads smaller than 20 AWG may be damaged if the solder temperature exceeds 260°C. Exposure time should be limited to 10 s or less.

Chemical-Related Failure. Aluminum electrolytic capacitors are susceptible to corrosion damage caused by halogenated hydrocarbon cleaning solvents. Solvents can penetrate the elastomer end seals of the capacitor, resulting in corrosion failure. The degree of degradation is dependent upon the operating time, temperature, and applied voltage. Excessive dc leakage, electrical open circuits, or internal gassing are the failure symptoms. Table 4.2 lists safe and unsafe solvents.

4.4.3   Electrolyte Failures

Failure of the electrolyte can result from application of a reverse bias to the component, or a drying of the electrolyte itself. Electrolyte vapor transmission through the end seals occurs on a continuous basis throughout the useful life of the capacitor. This loss has no appreciable effect on reliability during the useful-life period of the product cycle. When the electrolyte loss approaches 40 percent of the initial electrolyte content of the capacitor, however, the electrical parameters deteriorate, and the capacitor is considered to have worn out.

As a capacitor dries out, three failure modes may be experienced: leakage, a downward change in value, or *dielectric absorption.* Any one may cause a system to operate out of tolerance or fail altogether.

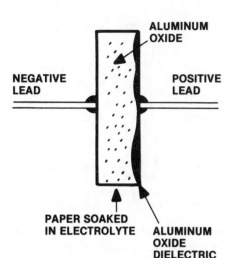

Figure 4.7   Failure mechanism of a leaky aluminum electrolytic capacitor. As the device ages, the aluminum oxide dissolves into the electrolyte, causing the capacitor to become leaky at high voltages. (Data from: Technical staff, *Sencore News,* issue 122, Sencore Corporation, Sioux Falls, S.D.)

The most severe failure mode for an electrolytic is increased leakage, illustrated in Figure 4.7. Leakage can cause loading of the power supply, or it can upset the dc bias of an amplifier. Loading of a supply line often causes additional current to flow through the capacitor, possibly resulting in dangerous overheating and catastrophic failure.

A change of device operating value has a less devastating effect on system performance. An aluminum electrolytic has a typical tolerance range of about 20 percent. A capacitor suffering from drying of the electrolyte can experience a drastic drop in value (to just 50 percent of its rated value, or less). This phenomenon occurs because, after the electrolyte has dried to an appreciable extent, the charge on the negative foil plate has no way of coming in contact with the aluminum oxide dielectric. This failure mode is illustrated in Figure 4.8. Remember, it is the aluminum oxide layer on the positive plate that gives the electrolytic capacitor its large rating. In effect, the dried-out paper spacer becomes a second dielectric, which significantly reduces the capacitance of the device.

The loss of capacitance in a circuit can result in increased ripple in power-supply applications, or a loss of low-frequency response in multistage circuits coupled using electrolytics.

Dielectric Absorption. Dielectric absorption is a side effect of electrolytic drying. Dielectric absorption refers to a capacitor's inability to completely discharge when its terminals are short-circuited. When a voltage is applied to an

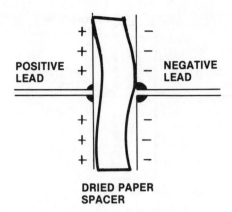

DRIED PAPER
SPACER

Figure 4.8 Failure mechanism of an electrolytic capacitor exhibiting a loss of capacitance. After the electrolyte dries, electrons can no longer come in contact with the aluminum oxide. The result is a decrease in capacitor value. (Data from: Technical staff, *Sencore News*, issue 122, Sencore Corporation, Sioux Falls, S.D.)

electrolytic capacitor, the dipoles in the dielectric become polarized and line up in an organized fashion. After the applied voltage is removed from the device, and the terminals are short-circuited, the dipoles should return to their random state, resulting in full discharge. A device experiencing the phenomenon of dielectric absorption, however, will return to a percentage of its original charge when measured with a high-impedance electrostatic voltmeter. This *battery effect* results from dipoles remaining in their polarized positions after the applied voltage has been removed and the capacitor has been discharged.

The effects of dielectric absorption are similar to those resulting from loss of capacitance value: poor filtering in power-supply applications and distorted waveforms in stage-coupling applications.

4.4.4 Capacitor Life Span

The life expectancy of a capacitor that is operating in an ideal circuit and environment will vary greatly, depending upon the grade of device selected. Typical operating life, according to capacitor manufacturer data sheets, ranges from a low of 3 to 5 years for inexpensive electrolytic devices to a high of greater than 10 years for computer-grade products. You get what you pay for.

Catastrophic failures aside, expected life is a function of the rate of electrolyte loss by means of vapor transmission through the end seals, and the operating or storage temperature. Properly matching the capacitor to the application is a key component in extending the life of an electrolytic capacitor. The primary operating parameters include:

- Rated voltage: The sum of the dc voltage and peak ac voltage that can be applied continuously to the capacitor. Derating of the applied voltage will decrease the failure rate of the device.
- Ripple current: The rms value of the maximum allowable ac current, specified by product type at 120 Hz and + 85C (unless otherwise noted). The ripple current may be increased when the component is operated at higher frequencies or lower ambient temperatures.
- Reverse voltage: The maximum voltage that may be applied to an electrolytic without damage. Electrolytic capacitors are polarized, and must be used accordingly.

4.4.5 Tantalum Capacitors

Tantalum electrolytic capacitors have become the preferred type of device where high reliability and long service life are primary considerations. The *tantalum pentoxide* compound possesses high dielectric strength and a high dielectric constant. As the components are being manufactured, a film of tantalum pentoxide is applied to the electrodes by means of an electrolytic process. The film is applied in various thicknesses. Figure 4.9 shows the internal construction of a typical tantalum capacitor. Because of the superior properties of tantalum pentoxide, tantalum capacitors tend to have as much as three times better capacitance per volume efficiency than an aluminum electrolytic capacitor. This, coupled with the fact that extremely thin films can be deposited during the electrolytic process, makes tantalum capacitors highly efficient with respect to the number of microfarads per unit volume.

The capacitance of any device is determined by the surface area of the conducting plates, the distance between the plates, and the dielectric constant of the insulating material between the plates. In the tantalum capacitor, the distance between the plates is small; it is just the thickness of the tantalum pentoxide film. Tantalum capacitors contain either liquid or solid electrolytes.

4.5 FAULT PROTECTORS

Fuses and circuit breakers are the two most common methods used in electronic equipment to prevent system damage in the event of a component failure. Although these devices are hardly new technology, many misconceptions exist about fuse and circuit-breaker ratings and operation.

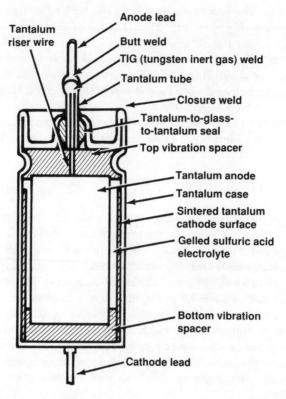

Figure 4.9   Basic construction of a tantalum capacitor. (Data from: Technical staff, "Introduction to Tantalum Capacitors," Sprague Applications Guide, Sprague Electric Company, Lansing, N.C., 1989)

4.5.1   Fuses

Fuses are rated according to the current they can pass safely. This may lead some to conclude, incorrectly, that excessive current will cause a fuse to blow. Actually, there is no amount of current that can cause a fuse to blow. It is, rather, power dissipation in the form of heat. In other words, it is the I^2R loss across the fuse element that causes the linkage to melt. The current rating of a given device, however, is not the brick-wall protection value that some may think it is. Consider the graph shown in Figure 4.10, which illustrates the relationship of rated current across a fuse to the blowing time of the device.

Fuse characteristics can be divided into three general categories: fast-acting, medium-acting, and slow-blow. Circuit protection for each type of device is a

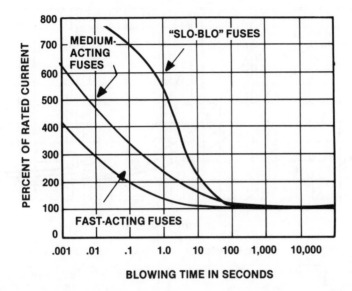

Figure 4.10   The relationship between the rated current of a fuse and its blowing time. Curves are given for three types of devices: fast-acting, medium-acting, and slow-blow. (Source: *Broadcast Engineering* magazine, Intertec Publishing, Overland Park, Kan.)

function of both current and time. For example, a slow-blow fuse will allow six times the rated current through a circuit for a full second before opening. Such delay characteristics have the benefit of offering protection against nuisance blowing because of high inrush currents during system start-up. This feature, however, comes with the price of possible exposure to system damage in the event of a component failure.

4.5.2   Circuit Breakers

Circuit breakers are subject to similar current let-through constraints. Figure 4.11 illustrates device load current as a percentage of breaker rating vs. time. The *A* and *B* curves refer to breaker load capacity product divisions. Note the variations possible in trip time for the two classifications. The minimum clearing time for the *A* group (the higher-classification devices) is 1 s for a 400 percent overload. As with fuses, these delays are designed to prevent nuisance tripping caused by normally occurring current surges from (primarily) inductive loads. Most circuit breakers are designed to carry 100 percent of their rated load continuously without tripping. They normally are specified to trip at between 101 and 135 percent of rated load after a period of time specified by the manufacturer. In this example, the must-trip point at 135 percent is 1 hour.

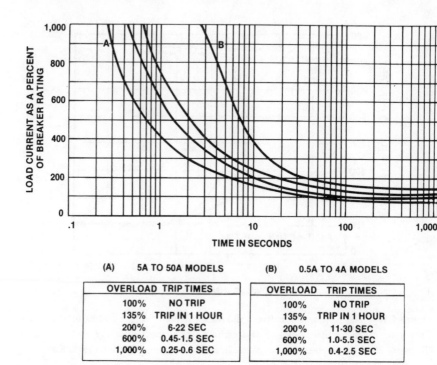

Figure 4.11 The relationship between the rated current of a circuit breaker and its blowing time. Curves (*a*) and (*b*) represent different product current ranges, as shown. (Source: *Broadcast Engineering* magazine, Intertec Publishing, Overland Park, Kan.)

Circuit breakers are available in both thermal and magnetic designs. Magnetic protectors offer the benefit of relative immunity to changes in ambient temperature. Typically, a magnetic breaker will operate over a temperature range of -40°C to +85°C without significant variation of the trip point. Time delays are usually provided for magnetic breakers to prevent nuisance tripping caused by start-up currents from inductive loads. Trip-time delay ratings range from instantaneous (under 100 ms) to slow (10 to 100 s).

4.5.3 Semiconductor Fuses

The need for a greater level of protection for semiconductor-based systems has led to the development of semiconductor fuses. Figure 4.12 shows the clearing characteristics of a typical fuse of this type. The total clearing time of the device, designed to be less than 8.3 ms, consists of two equal time segments: the *melting time* and the *arcing time*. The rate of current decrease during the latter period must be low enough that high induced voltages, which could destroy some semiconductor components, are not generated.

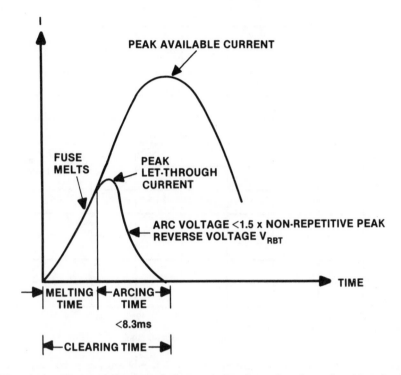

Figure 4.12   The current let-through characteristics of a semiconductor fuse. Note that the clearing time of the device is less than 8.3 ms. (Data from: *SCR Applications Handbook*, International Rectifier Corporation, El Segundo, Calif., 1977)

4.5.4   Application Considerations

Although fuses and circuit breakers are a key link in preventing equipment damage during the occurrence of a system fault, they are not without some built-in disadvantages. Lead alloy fuses work well and are the most common protection devices found, but because the current-interrupting mechanism is dependent on the melting of a metal link, their exact blow point is not constant. The interrupting current may vary depending on the type and size of fuse clip or holder, conductor size, physical condition of the fuse element, extent of vibration present, and ambient temperature. Figure 4.13 illustrates the effects of ambient temperature on blowing time and current-carrying capacity.

Transient Currents. Trip delays for fuses and circuit breakers are necessary because of the high start-up or inrush currents that occur when inductive loads or tungsten filament lamps are energized. The resistance of a tungsten lamp is high when hot, but low when cold. A current surge of as much as 15 times the rated

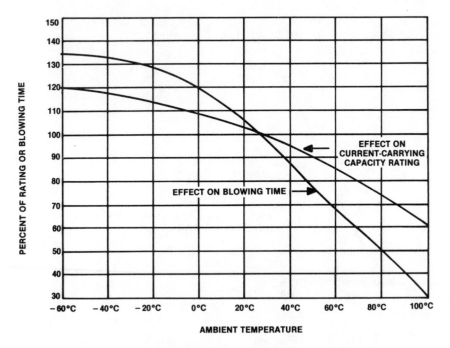

Figure 4.13 The effects of ambient temperature on fuse-blowing time and current-carrying capacity. (Data from: William Davis, "Selecting Circuit Protective Devices," *Plant Electrical Systems*, Intertec Publishing, Overland Park, Kan., March 1980)

steady-state value can occur when the lamp is first energized (pulse duration approximately 4 ms).

Transformer inrush currents can measure as high as 30 times the normal rated current for newer types of transformers that have grain-oriented, high-silicon steel cores. These transformer designs are becoming popular because of their favorable size and weight characteristics. Older transformers of more conventional design typically exhibit an inrush current approximately 18 times greater than the steady-state value. This transient current surge reaches its peak in the first half-cycle of applied ac power and decreases with each successive half-cycle. Such transients are relatively insensitive to the load placed on the secondary. The transient may, in fact, be smaller when the transformer is loaded than unloaded. A worst-case turn-on current surge will not occur every time the transformer is energized; rather, it will occur in a random fashion (perhaps one in every 5 or 10 turn-on events). Among the determining factors involved is the magnitude of the applied voltage at the instant the transformer is connected. Minimum transient energy occurs when the transformer is switched on at the zero-crossing point of the sine wave.

Table 4.3 Starting Surge Current Characteristics for Various Types of AC Motors Selected from a Random Test Group. Data from: William Davis, "Selecting Circuit Protective Devices," *Plant Electrical Systems*, Intertec Publishing, Overland Park, Kan., March 1980.

Motor type	Start current peak ampl. rms	Duration of start surge in sec	Load-sec %I•t sec
Shaded pole	150%	2.0	0.3
Split phase no. 1	600%	0.116	0.7
Split phase no. 2	425%	0.500	2.0
Capacitor (loaded) no. 1	400%	0.600	2.4
Capacitor (no load)	300%	0.100	0.3
Capacitor (loaded) no. 2	420%	0.500	2.1
Induction	700%	0.750	5.0
3 phase	350%	0.167	0.6
Cap. start split-phase run	290%	0.083	0.24

The magnitude of the turn-on current surge of an ac-to-dc power supply is determined mainly by the transformer transient and the capacitive load placed after the rectifier elements. Large filter capacitors commonly found in low-voltage, high-current supplies can place severe stress on the rectifiers and transformer. A fully discharged capacitor appears as a virtual short-circuit when power is applied. Some form of surge-limiting circuit often is provided for power supplies containing more than 10,000 µF total capacitance.

The surge current of an ac motor is spread over tenths of a second, rather than milliseconds, and varies considerably with the load connected to the unit. Table 4.3 lists typical motor surge currents for various types of devices. Note that the single-phase induction motor has the highest surge rating (seven times the running value for a period of 750 ms). Three-phase motors exhibit a relatively low surge current during start-up (350 percent for 167 ms).

Delay-Trip Considerations. The occurrence of turn-on surges for inductive loads, ac-to-dc power supplies, and tungsten filament lamps requires the installation of protective devices that exhibit delay-trip characteristics that match the given load. The problem is, however, that high surge currents of brief duration—not related to turn-on activities—can occur without tripping the circuit breaker or opening the fuse element. Subsequently, other circuit devices, such as semiconductors, can be damaged. To provide full protection to an electronic system, the overload-withstand characteristics of all components should match. This is not always an easy goal to accomplish.

For example, consider a simple SCR-controlled ac-to-dc power supply. The transformer will set the upper limit on surge current presented to the protective device and the SCR (assuming light capacitive loading). If that surge is 18 times the normal steady-state current for a period of 8 ms, then a protective device must be selected that will allow the surge to pass without tripping. An SCR must be selected for the circuit, therefore, that can withstand at least 18 times the normal rated current for 8 ms. If not, the SCR will become the weak link in the system, not the protective device.

4.6   BIBLIOGRAPHY

Anderson, Leonard R.: *Electric Machines and Transformers*, Reston Publishing Company, Reston, Va.

Davis, William: "Selecting Circuit Protective Devices," *Plant Electrical Systems*, Intertec Publishing, Overland Park, Kan., March 1980.

Fink, D., and D. Christiansen (eds.): *Electronics Engineers' Handbook*, 3rd ed., McGraw-Hill, New York, 1989.

Jordan, Edward C.: *Reference Data for Engineers: Radio, Electronics, Computer, and Communications*, 7th ed., Howard W. Sams Company, Indianapolis, Ind., 1985.

Lowdon, Eric: *Practical Transformer Design Handbook*, Howard W. Sams Company, Indianapolis, Ind., 1980.

Meeldijk, Victor: "Why Do Components Fail?," *Electronic Servicing & Technology* magazine, Intertec Publishing, Overland Park, Kan., November 1986.

Nenoff, Lucas:, "Effect of EMP Hardening on System R&M Parameters," *Proceedings of the 1986 Reliability and Maintainability Symposium*, IEEE, New York, 1986.

Pearman, Richard: *Power Electronics*, Reston Publishing Company, Reston, Va., 1980.

Wilson, Sam: "What Do You Know About Electronics?," *Electronic Servicing & Technology* magazine, Intertec Publishing, Overland Park, Kan., September 1985.

SCR Applications Handbook, International Rectifier Corporation, El Segundo, Calif., 1977.

SCR Manual, 5th ed., General Electric Company, Auburn, N.Y.

Technical staff: *Sencore News*, issue 122, Sencore Corporation, Sioux Falls, S.D.

Technical staff: "Aluminum Electrolytic Capacitors: Reliability, Expected Life, and Shelf Capability," *Sprague Applications Guide*, Sprague Electric Company, Lansing, N.C., 1989.

Technical staff: "Introduction to Tantalum Capacitors," *Sprague Applications Guide*, Sprague Electric Company, Lansing, N.C., 1989.

5

PLANNING THE WORKBENCH

5.1 INTRODUCTION

Some equipment problems still can be located with little more than a digital multimeter (DMM) and oscilloscope, given enough time and effort. But time costs money. Few technical managers are willing to make the trade. With current-technology equipment, the proper test equipment is a must. Table 5.1 lists the test instruments necessary for a well-equipped maintenance shop. This list does not include specialized instruments and gauges needed to maintain particular types of electromechanical hardware. The list shown can be expanded easily, depending upon the sophistication of the equipment at a particular facility. By and large, however, the recommendations given should provide the maintenance staff with most of the tools it needs.

Servicing certain types of hardware requires special test fixtures and/or test equipment. Maintenance of audio/video recording systems, for example, is nearly impossible without the proper gauges and test tapes. The purchase of alignment tools and detailed technical documentation is money well-spent.

5.1.1 Service Tools

The simplest maintenance task is next to impossible without the right tools. There is more to a well-stocked tool chest than meets the eye. Tools for maintenance may seem too basic to even warrant discussion, but an arsenal of proper tools is essential to efficient equipment repair. Table 5.2 lists common tools, fixtures, and chemicals that should be on hand in any shop. Deciding whether to invest in a customized tool kit, or to buy the tools individually is not always easy. At first glance, it might seem more economical to buy tools one at a time, as they are needed, but a careful examination of maintenance needs could point to a different conclusion. Manufac-

129

Table 5.1   Basic Test Equipment Required to Maintain Electronic Systems Today

Basic test equipment
_____ Oscilloscope
_____ Accurate bench DMM
_____ Isolation transformer
_____ Electrostatic discharge protection kit
   • Bench mat
   • Wrist strap
   • ESD protective packaging

Audio test equipment
_____ Audio signal generator
_____ Distortion analyzer
_____ AC voltmeter
_____ Wow and flutter meter
_____ Tape machine alignment gauges

Video test equipment
_____ Video signal/pattern generator
_____ Vectorscope
_____ Waveform monitor
_____ High-voltage probe
_____ CRT tester/restorer
_____ VCR/VTR test and alignment gauges

Digital hardware test equipment
_____ Logic probe
_____ Logic pulser
_____ Logic monitor clip (IC clip-on)
_____ Break-out box
_____ Logic comparator
_____ Logic state analyzer (optional)
_____ Logic timing analyzer (optional)

RF test equipment
_____ Transmitter dummy load
_____ Low-power dummy loads
_____ High-voltage probe
_____ Clamp-type current probes
_____ True-rms ac voltmeter
_____ Wattmeter with selection of slugs
_____ Assorted coaxial connectors/adapters
_____ Spectrum analyzer (optional)

Table 5.2   Basic Tool Complement for a Well-Equipped Maintenance Shop

Tools, fixtures, and chemicals

_____ Pliers and wire cutters
_____ Screwdriver set
_____ Nutdriver set
_____ Hex wrenches
_____ Wire stripper
_____ Crimping tool
_____ Wire wrap/unwrap tool
_____ Tweezers
_____ Vise
_____ PC board-holding fixtures
_____ Solder flux
_____ Flux remover
_____ Selection of clip-on heat sinks
_____ Heat sink grease
_____ Silicone lubricant
_____ Glass/plastic cleaner
_____ Antistatic spray
_____ Compressed air
_____ Adhesives
_____ Component coolant spray
_____ Tape head cleaner
_____ High-voltage insulating putty
_____ Spray-type contact cleaner

turers offer packages intended for special servicing applications. These kits include tools that may or may not be readily available from a local parts supplier.

Comprehensive, high-quality standard tool kits for professional engineers and technicians are available from a number of reputable suppliers. Special, off-the-shelf kits also can be found for telephone equipment, computers and peripherals, robotics, fiber optics, and other technologies. Because of the competitive nature of the tool kit market, buyers usually receive good value on any of these standard packages.

The tool kit should include portable, static-protective field-service gear consisting of a grounded wrist strap and static-dissipative work mat. The mat should be thin enough to fold neatly into the kit, and it should include one or more pockets to hold the wrist strap, ground cords, and other accessories.

Case style and construction are important points to consider when buying a tool kit. It does not makes sense to try to economize by purchasing a $50 case to carry expensive tools and instruments. If the case fails to provide adequate protection for its contents, it is of little or no value. The case is the backbone of the tool kit. Style

Table 5.3   Comparison of Common Tool Case Types. Data from: Jeff Richardson, "The Benefits of a Tool Kit Program," *Microservice Management* magazine, Intertec Publishing, Overland Park, Kan., May 1989.

Case Construction	Strength	Ability to Withstand abuse	Relative Cost
polyethylene	excellent	excellent	excellent
aluminum	excellent	excellent (might bend)	high
vinyl/wood walls reinforced with steel	excellent	excellent	high
vulcanized fiber reinforced with steel	excellent	very good	low
injection-molded polypropylene	good	moderate	moderate
molded ABS	good	poor	moderate
padded vinyl with wood frame	good	poor	moderate
vinyl zipper case	good	poor	poor
Cordura zipper case	excellent	excellent	low

and construction play a major role in its long-term functionality. When selecting a case, consider whether it will hold items other than tools, such as test instruments, electronic components, disc packs, documentation, and circuit boards. Consider also whether the case will be used in a single location or transported. A tool case that will be riding in the back of a pickup truck must be more substantial than one that seldom will be moved. With a properly selected case, tools and test equipment can be organized for easy access and safe transport to and from the job site. A good tool pallet does more than simply help keep track of tools; it allows placement of the most frequently used tools in the most accessible locations. Delicate tools should be protected, and the weight of the tools distributed evenly. Table 5.3 provides an overview of the types of tool cases designed to be used with tool pallets. Of the types described, the polyethylene, attaché-style case is the most popular.

5.1.2   Test Jigs and Fixtures

When organizing a maintenance shop, consider the need for jigs and fixtures for holding printed circuit boards and small components. Sometimes it takes more than two hands to repair a piece of electronic equipment. With modern hardware, components have become so small and their leads so closely spaced that all work must be performed with great care. Trying to work on a PWB that keeps sliding across the workbench is frustrating and senseless. The solution is to get "another pair of hands," either simple or elaborate, depending on the nature of the hardware being serviced.

One of the simpler devices for holding a small printed circuit board is a jig, which consists of either a weighted base or a bench clamp and a pair of alligator clips mounted so that they can be adjusted into various positions. Larger, heavier boards call for something more elaborate. A number of board-holders are available for securing the PWB in a convenient position for servicing. A technician working on a large piece of equipment should use some type of holding fixture.

Test-equipment manufacturers have devised a number of tools to aid in troubleshooting of circuit boards crowded with components. These are among the most popular:

- *Spring-loaded hook probe.* Allows the technician to grab one pin out of a group of closely spaced pins without contacting other conductors. With care, it may be possible to leave the power on and move this type of hook from pin to pin to trace a fault. Conventional wisdom, however, suggests that when pins are packed tightly enough to warrant using one of these hooks, the technician should avoid the possibility of damage by disconnecting power before moving to a new test point.
- *Test clip.* Operates similar to a spring clothespin. Each side of the clip has a number of conductors that are spaced like the leads on a DIP IC package. The test points at the top of the clip provide easy access to the IC leads, away from the crowded surface of the PWB.
- *Cable interface adapter.* Consists of a short extender plug that provides test points for each pin of a cable connector. This permits test measurements to be taken at the point where the cable connects to a device. A variation on this theme provides not only the cable interface and test points, but also two test pins with a switch between them for each conductor of the cable. With this type of adapter, it is possible to isolate a single conductor, or several, and probe on the *device side* as well as the *cable side* of the open switch.

5.2   SOLDERING EQUIPMENT

With the advent of surface-mount technology and high-density, multilayer PWBs, soldering and board repair has become as much science as art. Whatever piece of

Table 5.4   Checklist of Solder Equipment Necessary for Equipment Maintenance

Soldering/desoldering products

_____	Fine- and medium-grade solder
_____	Soldering irons (35 and 75 W)
_____	Temperature-controlled soldering iron
_____	Solder sucker
_____	Heat sinks of various sizes
_____	Desoldering braid
_____	IC insertion/extraction tools
_____	Selection of chip-mount solder heads
_____	Selection of SMC solder heads

equipment may be ready for servicing, the technician always should have high-quality soldering and desoldering equipment on hand. Many facilities unwisely skimp on this type of hardware, perhaps because, in the past, soldering seemed elementary. However, with the widespread use of surface-mount components (SMCs) and VLSI chips, all that has changed. Table 5.4 lists the recommended solder rework items.

When it comes to soldering and desoldering current technology products, the rules have changed dramatically. For circuit-board work, product manufacturers recommend the use of a 10 to 50 W temperature-controlled soldering pencil. For desoldering an IC, either use a soldering iron with a head that will heat up all the leads at one time, or find some method of removing the solder from each lead in turn. A number of products may be used to remove solder from component leads, including:

- _Desoldering braid:_ Fine copper formed into a braided wire, impregnated with rosin flux and rolled into a coil for easy handling.
- _Suction devices:_ Add-on hardware for a soldering iron. The tip is applied to the joint while the solder is molten. A squeeze bulb or spring-loaded plunger is used to suck out the solder.
- _Solder/desolder stations:_ PWB rework equipment for sophisticated users. These units start at a few hundred dollars, but they do everything from soldering and desoldering surface-mount components to repairing PWB conductor traces. Some of the more sophisticated soldering stations have two or more heat-producing elements on the same unit so that, without changing tips, the technician can vacuum desolder with one while soldering with the other.

The use of sensitive integrated circuits in many products today requires attention to electrostatic discharge, and this caution extends to soldering equipment. Soldering tools usually are grounded through a resistance of 250 kΩ to 1 MΩ to drain away static buildup without posing a shock hazard to the user.

Any solder rework station also should include a quality magnifying lamp so that soldering work and circuit-board traces can be inspected closely.

5.2.1   Approach to Servicing

A technician who attempts to troubleshoot a problem by removing components from a crowded PWB on a trial-and-error basis will simply create a whole new problem. Under the best of conditions, a trained technician with the proper tools causes stress to both components and the board. Replacing components one at a time in an attempt to find the cause of a fault may result in a board that is beyond repair.

Soldering skill always has been a determining factor in the successful repair of electronic circuits. Poor soldering technique can lead to cold solder joints or voids that, in time, result in equipment failure. Anyone who will be doing a significant amount of PWB rework should take advantage of the instructional videotapes offered by many consumer electronics companies, soldering equipment manufacturers, and industry organizations such as the EIA (Electronic Industries Association).

Component Removal. The removal of faulty components is the most frequently performed physical repair task. The simplest and most common method involves a continuous-vacuum desoldering iron. Such a tool typically has a handpiece with a hollow, heated tip. The heated tip melts the solder, and a continuous vacuum—applied as soon as the solder has melted—extracts the molten solder and deposits it in a solder-collection chamber, where it solidifies. Such a system may come with its own vacuum pump, or it may connect to a building compressed-air system. For the desoldering action to work properly, the desoldering system must have a rapid vacuum rise time. Sufficient vacuum must develop quickly for all the molten solder to be withdrawn in one continuous slug, leaving the joint free of solder.

The solder should melt completely within about 3 s. Prolonged exposure of a PWB to high temperatures may damage the board traces or components mounted on the board. Advanced soldering irons include thermostatically controlled heating elements. Two types of systems are common:

- Open-loop iron. The temperature of the soldering tip is adjustable, but there is no sensor at the tip to set the current to the heating element to maintain the desired temperature when the tip is under a heavy thermal load.
- Closed-loop iron. A sensor is located at the tip to keep the temperature constant, regardless of the thermal load.

An important benefit of the closed-loop iron is that solder melting time can be kept reasonably constant, even if the work involves thermally massive multilayer boards. If desoldering or soldering operations are performed on a portion of a multilayer board in rapid succession, a high-quality, closed-loop-controlled system is critical so that the proper operating-tip temperature is sustained from one joint to the next. It follows that the rate at which the temperature may be adjusted is an important factor in the usefulness of the closed-loop design. Many advanced closed-loop soldering stations include digital readouts of both the set temperature and the actual temperature at the workpiece.

Experience has shown that on some thermally massive boards, even a high-quality soldering station cannot heat the joint completely or rapidly enough. In such cases, *thermal soaking* (preheating) of the joint and/or auxiliary heating may be required.

Repairing Damaged PWBs. Damaged conductors on a circuit board usually can be repaired by replacing a portion of the conductor with a new, flat conductive material that has equal or greater current-carrying capacity. The physical dimensions of the patch should be similar to those of the original conductor. Solder equipment manufacturers offer patch kits that include pretinned replacement conductors and a selection of eyelets or funnelets for repairing plated through-holes.

5.2.2   Solder Types

Reliable PWB rework requires the use of quality solder. The best type for most equipment is 0.028-in (22-gauge) solder with a rosin core. Small solder strands melt fast and lose less heat, which allows better control over the amount of solder applied to the joint. The solder should be made of virgin tin and desilvered lead, and be free from impurities such as zinc, aluminum, iron, copper, and cadmium. Do not try to save money on solder. Solder with the lowest melting point, and of the type that yields the strongest bond, is made with a ratio of 63 percent tin to 37 percent lead, but 60-tin/40-lead is nearly as good.

For certain jobs, some manufacturers recommend *silver solder.* Silver solder contains about 3 percent silver along with the lead and tin. Silver solder is used for soldering components such as ceramic capacitors, which have silver-palladium fired onto the conductive surfaces. If common tin-lead solder were used, it might absorb some of the silver from the component, causing a weak joint and poor adhesion. The small amount of silver in the solder reduces migration of the silver from the component connections. Silver solder is used in the same manner as ordinary solder and performs essentially the same, except for a slightly higher melting point. Before replacing a component, always check the manufacturer's service manual for comments relating to soldering. If the literature calls for silver solder, use it.

Solder Flux. The rosin core of the solder usually is adequate to clean oxide from the joint during soldering. In some cases, however, additional flux may be required. It is good practice to have some high-quality liquid rosin flux on hand. Rosin is a nonconductive, noncorrosive flux that is recommended for work on electronic circuits. Rosin flux is sticky, however, and it will collect dust if allowed to remain on the solder joint. After soldering, clean off any excess flux.

5.2.3   Moving Heat to the Joint

Most soldering irons are sold on the basis of heating-element wattage. This rating is, unfortunately, a common source of misunderstanding. The wattage indicates only the potential amount of heat an iron can produce. The amount of heat that actually reaches the tip will be considerably less than the iron's rating. The amount of heat delivered to the work point is determined by the heat-transfer efficiency of the iron, the shape of the soldering tip, and the distance between the heating element and the work.

In most electronics applications, 650°F is the minimum amount of heat required to *reflow* a solder joint. However, this does not take into consideration heat lost from the tip. A heat reserve also must be figured into the choice of an iron. Be careful; too much heat can ruin components and carbonize the flux before it has a chance to do its job. In general, most connections require about 800°F. Large connections, such as braided-wire grounds or heavy-gauge wire, require about 1000°F.

Often, it is difficult to solder or desolder ICs from a multilayer PWB. Moderate heat (800°F or so) is adequate for nearly all connections, except for pins tied to the ground plane or power-supply rails. The large mass of copper used to conduct power throughout a board also can conduct a fair amount of heat. When faced with these various heat requirements, use a closed-loop iron or two different soldering irons. Do not use a 1000°F iron on connections that can be soldered or desoldered with an 800°F iron.

Soldering iron tips available today usually are made of copper with a thick iron or nickel plating for long life. They must never be *redressed*; filing or grinding will destroy the tip. Coated tips also need to be retinned less often than traditional copper tips. Corrosion is the worst enemy of a soldering tip because it prevents efficient transfer of heat to the work point. When a clad tip becomes corroded, sand it lightly with a piece of emery cloth, and retin. Never retin a soldering iron while it is hot. Let the iron cool, then warm it for about 1 minutes, and apply flux-core solder. This procedure is recommended because corrosion is faster at higher temperatures. Tinning while the tip is cooler will provide the best soldering surface. Replace a clad tip when the plating becomes corroded.

The dangers of using too much heat to solder or desolder a component are obvious, but what about using inadequate heat? If the solder connection is not heated sufficiently, a *rosin joint* will result. Although a rosin joint looks nearly identical to a good solder joint, the flux resins in a rosin joint insulate the component lead and prevent reliable contact with the PWB trace.

After the circuit board has been soldered, clean all flux residues and inspect for rosin joints. A good solder joint is smooth and shiny; a rosin joint is dull gray and full of pinholes.

5.2.4 Dealing with Chip Components

The biggest challenge facing maintenance engineers today in PWB rework is the soldering and unsoldering of chip capacitors, transistors, integrated circuits, and other surface-mounted components. Replacing one of these parts, without destroying it or the PWB, requires the right tools and procedures. There is little doubt that surface-mount technology will continue to replace leaded-through-hole technology. As more components become available in SMC packages, more PWBs will be designed to use them.

Chip components are tiny and are soldered on the same side of the PWB on which they are mounted. To remove a 2-terminal device, grasp the failed component with pliers or tweezers and melt the solder at both ends, using a dual tip designed for that purpose. If such a tip is unavailable, melt the solder at one end of the device, then quickly apply the soldering iron to the other end. The dual-tip procedure is, not surprisingly, easier than the latter.

While heat is being applied to the chip component, use a gentle twisting motion to free the device. In many cases, manufacturers apply a drop of adhesive to the component and board to hold it in place while the board is mass-soldered. This glue is formulated so that once it has cured, it will shatter if a twisting force is applied, thereby freeing the device. The adhesive also may be heat-sensitive. Application of the soldering iron may soften the adhesive enough to allow removal of the component. Not all SMC parts on a board are glued down, however. In many cases, certain ICs or other parts are soldered manually after the mass soldering of all other components is complete.

Another approach to chip component removal involves the use of two soldering irons, one placed at each joint. Once the solder has softened, the part can be removed. This usually requires the assistance of another technician.

Using any of the methods described probably will result in excess solder on the PWB *lands* (the points to which the chip components connect). Application of a clean soldering iron may be enough to remove the excess. If not, use solder wick or a vacuum device along with heat from the iron to get rid of the excess.

When replacing a chip device, load one land with solder, and bond the chip to the board using the soldering iron and a small amount of additional solder. After the first bond has solidified, apply heat and solder to the other end to complete the job.

Chip transistors can be handled in much the same way as 2-terminal chip capacitors and resistors. Because three leads are involved, however, use a tool designed for desoldering chip transistors.

Never reuse a chip device that has been removed from a PWB. If the part was not destroyed by repeated soldering and desoldering, its life may have been significantly shortened.

5.2.5   Dealing with Flat-Pack ICs

The technician should never remove a *flat-pack* IC unless dead-sure that it is defective. As with other components, there are several methods to remove a flat-pack device. Some are better than others, and all require a good deal of skill and patience. The most common approach involves use of a special desoldering head that heats all the terminals simultaneously. Make sure the head matches the flat-pack device being replaced.

After the defective component has been removed, use soldering braid to clean the lands on the PWB as necessary. Inspect all connection points to be sure no solder bridges were formed during the removal process. To replace the device, brush the lands with a small amount of liquid flux. This is recommended for two reasons: First, because very little solder is needed for each connection, the flux built into the core solder might not be enough to clean the joint properly. Second, the flux is sticky, and it will help hold the device in place while it is being soldered. Do not let the flux evaporate or thicken before soldering the component.

After the IC is in place and aligned properly with the PWB lands, solder down a couple of leads, then carefully solder the rest, one at a time. Apply enough solder to form a good bond, but not so much that a solder bridge can form. After soldering, carefully inspect the device for solder bridges. If a bridge is found, a soldering-iron tip (free of excess solder) can be drawn along the length of the gap between the two leads. The tip should pick up the excess solder forming the bridge. As an alternative, apply solder wick and heat from the iron at the solder bridge, absorbing the excess solder with the wick. This procedure should leave enough solder between the leads and the lands to ensure a good connection.

Pulse-Heat Systems. A new generation of desoldering tools is available for removing SMCs with a minimum of risk to the component and the PWB. Pulse-heat SMC reflow systems can be used to install and remove a wide range of surface-mount components. Unlike a continuously heated soldering iron, a pulse-solder system is not hot when idle. The operator applies the tip to the joint to be soldered (or unsoldered), then activates a foot switch. The handpiece provides a controlled temperature rise to the solder-reflow point. This approach eliminates problems associated with thermal shock. Small outline ICs (SOICs) and plastic leaded chip carriers (PLCCs) may be damaged by the application of a thermally massive soldering head to the workpiece. Because the tip is extremely hot when applied to the device leads, the sudden temperature rise may damage the device. A selection of handpiece types and tips is available to accommodate a range of standard lead configurations.

When a pulse-heat iron is used to solder a device into place, the handpiece is applied to the SMC leads, and the foot switch is depressed. After the solder has melted, the switch is released, and the soldering tip cools down. (The tip is made of a material that prevents solder adhesion.) This ensures good alignment of the component with the PWB lands. When solder paste is being used, the controlled temperature rise tends to drive off volatile materials in the paste before the solder melts. This minimizes splattering and the formation of solder balls.

Hot-Air Reflow Stations. High-volume PWB rework requires more efficient methods of removing defective components. The hot-air reflow approach directs heated air into the lead attachment area of the SMC being installed or removed. Advanced reflow workstations provide the operator with a choice of vision systems to align the component to the substrate. Precise air/gas temperatures of 500°F to 800°F are available to the operator. Hot-air reflow workstations are ideal for PWB rework involving 4-sided SMC components.

5.2.6   Removing Conformal Coatings

A PWB with a conformal coating may present special rework problems unless the proper tools are available. If the coating covers the solder joints, whether partially or completely, it must be removed. Coatings can create heat barriers that make it difficult to melt and remove solder from a joint. Before attempting rework, break the coating seal around each of the leads on the component side of the board to ensure positive airflow through the solder joint. Three methods commonly are used to remove conformal coatings:

- Thermal parting. A controlled, low-temperature localized heating probe is used to remove thick coatings with an *overcuring* or thermal-degradation process. A variety of temperature-controlled tips are available to allow access to the workpiece.
- *Abrasion.* A low-RPM, high-torque grinding tool is used to slice away the coating.
- *Hot-air jet.* Hot air, directed at the coating, is used to break down the adhesives of the coating.

5.3   USING CHEMICALS IN THE SHOP

Chemicals are an important element in servicing professional electronic products. The right chemical applied to the right point often will help identify a problem or restore a system to proper operation. Every maintenance technician knows that periodic cleaning of electronic hardware is essential to its reliability. Cleaning removes oxides, dust, grease, and other environmental contaminants that can reduce electrical conductivity and trap heat. Cleaning also helps protect against frictional wear, corrosion, and static buildup.

5.3.1   Chemicals For Troubleshooting

A number of solvents and cleaning agents are available in both aerosol and liquid form to remove performance-inhibiting contaminants from electronic equipment. The principal applications of these chemicals include:

- Removing oxides from tape heads and drive components
- Cleaning printer mechanisms
- Degreasing contacts and PWB connectors
- Removing organic flux after soldering

Although personal preference and habit often determine which cleaning solvents technicians use in the shop, more objective factors such as safety, effectiveness, and convenience should be considered. Electronic cleaning solvents can be divided into four major categories based on chemical composition:

1. Chlorofluorocarbons
2. Chlorinated solvents
3. Alcohols
4. Blends of chemicals

Aerosols are preferred by many technicians because they deliver a continually fresh supply of uncontaminated solvent with sufficient pressure to dislodge and remove even encrusted grease without scrubbing. Aerosol solvents can be applied effectively using a lint-free cloth for catching overspray and wiping the PWB and/or component. Carbon-dioxide-propelled aerosol systems provide the greatest initial spray pressure.

Chlorinated solvents, such as 1,1,1-trichloroethane, are the strongest electronic cleaners used today. They are nonflammable and are available as both liquids and high-pressure aerosols. Chlorinated solvents, however, are more toxic and slightly more unstable than chlorofluorocarbons. They also have a tendency to cause swelling in certain types of plastics. Alcohols, such as isopropanol, commonly are used for electronic field service work. Although they are good general-purpose solvents, alcohols have several notable disadvantages: They are flammable; unavailable as aerosols; and liable to react with some plastic, polycarbonate, and polystyrene materials.

A number of excellent chemical blends also are available. Each blend possesses a specific measure of solvency, reactivity, and flammability. When using these products, check the package label for contents and capabilities, as well as precautions against adverse reactions with materials to be cleaned.

Applicators. For many years, household cotton swabs have been used routinely to apply solvents and remove contaminants from electronic hardware. Although inexpensive and readily available, everyday cotton swabs produce lint and are not correctly sized or textured for many cleaning jobs. A superior alternative is found in the large variety of precision specialized swabs and applicators with tips of polyurethane foam, polyester cloth, and other advanced synthetic materials. These products are designed to be highly absorbent, thoroughly clean, and completely free of particles and extractables. Available in a wide assortment of sizes and shapes, foam-tip swabs are thermally mounted (without contaminating adhesives) on handles of varying lengths and flexibility for a variety of cleaning jobs.

5.3.2 Presaturated Pads and Swabs

Rapidly gaining in popularity with maintenance technicians, disposable presaturated pads and swabs offer the ultimate in field maintenance convenience. These products combine a premeasured amount of liquid solvent with an application device—either a cloth pad or a foam-tip swab—sealed in individual foil packets. The user simply tears open a packet, cleans the surfaces, and discards the used cleaning item. Premoistened pads and swabs are easy to store and use. They are used only once, preventing exposure of sensitive surfaces to contamination. Premoistened pads are highly recommended for cleaning PWB edge connectors and equipment housings.

For cleaning tape heads and electromechanical components, presaturated, lint-free pads containing either solvent or isopropanol alcohol are available. Presaturated, nonresidual isopropanol swabs serve the same purpose and are especially useful for hard-to-reach areas. In any event, follow the recommendations of the equipment manufacturer.

5.3.3 Compressed-Gas Dusters

Made from a microscopically clean, moisture-free inert fluorocarbon (dichlorodifluoromethane), compressed-gas dusters deliver powerful jet action to instantly remove dust and particulates from even the most inaccessible areas of electronic equipment. These products are especially useful for:

- Removing abrasive contaminants from magnetic tape heads before cleaning
- Cleaning dusty PWBs, switches, and computer keyboards
- Blowing paper dust out of printers

Compressed-gas dusters are invaluable for cleaning delicate components in applications where liquid solvents are inappropriate or inconvenient. Any surface, no matter how sensitive, can be cleaned safely with a compressed-gas duster. There is no risk of abrasive damage or microscopic contamination. Packaged in convenient cans, many of these dusters come with plastic tubes for pinpoint application and variable-control trigger valves for one-hand operation.

Exercise care, however, when using compressed-gas duster products. Make sure the stream of air is used to move dust and dirt *out of* the equipment being serviced, not simply to another part of the hardware. Be careful not to simply move dirt from where it can be seen to where it cannot. The contaminants may continue to cause problems for the equipment, regardless of their visibility.

5.3.4   Contact Cleaner/Lubricants

New chemical products that combine precision cleaning agents with fine-grade lubricants provide a convenient method to clean, restore and protect fragile electronic contacts. Available as aerosol sprays and premoistened pads, contact cleaner/lubricants are especially useful for protecting microthin, precious metal surfaces against frictional wear, oxidation, and corrosion. Applied periodically to gold PWB edge connectors and other electronic contacts, cleaner/lubricants prolong the life and preserve the electrical continuity of metallic surfaces. When choosing a product of this type, look for the following important properties:

- High-temperature lubricity
- Low volatility
- Controlled *creep* onto neighboring surfaces
- Oxidation stability (to prevent the formation of gummy by-products)
- Compatibility with adjacent materials

Consult the equipment manufacturer's service department for suggestions on the types of cleaner/lubricants to use.

5.4   FINDING REPLACEMENT PARTS

Although finding the right replacement part sometimes can seem to be an impossible task, the more suppliers a technician becomes familiar with, the greater the likelihood of finding a necessary component. Some suppliers sell through traditional distributors; others sell via mail order. Check out all of them. An original equipment manufacturer, if unable to provide the replacement part, may suggest other parts houses that can help.

Many popular semiconductors now are available from alternative parts sources, frequently at attractive prices. Most of these vendors are catalog operations. Look at the catalog, review the sales terms and conditions, and talk to customer service representatives on the phone to learn about the operation before deciding whether to do business with the company. Choose companies that offer such services as fast delivery and on-line information regarding in-stock and out-of-stock items. Buy from companies that offer easy return privileges for unacceptable or defective parts. After receiving an order from the supplier, consider the following questions:

- How long did it take for the order to arrive? If the parts are cheap, but it takes more than a week to receive a "rush" order, then *the technician* is the loser.
- How were the products packaged? The use of paper bags, newspaper, and other cheap materials can allow the goods to be damaged.
- How much of what was ordered actually arrived? It should not be less than 70 percent.
- Is the invoice clear? Are all charges reasonable? Avoid buying from companies that charge for handling or assess surcharges. Reputable companies bill freight only in the amount charged to them by the carrier.
- Is the order accurate? Mistakes do happen, but an excessive number of mistakes is a clear sign of a problem company.

5.4.1   Matching Replacement Devices

The suitability of replacement semiconductors—usually the biggest parts headache—varies, depending on the type of component and the circuit in which it is used.

Differing IC Packages. Occasionally, a situation arises in which a so-called direct replacement does not look like the original. Consider the following examples:

- A 14-lead linear DIP IC was removed, but the replacement component has only eight leads. In this case, pins 1, 2, 7, 8, 13, and 14 are not connected to anything internally in the original package. The 8-pin replacement is, more or less, a direct replacement, and it should be installed as shown in Figure 5.1.
- The TO-220 outline is a direct retrofit for the JEDEC metal TO-66 case. Likewise, the CP-3 plastic device can be used to replace types in the TO-3 metal case. In either situation, the replacement is made by cutting the center lead of the plastic device and bending the two outside leads to install the new component, as shown in Figure 5.2.

Stud-Mounted Devices. When changing a stud-mounted semiconductor, it is sometimes impossible to find a replacement with the same thread and stud size as the original. In this case, the device may be replaced with a component that has a larger stud. Do not drill the mounting hole any more than 1/64-in larger than the size of the replacement stud. Remove any burrs and surface irregularities from the mounting plate. Make sure to drill the hole perpendicular to the mounting surface.

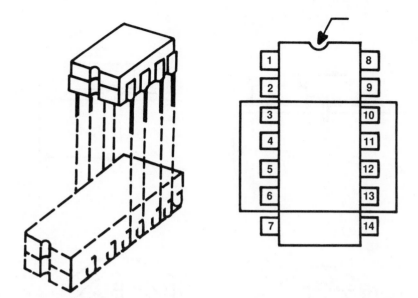

Figure 5.1 Some linear ICs that originally were supplied in a 14-lead DIP case now are available in an 8-pin mini-DIP case. Pins 1, 2, 7, 8, 13, and 14 were not connected in the original version, so the 8-lead package can be installed as shown. (Source: *Broadcast Engineering* magazine, Intertec Publishing, Overland Park, Kan.)

Use heat sink compound between the heat sink and the replacement device. Apply the proper torque to mount the device in place. Mounting hardware usually consists of an insulating washer and sleeve, a nut, and a lockwasher.

High-Frequency Circuits. Equipment operating at VHF or UHF frequencies usually is sensitive to changes in semiconductor operating parameters. When installing a universal replacement, do not change any of the mechanical details of the original circuit. Before removing the original transistor (or other component), carefully note its position with respect to other components. Also note the length and placement of the leads; duplicate them as precisely as possible. Failure to do this could result in improper tuning or circuit instability, particularly in circuits operating at UHF frequencies.

When using a universal replacement device in an untuned RF amplifier stage operating at a low signal level, it usually is unnecessary to make adjustments to ensure proper performance. However, when a replacement is made in a tuned RF amplifier, oscillator, or converter, check alignment of the associated circuits to ensure proper tracking and to achieve the required gain without loss of stability.

When replacing a transistor in an amplifier stage that operates at relatively high power levels (10 to 100 W), give extra attention to mounting of the device on its

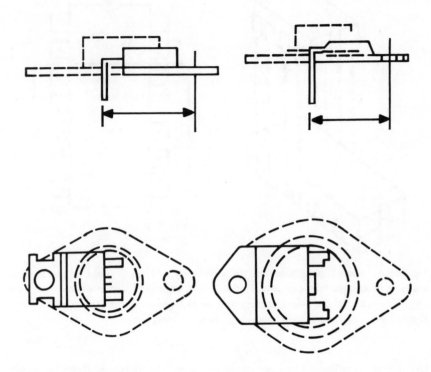

Figure 5.2   Some plastic-case devices are direct retrofits for metal-case types. Cut the center lead of the plastic component, and bend the two outside leads. Check the pinouts of the original device and the replacement to make sure the leads are going where they should. (Source: *Broadcast Engineering* magazine, Intertec Publishing, Overland Park, Kan.)

heat sink. (See Chapter 2, Sec. 2.6.1.) Failure to provide a good thermal bond between the device and the heat sink could result in shortened life of the component. Indeed, a poor thermal bond could have caused the problem in the first place.

5.4.2   Leasing Replacement Parts

Because of the diversity of hardware and software options available today, maintenance department managers often are forced to stock large inventories of equipment to meet customer needs. This practice can dramatically affect operating costs, especially when a large portion of the replacement parts inventory sits idle. Leasing replacement parts can be an efficient and cost-effective alternative to purchasing. Used correctly, leasing is a tool by which the maintenance department can better manage its parts costs and usage. Leasing high-tech replacement parts can help achieve the following:

- *Spread out acquisition costs.* Leasing allows the costs of acquiring replacement parts to be spread out over the term of the lease. Sudden and sporadic lump-sum payments are not encountered, as when parts are purchased on an as-needed basis.
- *Gain control of monthly expenses.* Leasing provides consistent and predictable monthly expenses, eliminating budgetary spikes that occur with conventional purchasing practices.
- *Reduce downtime and service delays.* By leasing "kits" of parts for major product groups, the maintenance technician is assured of having sufficient quantities of components needed to service the equipment at a given facility. There is no downtime in which the technician must wait for parts orders to be filled by the manufacturer or another supplier.
- *Reduce inventory and overhead.* The need to purchase large quantities of replacement parts and stock them in inventory is eliminated. Available kits usually meet all short-term requirements.
- *Track costs and usage for multiple locations.* Leasing can be especially advantageous for technicians that service multiple locations. One master lease agreement can be written to cover all satellite offices. Management then can have access to billing and tracking data to help monitor companywide parts usage and costs, assist in planning and managing budgets, and facilitate internal chargeback by department or location.
- *Preserve working capital for other necessities.* The decision to make a one-time $75,000 payment to purchase needed replacement parts may be prohibitively expensive. The ability to lease those same parts at a fixed monthly cost to the operating budget, however, may offer significant benefits.
- *Maintain flexibility to meet changing requirements.* Equipment and service needs constantly are evolving. Leasing replacement parts provides the flexibility to make changes during the lease term as dictated by hardware upgrades or business conditions.

Most hardware lease agreements span 3 to 5 years, depending on the specific needs of the customer. Parts leases typically are written separately from these leases, covering a shorter period of time. When a company enters into a lease agreement, the parts supplier maintains ownership of the parts and charges a monthly usage fee. This arrangement works because, in most cases, the supplier assumes that the parts will have a residual value when they come off lease. The supplier can refurbish the components, and then sell or lease them to another customer. Suppliers typically recommend the lease of parts kits — customized groupings of parts tailored to meet specific needs. Individual components can be leased as well. Kits can be customized as required.

When the lease term has ended, the maintenance department has several options, including:

- Returning the components to the lessor and replacing them with other parts that meet current needs, in which case a new lease agreement is written.
- Renewing the lease for an additional period of time (probably at reduced rates).
- Purchasing the parts outright at their fair market value.
- Returning the parts to the lessor with no further obligation.

New Vs. Used Parts. As the costs of new technologies continue to rise, and as product life cycles continue to shorten, maintenance departments must constantly seek more cost-effective ways to service their end-users. As a result of these market conditions, many manufacturers are stocking smaller inventories of new parts in their regional facilities. The result is that customers may experience delays in receiving parts orders. The use of refurbished replacement parts is one cost-effective solution. Refurbished parts are typically of high quality, and many feature the latest engineering changes and updates. Reputable refurbishers put their components through full diagnostics and on-line testing, and back them with standard repair/replacement warranties. Like new replacement parts, refurbished parts can be leased or purchased.

5.5   BIBLIOGRAPHY

Bausel, James: "Focus on Soldering and Desoldering," *Electronic Servicing & Technology* magazine, Intertec Publishing, Overland Park, Kan., November 1989.

Cooper, Gershon: "Choosing an Alternative Vendor," *Electronic Servicing & Technology* magazine, Intertec Publishing, Overland Park, Kan., December 1987.

Fenton, Christopher: "Choosing a Soldering Iron," *Electronic Servicing & Technology* magazine, Intertec Publishing, Overland Park, Kan., May 1988.

Graham, Edward S.: "Designing a Working Service Kit," *Microservice Management* magazine, Intertec Publishing, Overland Park, Kan., May 1989.

Jensen, Sherman: "When to Buy the Custom Tool Kit," *Microservice Management* magazine, Intertec Publishing, Overland Park, Kan., February 1988.

O'Brien, Gil: "Cleaning Supplies for Computers," *Microservice Management* magazine, Intertec Publishing, Overland Park, Kan., August 1988.

Persson, Conrad: "Solder: The Tin that Binds," *Electronic Servicing & Technology* magazine, Intertec Publishing, Overland Park, Kan., February 1986.

_____: "I Only Have Two Hands!," *Electronic Servicing & Technology* magazine, Intertec Publishing, Overland Park, Kan., July 1986.

_____: "Setting up a Test Bench," *Electronic Servicing & Technology* magazine, Intertec Publishing, Overland Park, Kan., March 1987.

_____: "Locating Replacement Parts," *Electronic Servicing & Technology* magazine, Intertec Publishing, Overland Park, Kan., December 1987.

Richardson, Jeff: "The Benefits of a Tool Kit Program," *Microservice Management* magazine, Intertec Publishing, Overland Park, Kan., May 1989.

Romano, Joseph: "Leasing Replacement Parts: When is it Right for You?," *Microservice Management* magazine, Intertec Publishing, Overland Park, Kan., December 1990.

Schmerbauch, James: "PCB Rework," *Microservice Management* magazine, Intertec Publishing, Overland Park, Kan., August 1987.

6

DIGITAL TEST INSTRUMENTS

6.1 INTRODUCTION

As the equipment used by consumers and industry becomes more complex, the requirements for highly skilled maintenance technicians also increase. Maintenance personnel today require advanced test equipment and must think in a "systems mode" to troubleshoot much of the hardware now in the field. New technologies and changing economic conditions have reshaped the way maintenance professionals view their jobs. As technology drives equipment design forward, maintenance difficulties will continue to increase. Such problems can be solved only through improved test equipment and increased technician training.

Servicing computer-based professional equipment typically involves isolating the problem to the board level, then replacing the defective PWB. Taken on a case-by-case basis, this approach seems efficient. What is inefficient about the approach, however (and readily apparent), is the investment required to keep a stock of spare boards on hand. Furthermore, because of the complex interrelation of circuits today, a PWB that appears to be faulty may actually turn out to be perfect. The ideal solution is to troubleshoot down to the component level and replace the faulty device instead of swapping boards. In many cases, this approach requires sophisticated and expensive test equipment. In other cases, however, simple test instruments will do the job.

Although the cost of most professional equipment has been going up in recent years, maintenance technicians have seen a buyer's market in test instruments. The semiconductor revolution has done more than simply given consumers low-cost computers and disposable calculators. It also has helped to spawn a broad variety of inexpensive test instruments.

Servicing computer-based systems can be done on just about any level. The technician can swap boards (where the product is modular) or troubleshoot to the component level. If the choice is made to troubleshoot to the component level, a wide variety of test instruments—from simple to complex—is available. Regardless of the method selected, the most important thing the maintenance professional can bring to the test bench is good, solid information about the equipment being serviced and sharply honed troubleshooting skills.

6.2 DIGITAL MULTIMETERS

The most basic multimeter in today's arsenal of electronic test instruments is the analog volt-ohm-milliammeter (VOM). These units are available for well under $100. In addition to the basic voltage, resistance, and current measurements, semiconductor junction testing is possible, to a degree, with a VOM. Although useful for many applications, the analog VOM has its limitations, primarily frequency response and measurement accuracy.

The analog VOM has been replaced largely by the digital multimeter (DMM). When it was introduced more than a decade ago, its price tag was high and its reliability marginal (compared with the analog VOM). Advanced semiconductor technology has changed the equation, however. Beyond the basic voltage-resistance-current measurements, sophisticated digital multimeters provide a number of unique functions. Large-scale integration has made it possible to pack more features into the limited space inside a DMM, and to do so without significantly increasing the cost of the instrument. In addition, because of competition among manufacturers, the functionality of DMM instruments is continually improving. The result: increasingly sophisticated products.

6.2.1 DMM Features

Digital multimeters can be divided into two basic classifications:

1. *Portable:* Hand-held instruments intended for field applications offer extensive capabilities in a portable, rugged package. Some hand-held meters include true-rms measurements, 1 ms response times, capacitance and frequency modes, and recording capabilities.
2. *Benchtop:* Instruments intended for fixed applications usually offer a wide range of features and interface to other instruments. Options include dual display meters that have two fully functional digital readouts, and built-in communications capabilities (RS-232 and IEEE-488). A communications interface permits the meter to be controlled over standard computer ports for data storage, analysis, and remote operation.

Evolutionary developments in DMM design have provided a number of helpful features to maintenance technicians, including:

- *Audible continuity check.* The instrument emits a beep when the measured resistance value is less than 5 Ω or so. This permits the technician to perform point-to-point continuity tests without actually looking at the DMM.
- *Audible high-voltage warning.* The instrument emits a characteristic beep when the measured voltage is dangerously high (100 V or more). This safety feature alerts the technician to use extra care when making measurements.
- *Sample and hold.* The instrument measures and stores the sampled value after it has stabilized. Completion of the sample function is signaled by an audible beep. This feature permits the technician to concentrate on probe placement rather than on trying to read the instrument.
- *Continuously variable tone.* The instrument produces a tone that varies in frequency with the measured parameter. This feature permits the technician to adjust a circuit for a peak or null value without actually viewing the DMM. It also may be used to check for intermittent continuity.
- *Ruggedized case.* Durable cases for portable instruments permit normal physical abuse to be tolerated. Many DMMs in use today may be dropped or exposed to water without failing.
- *Input overload hardening.* Protection devices are placed at the input circuitry to prevent damage in the event of an accidental overvoltage. Fuses and semiconductor-based protection components are used.
- *Logic probe functions.* Signal high/low and pulse detection circuitry built into a DMM probe permits the instrument to function as both a conventional multimeter and a logic probe. This eliminates the necessity to change instruments in the middle of a troubleshooting sequence.
- *Automatic shutoff.* A built-in timer removes power to the DMM after a preset period of nonoperation. This feature extends the life of the internal batteries.

DMM functions are further expanded by the wide variety of probes and accessories available for many instruments. Specialized probes include:

- *High voltage:* Used for measuring voltages above 500 V. Many probes permit measurement of voltages of 50 kV or more.
- *Clamp-type current:* Permits current in a conductor to be measured without breaking the circuit. Probes commonly are available in current ranges of 1 to 2 A and up.
- *Demodulator:* Converts a radio frequency signal into a dc voltage for display on the DMM.
- *Temperature:* Permits the DMM to be used for displaying the temperature of a probe device. Probes are available for surface measurement of solids and for immersion into liquid or gas.

Advances in DMM technology have resulted in increased reliability. DMMs are engineered to include a minimum of moving parts. Auto-ranging eliminates much of the mechanical switch contacts and movement common with older-technology instruments. This approach reduces the number of potential sources of failure, such as corrosion and intermittent switch contacts. Gold plating of the remaining contacts in the DMM further reduces the chance of failure.

6.2.2 Voltage Conversion

Figure 6.1 shows a generalized block diagram of a DMM. Conversion of an input voltage to a digital equivalent can be accomplished in one of several ways. The *dual-slope conversion* method (also know as *double-integration*) is one of the more popular techniques. Figure 6.2 illustrates the dual-slope conversion process. V_{in} represents the input voltage (the voltage to be measured). V_{REF} is a reference voltage with a polarity opposite that of the measured voltage, supplied by the digital conversion circuit. Capacitor C1 and operational amplifier U1 constitute an integrator. S1 is an electronic switch that is initially in the position shown. When the meter probes are connected to the circuit, the sample voltage is applied to the input of the integrator for a specified length of time, called the *integration period*. The integration period usually is related to the 60 Hz line frequency; integration periods of 1/60 s and 1/10 s are common. The output signal of the integrator is a voltage determined by the RC time constant of R1 and C1. Because of the nature of the integrator, the maximum voltage (the voltage at the end of the integration period) is proportional to the voltage being measured. At the end of the integration period, switch S1 is moved to the other position (V_{REF}), and a voltage of opposite polarity to the measured voltage is applied. The capacitor is then discharged to zero. As shown in the figure, the discharge interval is directly proportional to the maximum voltage, which in turn is proportional to the applied voltage.

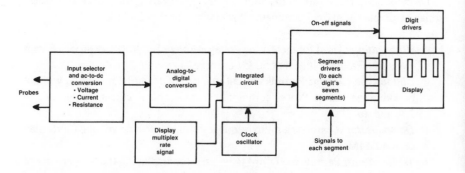

Figure 6.1   Block diagram of a basic DMM. LSI chip technology has reduced most of the individual elements shown to a single IC. (Data from: Conrad Persson, "The New Breed of Test Instruments," *Broadcast Engineering* magazine, Intertec Publishing, Overland Park, Kan., November 1989)

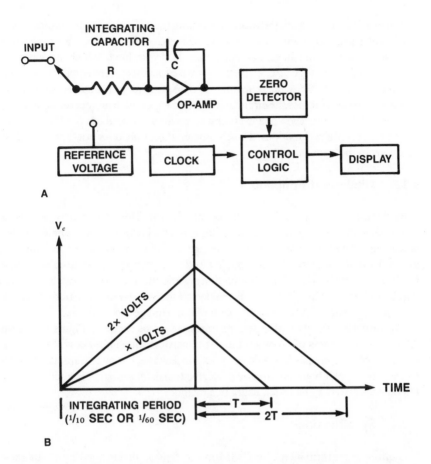

Figure 6.2 The analog-to-digital conversion process: (a–top) The basic circuitry involved. (b–bottom) Graph of V_c vs. time. Because the slope of the discharge curve is identical regardless of the ultimate voltage reached by the integrator, the discharge period and, therefore, the number of counts recorded by the logic circuitry, are proportional to the input voltage. (Data from: Conrad Persson, "The New Breed of Test Instruments," *Broadcast Engineering* magazine, Intertec Publishing, Overland Park, Kan., November 1989)

At the same time that the integration interval ends and the discharge interval begins, a counter in the meter begins counting pulses generated by a clock circuit. When the voltage reaches zero, the counter stops. The number of pulses counted is, therefore, proportional to the discharge period. This count is converted to a digital number and displayed as the measured voltage. Although this method works well, it is somewhat slow, so many microcomputer-based meters use a variation called *multislope integration*.

When a DMM is used in the resistance testing mode, it places a low voltage with a constant-current characteristic across the test leads. After the leads are connected, the voltage across the measurement points is determined, which provides the resistance value (voltage divided by the known constant current equals resistance). The meter converts the voltage reading into an equivalent resistance for display. The voltage placed across the circuit under test is kept low to protect semiconductor devices in the circuit. The voltage at the ohmmeter probe is about 0.1 V or less, too low to turn on silicon junctions. The ohmmeter, therefore, does not "see" transistor junctions in the circuit.

6.2.3   Diode Test Function

A diode test feature typically is included in a DMM. This feature may be used to test diodes, transistor junctions, and other semiconductor devices. The diode test is accomplished by placing a higher voltage across the meter's test leads than is available for most resistance ranges, but limiting the current to a smaller value than is allowed for resistance. This combination ensures that the semiconductor junction will be turned on when the meter is connected in the forward direction, while the current is limited to a safe amount that will not damage the junction.

The diode test function also permits transistors to be checked. Figure 6.3 shows the diode equivalents of common bipolar transistors. Such checks will not reveal subtle problems, such as low gain or excessive leakage, but most transistor failures are catastrophic. If both junctions test OK, the transistor probably is good. Darlington devices also may be tested using this method.

6.2.4   Specifications

The ultimate performance of a DMM in the field is determined by its inherent accuracy. Key specifications include:

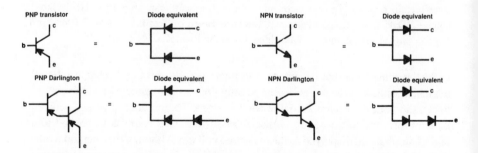

Figure 6.3   The diode equivalent of common transistors. (Data from: William L. Detwiler, "Troubleshooting with a Digital Multimeter," *Mobile Radio Technology* magazine, Intertec Publishing, Overland Park, Kan., September 1988)

- *Accuracy:* How closely the meter indicates the actual value of a measured signal, specified in percent of error. Zero percent of error indicates a perfect meter.
- *Frequency response:* The range of frequencies that can be measured by the meter without exceeding a specified amount of error.
- *Input impedance:* The combined ac and dc resistance at the input terminals of the multimeter. The impedance, if too low, can load critical circuits and result in measurement errors. An input impedance of 10 MΩ or greater will prevent loading.
- *Precision:* The degree of accuracy to which a measurement is carried out. As a rule, the more digits a DMM can display, the more precise the measurement will be.

6.2.5 Conventional Test Instruments

Troubleshooting a digital system requires a wide range of test instruments, from simple to complex. Following are descriptions of some of the more common conventional test instruments:

- *Component tester.* Component checkers apply an ac voltage across a device under test and measure the resulting ac current. Each value and type of component produces an individual *signature* that is displayed as a graph of voltage vs. current on a CRT (cathode-ray tube). Components that can be tested with such instruments include resistors, inductors, diodes, transistors, and linear integrated circuits. Both in-circuit and out-of-circuit measurements can be performed.
- *Transistor checker.* Simple semiconductor test instruments are used to check the in- or out-of-circuit performance of diodes and transistors. Measurements include leakage and gain. Many transistor checkers also can be used to identify the leads of a device.
- *L/C tester.* Inductor and capacitor (*L/C*) testers are dedicated to measuring the value of out-of-circuit components. Such instruments can be used to check device tolerance, sort values, select precision values, and measure cable or switch stray capacitance/inductance.

Frequency Counter. A frequency counter provides an accurate measure of signal cycles or pulses over a standard period of time. It is used to totalize or to measure frequency, period, frequency ratio, and time intervals. Key specifications for a counter include:

- *Frequency range:* The maximum frequency that the counter can resolve.
- *Resolution:* The smallest increment of change that can be displayed. The degree of resolution is selectable on many counters. Higher resolution usually requires a longer acquisition time.

- *Sensitivity:* The lowest amplitude signal that the instrument will count (measured in fractions of a volt).
- *Time base accuracy:* A measure of the stability of the time base. Stability is measured in parts per million (ppm) while the instrument is subjected to temperature and operating voltage variations.

Multifunction Hand-Held Instrument. Although specialized instruments such as a separate frequency counter or capacitance meter may be the best choice for a specific range of tasks, combinations of instruments are available for the convenience of maintenance personnel. Separate analyzers usually offer a broader range of measurement capabilities, but a combined instrument often makes field servicing easier.

6.3 LOGIC INSTRUMENTS

The simplest of all logic instruments is the logic probe. A basic logic probe tells the technician whether the logic state of the point being checked is high, low, or pulsed. Most probes include a pulse stretcher that will show the presence of a 1-shot pulse. The indicators usually are LEDs (light-emitting diodes). In some cases, a single LED is used to indicate any of the conditions (high, low, or pulsing); generally, individual LEDs are used to indicate each state. Probes usually include a switch to match the level sense circuitry of the unit to the type of logic being checked (TTL or CMOS). The probe receives its power from the circuit under test. Connecting the probe in this manner sets the approximate value of signal voltage that constitutes a logic low or high. For example, if the power-supply voltage for a CMOS logic circuit is 18 V, a logic low would be about 30 percent of 18 V (5.4 V), and a logic high would be about 70 percent of 18 V (12.6 V).

The logic pulser is the active counterpart to the logic probe. The logic pulser generates a single pulse, or train of pulses, into a circuit to check the response.

6.3.1 Logic Current Tracer

The close spacing of pins and conductors on a crowded PWB invites short circuits. On a large board, locating a short can be a difficult and time-consuming task. The logic current tracer can be a valuable tool in troubleshooting such failures.

Even in an extremely thin printed circuit conductor, a change in current induces a magnetic field. The probe of a logic current tracer is designed to sense changes in the minute magnetic field of a PWB trace. An LED indicates the presence of the field. To locate a short circuit, the current tracer is used in conjunction with the logic pulser. As illustrated in Figure 6.4, pulses are injected at an identifiable failure point. The tracer then is moved slowly along the circuit trace until the LED goes off, indicating that the point of the fault has been passed.

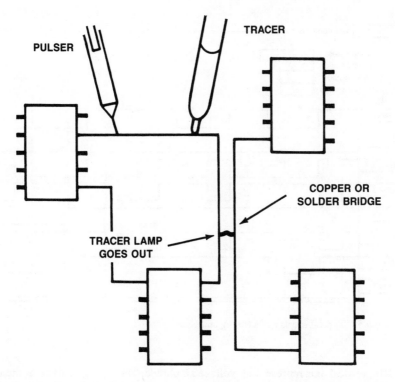

Figure 6.4 Use of a logic pulser and current tracer to locate a short circuit on a PWB.

6.3.2 Logic Test Clip

A logic test clip is a device that clips directly over an IC package to display the logic status of all pins on the chip. Each probe of the clip contacts one pin of the IC. The probes are connected to LEDs that indicate whether that point in the circuit is at a logic high or low, or pulsed.

The IC comparator is an extension of the logic test clip. The comparator allows the operation of an in-circuit IC to be compared with a known-good IC of the same type placed in the comparator's *reference* socket. While the circuit containing the IC under test is operating, the comparator matches the responses of the reference IC to the test device.

6.3.3 Logic Analyzer

The logic analyzer is really two instruments in one: a *timing analyzer* and a *state analyzer*. The timing analyzer is analogous to an oscilloscope. It displays information in the same general form as a scope, with the horizontal axis representing time

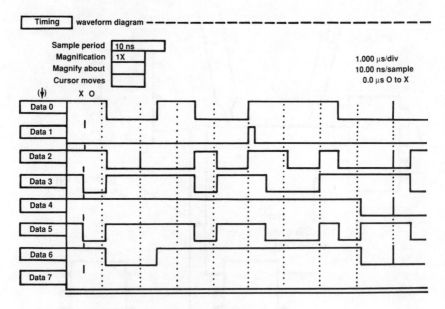

Figure 6.5 Typical display of a timing analyzer.

and the vertical axis representing voltage amplitude. The timing analyzer samples the input waveform to determine whether it is high or low. The instrument cares about only one voltage threshold. If the signal is above the threshold when it is sampled, it will be displayed as a 1 (or high); any signal below the threshold will be displayed as a 0 (or low). From these sample points, a list of 1s and 0s is generated to represent a picture of the input waveform. This data is stored in memory and used to reconstruct the input waveform, as shown in Figure 6.5. A block diagram of a typical timing analyzer is shown in Figure 6.6.

Sample points for the timing analyzer are developed by an internal clock. The period of the sample can be selected by the user. Because the analyzer samples asynchronously to the unit under test (UUT) under the direction of the internal clock, a long sample period results in a more accurate picture of the data bus. A sample period should be selected that permits an accurate view of data activity, but does not fill up the instrument's memory with unnecessary data.

Accurate sampling of data lines requires a trigger source to begin data acquisition. A fixed delay may be inserted between the trigger point and the *trace point* to allow for bus settling. This concept is illustrated in Figure 6.7. Various trigger modes commonly are available on a timing analyzer, including:

- Level triggering. Data acquisition and/or display begins when a logic high or low is detected.

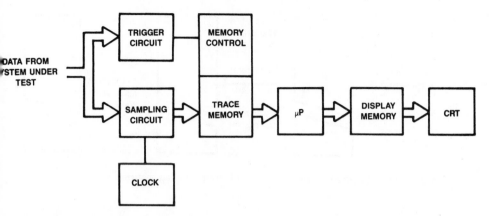

Figure 6.6 Block diagram of a timing analyzer.

- Edge triggering. Data acquisition/display begins on the rising or falling edge of a selected signal. Although many logic devices are level-dependent, clock and control signals are often edge-sensitive.
- Bus state triggering. Data acquisition/display is triggered when a specific code is detected (specified in binary or hexadecimal).

The timing analyzer constantly is taking data from the monitored bus. Triggering, and subsequent generation of the trace point, controls the *data window* displayed to the technician. It is possible, therefore, to configure the analyzer to display data that precedes the trace point. (See Figure 6.8.) This feature can be a powerful troubleshooting and developmental tool.

The timing analyzer is probably the best method of detecting *glitches* in computer-based equipment. A glitch is any transition that crosses a logic threshold

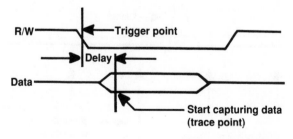

Trace point = trigger + delay

Figure 6.7 Use of a delay period between the trigger point and the trace point of a timing analyzer.

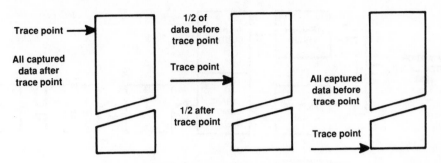

Figure 6.8   Data-capturing options available from a timing analyzer.

more than once between clock periods. (See Figure 6.9.) The triggering input of the analyzer is set to the bus line that is experiencing random glitches. When the analyzer detects a glitch, it displays the bus state before, during, or after the disturbance.

The state analyzer is the second half of a logic analyzer. It is used most often to trace the execution of instructions through a microprocessor system. Data, address, and status codes are captured and displayed as they occur on the microprocessor bus. A *state*—a sample of the bus when the data is valid—usually is displayed in a tabular format of hexadecimal, binary, octal, or assembly language. Because some microprocessors multiplex data and addresses on the same lines, the analyzer must be able to clock in information at different clock rates. The analyzer, in essence, acts as a demultiplexer to capture an address at the proper time, then to capture data present on the same bus at a different point in time. A state analyzer also enables the operator to *qualify* the data stored. Operation of the instrument may be triggered by a specific logic pattern on the bus. State analyzers usually offer

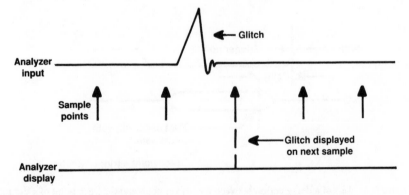

Figure 6.9   Use of a timing analyzer for detecting glitches on a monitored line.

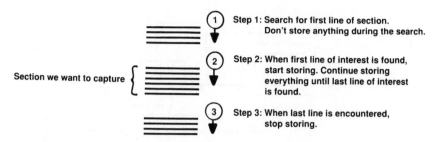

Figure 6.10   Illustration of the selective storage capability of a state analyzer.

a *sequence term* feature that aids in triggering. A sequence term allows the operator to qualify data storage more accurately than would be possible with a single trigger point. A sequence term usually takes the following form:

Find	xxxx
Then find	yyyy
Start on	zzzz

A sequence term is useful for probing a subroutine from a specific point in a program. It also makes possible selective storage of data, as shown in Figure 6.10. To make the acquired data easier to understand, most state analyzers include software packages that interpret the information. Such *disassemblers* (also known as *inverse assemblers*) translate hex, binary, or octal codes into assembly code to make them easier to read.

The logic analyzer is used routinely by design engineers desiring an in-depth look at signals within digital circuits. A logic analyzer can operate at 100 MHz and beyond, making the instrument ideal for detecting glitches resulting from timing problems. Such faults usually are associated with design flaws, not with manufacturing defects or failures in the field.

Although the logic analyzer has benefits for the service technician, its use is limited. Designers must be able to verify hardware and software implementations with test equipment; service technicians simply need to isolate a fault quickly. It is difficult and costly to automate test procedures for a given PWB using a logic analyzer. The technician must examine a long data stream and decide whether the data is good or bad. Writing programs to validate state analysis data is possible, but time-consuming.

6.3.4   Signature Analyzer

Signature analysis is a common troubleshooting tool based on the old analog method of signal tracing. In analog troubleshooting, the technician followed an

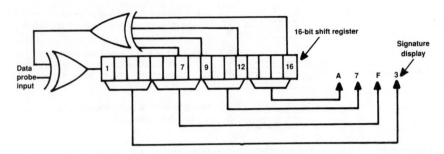

Figure 6.11 Simplified block diagram of a signature analyzer. (Data from: "Signature Analyzers," Mike Wickstead, *Microservice Management* magazine, Intertec Publishing, Overland Park, Kan., October, 1985)

annotated schematic that depicted waveforms and voltages that should be observed with an oscilloscope and voltmeter. To accomplish the same step-by-step troubleshooting in a digital system would be extremely difficult. Even if there were room to annotate a data stream on a schematic, comparing it with what is seen on the oscilloscope or timing analyzer would be almost impossible. Missing only 1 data bit could be catastrophic in a digital system, and there can be thousands or millions of data bits to examine. The signature analyzer overcomes these problems by taking a complex data stream from a test point on the PWB and converting it into a 4-digit hexadecimal signature on an LED or LCD display. These hexadecimal signatures are easy to annotate on schematics, permitting their use as references for comparison against signatures obtained from a UUT. The signature by itself means nothing; the value of the signature analyzer is in the ability to compare a known-good signature with the UUT.

A block diagram of a simplified signature analyzer is shown in Figure 6.11. A digital bit stream, accepted through a data probe during a specified measurement window, is marched through a 16-bit linear feedback shift register. Whatever data is left over in the register after the specified measurement window has closed is converted into a 4-bit hexadecimal readout. The feedback portion of the shift register allows a single faulty bit in a digital bit stream to create an entirely different signature than would be expected in a properly operating system. Hewlett-Packard, which developed this technique, terms it a *pseudorandom binary sequence* (PRBS) generator. This method allows the maintenance technician to identify a single bit error in a digital bit stream with 99.998 percent certainty, even when picking up errors that are timing-related.

To function properly, the signature analyzer requires start and stop signals, a clock input, and the data input. The start and stop inputs are derived from the circuit being tested and are used to bracket the beginning and end of the measurement window. During this gate period, data is input through the data probe. The clock input controls the sample rate of data entering the analyzer. The clock is most often taken from the clock input pin of the microprocessor. The start, stop, and clock inputs may be configured by the technician to trigger on the rising or falling edges of the input signals.

Signature analysis on a microprocessor-based system is possible only when the address and data lines are active. Any number of hardware failures can cause a microprocessor to shut down, preventing the use of a signature analyzer. One solution is the purchase of a *no-operation* (NOP) adapter for the microprocessor chip. A NOP adapter is a simple device that plugs into the microprocessor socket after the device has been removed. The chip then is inserted into the socket on the NOP fixture, which is configured to sever the connections between the data and interrupt lines of the processor and the board being tested.

The fixture hard-wires a no-operation instruction to the microprocessor, which essentially turns it into a 64K counter (on a 16-bit microprocessor system), counting off numbers on its address lines. This supplies the technician with predictable, repeatable activity for taking reliable signatures.

Application Example. The output of a pair of ROM chips needs to be verified. The analyzer clock input is connected to the microprocessor clock, and both the start and stop inputs are attached to the chip enable input of the device under test. The start input is set to trigger on the falling edge, and the stop input is set for a rising edge trigger. This arrangement will ensure a valid measurement window because a typical ROM will show valid data only when its chip enable pin is at a low logic level. The data probe is placed at one of the data outputs of the ROM. The analyzer then will display a stable, predictable signature for the valid data activity on the data output line.

6.3.5   Manual Probe Diagnosis

Manual probe diagnosis integrates the logic probe, logic analyzer, and signature analyzer to troubleshoot computer-based hardware. The manual probe technique employs a database consisting of nodal-level signature measurements. Each node (primary junction or branch point) of a known-good unit under test is probed and the signature information saved. This known-good data then is compared to the data from a faulty UUT. Efficient use of the manual probe technique requires a skilled operator to determine the proper order for probing circuit nodes.

6.3.6   Checking Integrated Circuits

An IC tester identifies chip faults by generating test patterns that exercise all possible input-state combinations of the unit under test. For example, the NAND gate shown in Figure 6.12 has a maximum of four input patterns that the instrument must generate to completely test the device. The input test patterns and the resulting

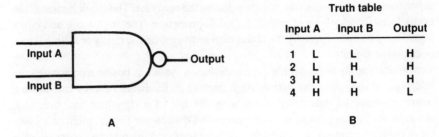

Truth table		
Input A	Input B	Output
1 L	L	H
2 L	H	H
3 H	L	H
4 H	H	L

A   B

Figure 6.12   Truth table for a NAND gate.

device outputs are arranged in a truth table. The table represents the response that should be obtained with specified input stimulus. The operator inputs the device type number, which configures the instrument to the proper pattern and truth table. As illustrated in Figure 6.13, devices that match the table are certified as good; those failing are rejected.

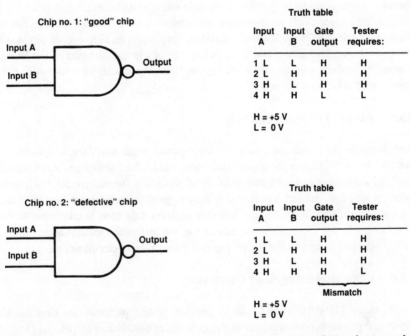

Chip no. 1: "good" chip

Truth table			
Input A	Input B	Gate output	Tester requires:
1 L	L	H	H
2 L	H	H	H
3 H	L	H	H
4 H	H	L	L

H = +5 V
L = 0 V

Chip no. 2: "defective" chip

Truth table			
Input A	Input B	Gate output	Tester requires:
1 L	L	H	H
2 L	H	H	H
3 H	L	H	H
4 H	H	H	L

Mismatch

H = +5 V
L = 0 V

Figure 6.13   Component accept/reject process based on comparing observed states and an integrated circuit logic truth table.

Such conventional functional tests are well-suited for out-of-circuit device qualification. Under out-of-circuit conditions, the instrument is free to toggle each input high or low, thereby executing all patterns of the truth table. For many in-circuit applications, however, inputs are wired in ways that prevent the instrument from executing the predetermined test pattern. Some instruments are able to circumvent the problem of nonstandard (in-circuit) configurations by including a "learning mode" in the system. Using this mode, the IC tester exercises the input of a gate that is known to be operating properly in a nonstandard configuration with the normal set of input signals. The resulting truth table is stored in memory. The instrument then is connected to an in-circuit device to be checked.

6.3.7 Emulative Tester

Guided diagnostics can be used on certain types of hardware to facilitate semiautomated circuit testing. The test connector typically taps into one of the system board input/output ports and, armed with the proper software, the instrument goes through a step-by-step test routine that exercises the computer circuits. When the system encounters a fault condition, it outputs an appropriate message. The *emulative tester* is a variation on the general guided diagnostics theme. Such an instrument plugs into the microprocessor socket of the computer and runs a series of diagnostic tests to identify the cause of a fault condition. An emulative tester imitates, or emulates, the board's microprocessor while verifying circuit operation. Testing the board "inside-out" allows easy access to all parts of the circuit; synchronization with the various data and address cycles of the PWB is automatic. Test procedures, such as read/write cycles from the microprocessor are generic, so that high-quality functional tests can be created quickly for any board. Signature analysis is used to verify that circuits are operating correctly. Even with long streams of data, there is no maximum memory depth.

Several different types of emulative testers are available. Some are designed to check only one brand of computer, or only computers based on one type of microprocessor. Other instruments can check a variety of microcomputer systems, using *personality* modules that adapt the instrument to the system being serviced.

Guided-fault isolation (GFI) is practical with an emulative tester because the instrument maintains control over the entire system. Automated tests can isolate faults to the node level. All board information and test procedures are resident within the emulative test instrument, including prompts to the operator on what to do next. Using the microprocessor test connection combined with movable probes or clips allows a closed-loop test of any part of a circuit. Input/output capabilities of emulative testers range from single-point probes to more than a hundred I/O lines. These lines can provide stimulus as well as measurement capabilities.

The principal benefit of the guided probe over the manual probe (discussed in Sec. 6.3.5) is derived from the creation of a *topology database* for the UUT. The data base describes devices on the UUT, their internal fault-propagating characteristics, and their interdevice connections. In this way, the tester can guide the operator down the proper logic path to isolate the fault. A guided probe accomplishes the same analysis as a conventional fault tree, but differs in that troubleshooting is based on a generic algorithm that uses the circuit model as its input. First, the system compares measurements taken from the failed UUT against the expected results. Next, it searches for a database representation of the UUT to determine the next logical place to take a measurement. The guided probe system automatically determines which nodes impact other nodes. The algorithm tracks its way through the logic system of the board to the source of the fault. Programming of a guided probe system requires the following steps:

- Development of a stimulus routine to exercise and verify the operating condition of the UUT (go/no-go status).
- Acquisition and storage of measurements for each node affected by the stimulus routine. The programmer divides the system or board into measurement sets (MSETs) of nodes having a common time base. In this way, the user can take maximum advantage of the time-domain feature of the algorithm.
- Programming a representation of the UUT that depicts the interconnection between devices, and the manner in which signals can propagate through the system. Common *connectivity libraries* are developed and reused from one application to another.
- Programming a measurement set cross-reference database. This allows the guided probe instrument to cross MSET boundaries automatically without operator intervention.
- Implementation of a test program control routine that executes the stimulus routines. When a failure is detected, the guided probe database is invoked to 'determine the next step in the troubleshooting process.

6.3.8   Protocol Analyzer

Testing a local area network (LAN) presents special challenges to the maintenance technician. The protocol analyzer commonly is used to service LANs and other communications systems. The protocol analyzer performs data monitoring, terminal simulation, and bit error rate tests (BERTs). Sophisticated analyzers provide *high-level decide* capability, which refers to the open system interconnection (OSI) network 7-layer model. (OSI networks are discussed in Chapter 8.) Typically, a sophisticated analyzer based on a personal computer can decide to level 3. PC-based devices also may incorporate features that permit performance analysis of wide area networks (WANs). Network statistics, including response time and use, are measured and displayed graphically. The characteristics of a line can be viewed and measured easily. These functions enable the technician to observe activity of a communications link and exercise the facility to verify proper operation. Simulation capability usually is available to emulate almost any data terminal or data communications equipment.

LAN test equipment functions include monitoring of protocols and emulation of a network node, bridge, or terminal. Statistical information is provided to isolate problems to particular devices, or to high periods of activity. Statistical data includes use, packet rates, error rates, and collision rates.

6.4   AUTOMATED TEST INSTRUMENTS[1]

Computers can function as powerful troubleshooting tools. Advanced-technology personal computers, coupled with add-on interface cards and applications software, provide a wide range of testing capabilities at a reasonable price. Computers also can be connected to many different instruments to provide for automated testing. The two basic types of stand-alone automated test instruments (ATEs) are *functional* and *in-circuit*.

Functional testing exercises the unit under test to identify faults. Individual PWBs or subassemblies may be checked using this approach. Functional testing provides a fast go/no-go qualification check. Dynamic faults are best discovered through this approach. Functional test instruments are well-suited to high-volume testing of PWB subassemblies, but programming is a major undertaking. An intimate knowledge of the subassembly is needed to generate the required test patterns. An in-circuit emulator is one type of functional tester.

In-circuit testing is primarily a diagnostic tool. It verifies the functionality of the individual components of a subassembly. Each device is checked, and failing parts are identified. In-circuit testing, although valuable for detailed component checking, does not operate at the clock rate of the subassembly. Propagation delays, race conditions, and other abnormalities may go undetected. Access to key points on the PWB may be accomplished in one of two ways:

- Bed of nails. The subassembly is placed on a dedicated test fixture and held in place by a vacuum or through mechanical means. Probes access key electrical traces on the subassembly to check individual devices. Through a technique known as *backdriving*, the inputs of each device are isolated from the associated circuitry, and the components are tested for functionality. This

---

1 Portions of Section 6.4 were adapted from: Gregory Carey, "Automated Test Instruments," *Broadcast Engineering* magazine, Intertec Publishing, Overland Park, Kan., November 1989.

type of testing, which is expensive, is practical only for high-volume subassembly qualification testing and rework.

- *PWB clips.* Intended for lower-volume applications than a bed-of-nails instrument, PWB clips replace the dedicated test fixture. The operator places the clips on the board as directed by the instrument. Because simultaneous access to all components is not required, the tester is less hardware-intensive, and programming is simplified. Many systems include a library of software routines designed to test various classes of devices. Test instruments using PWB clips tend to be slow. Test times for an average PWB range from 8 to 20 minutes, compared with 1 minute or less for a bed of nails.

6.4.1 Computer-Instrument Interface

Two types of computer interface systems are used to connect test instruments and computers: IEEE-488 and RS-232. The IEEE-488 format also is called the *general-purpose interface bus* (GPIB) or the *Hewlett-Packard interface bus* (HPIB). RS-232 is the standard serial interface used on many computers. Both of the interface systems have their own advantages. Both provide for connection of a computer to one or more measuring instruments. Both are bidirectional, which allows the computer to either send information or receive it from the outside world. Some systems provide both interfaces, but most have one or the other. Figure 6.14 illustrates the differences between RS-232 and GPIB.

Test instruments utilizing the GPIB interface greatly outnumber those with RS-232. Several thousand test instruments are available with GPIB interfacing as an option. Some plotters and printers also accept a GPIB input.

By comparison, RS-232 is more common than GPIB in computer applications. Printers, plotters, scanners, and modems often use the *standard serial interface*, which is another name for RS-232. Some printers use a third protocol known as the *Centronics parallel* standard. This parallel format is used when data flows in only one direction, from the computer to the printer. Test instruments incorporating RS-232 typically are those used for remote sensing, such as RF signal-strength meters or thermometers.

Neither RS-232 nor GPIB is ideal for every application. Each protocol works well in some uses, marginally in others, and poorly in still others. Table 6.1 lists the relative advantages and disadvantages of the two protocols. Notice that the only advantage common to both is the capability to move data in both directions. Beyond that, the two are quite different. The decision of which protocol to use for a particular application must be based on what the system needs to do. Because RS-232 is already built into personal computers, many users want to employ it for automation. Yet, GPIB is the preferred protocol for most test-equipment applications.

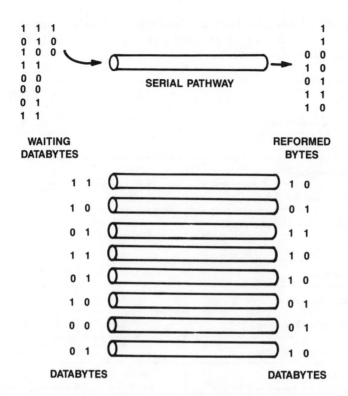

Figure 6.14 Comparison of data formats: (a) Serial RS-232 protocol. (b) Parallel GPIB protocol. (Data from: Gregory D. Carey, "Automated Test Instruments," *Broadcast Engineering* magazine, Intertec Publishing, Overland Park, Kan., November 1989)

Format Differences. RS-232's single biggest advantage is that it can send signals over longer distances easily. It can send data directly about 1000 feet in one run of cable. A *line extender* lets the run go even farther. For example, many mainframe computers use RS-232 to send data to printers located at different sites of a business. Inexpensive twisted-pair cables interconnect the computer and printer. If data must be sent over longer distances, a modem can be used to convert RS-232 signals into a form that can be fed over a standard telephone line. GPIB signals first must be converted to RS-232 if a modem is needed.

By contrast, GPIB's single biggest advantage is that it can work with several instruments simultaneously. This capability is essential when an automated test requires more than one item to be under computer control. For example, a manufacturer might use GPIB to automate several pieces of test equipment at the end of a production line. Up to 15 different units can be connected simultaneously when GPIB is used.

Table 6.1   Comparison of the RS-232 and GPIB Standards. Data from: Gregory D. Carey, "Automated Test Instruments," *Broadcast Engineering* magazine, Intertec Publishing, Overland Park, Kan., November 1989.

Advantages	RS-232	GPIB
Bidirectional data transfer	X	X
Works with long cables	X	
Sends data by phone	X	
Included in most computers	X	
Inexpensive cables and connectors	X	
Controls one to 15 units		X
Fast data transfer		X
Standard to most test equipment		X
Automatically adjusts speed		X
Plug-together compatibility		X
One standard connector		X
Advanced software available		X
Disadvantages		
Controls only one unit	X	
Speed of computer must match controlled unit	X	
Many data formats	X	
Many wiring variations	X	
Several connector styles	X	
Higher cost to add		X
Short cable runs only		X
Expensive multiconductor cable		X

The reason for these differences lies in the way the signals are fed to and from the computer. RS-232 is a bidirectional serial system. GPIB is a parallel format. Eight separate wires carry the GPIB data into or out of the computer, allowing an entire byte to move at one time. If all things were equal, this would make GPIB eight times faster than RS-232. GPIB actually can transfer data about 260 times faster than the fastest RS-232 data rate because of other electrical differences in GPIB. The parallel structure of GPIB permits external instruments to be addressed one at a time or in groups, allowing connection of several units. RS-232 requires complicated mechanical or electrical signal switching to work with multiple instruments.

6.4.2   Availability of Interfaces

RS-232 interface ports are either included as part of a personal computer or easily added with a low-cost accessory board. RS-232 will interface with many printers. Third-party manufacturers make accessories to add GPIB to IBM (and IBM-compatible) and Apple/Macintosh computers. (IBM is a registered trademark of International Business Machines. Apple and Macintosh are registered trademarks of Apple Computers.) Specialized computers are available, designed as instrument controllers and using GPIB as their main input/output port.

Because RS-232 is common on computers but GPIB is often needed for test equipment automation, several manufacturers make protocol converters that translate RS-232 signals to GPIB. These converters permit the benefits of both communications protocols to be employed. Both RS-232 and GPIB are based on industrywide standards, but only GPIB is a *true standard*. RS-232 has numerous variations, making direct connection more difficult.

RS-232 Formats. The RS-232 standard specifies voltage levels and polarity so that one RS-232 feeds another directly. However, the many variations in RS-232 make it difficult to work with. Most initial problems result from variations in data-transfer rates, data formats, and electrical connectors. After an RS-232 link is up and running, it usually works well.

RS-232 encompasses 15 different data-transfer rates. The rate of data transfer is measured in *baud*. One baud is the transfer of one data bit per second (b/s). A 300-baud device transfers data at 300 b/s. It takes about 10 bits (7 or 8 data bits plus 2 or 3 control bits) to form 1 character (byte), so data transfers at one-tenth the baud rate. A baud rate of 300 yields about 30 characters per second (c/s), and the fastest RS-232 baud rate of 19,200 sends data at approximately 1900 c/s.

The computer and the external device must use the same baud rate to communicate. If the two rates are different, each character is garbled, and all data is lost. Most RS-232 devices have configuration switches for matching the baud rate to the computer system. Aside from speed variations, there are nearly a dozen data format variations. Data bytes can be either 7 or 8 bits long. RS-232 adds stop bits

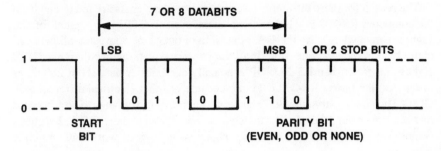

Figure 6.15 Characteristic data format for RS-232. (Data from: Gregory D. Carey, "Automated Test Instruments," *Broadcast Engineering* magazine, Intertec Publishing, Overland Park, Kan., November 1989)

and parity bits, which help ensure accurate data transfer. (See Figure 6.15.) There can be 1 or 2 stop bits, and parity can be none, even, odd, mark, or space. The number of stop and parity bits also must match in order for data to move from one device to the other. Again, switches on the unit let the user match an instrument to the computer system.

RS-232 also uses at least four different physical connectors with 5, 8, 9, or 25 pins. The 25-pin version is the most common, but it has dozens of different wiring variations. Luckily, the main data and ground pins are always the same. Table 6.2 shows the four pins that are always the same.

The data-out and data-in pins interchange, depending on whether the computer end or the instrument end of the cable is referenced. The "out" pin of the computer must feed the "in" pin of the external device, and vice versa. There is a 50-50 chance that an RS-232 cable will connect the inputs and the outputs correctly. Some devices have switches or jumpers to let the user exchange the wiring of the "in" and "out" pins. If there is no way to switch the pins internally, there are two solutions: custom-wired cables or an adapter plug that reverses the connectors (a null-modem adapter, modem eliminator, cable switcher, or line reverser). The null-modem has a female connector on one side and a male connector on the other with the data wires exchanged between them (pin 2 on one side connects to pin 3 on the other, and vice versa). The null-modem also exchanges the pins used for handshaking functions, which will be discussed in more detail in Sec. 6.4.3.

GPIB Format. All GPIB connections and signals are the same. Any GPIB device may be connected to any GPIB computer. Because of its standard format, GPIB does not require settings for baud rate, stop bits, and parity bits. The format uses the same pins for all data going to or coming from the computer. In addition, the system automatically adjusts the data-transfer speed. GPIB can transfer data at any speed from less than 1 c/s to 500,000 c/s, making it up to 260 times faster than

Table 6.2   The Four Standard Pins of the RS-232 Connector. Data from: Gregory D. Carey, "Automated Test Instruments," *Broadcast Engineering* magazine, Intertec Publishing, Overland Park, Kan., November 1989.

Function	Pin
Safety ground	1
Data out	2
Data in	3
Data ground	7

an RS-232 system operating at 19,200 baud. Few instruments supply data that fast, but the system is capable of this speed without modification.

The GPIB connector, cable, and signals are always the same. A standard 24-pin connector hooks one piece to another. Each connector has both a male and a female connector, allowing them to be stacked on top of each other. Some systems are built in a "star" arrangement, with each cable terminating in a single point; others are configured as a "daisy chain," looping from one instrument to the next. It is possible to mix stacking and chaining for any connection scheme required. There is a limit of about 2 meters per connection, however, which limits GPIB to short runs. Longer runs cause capacitive loading, which may distort high-speed data.

6.4.3   Handshaking

Handshaking is the method by which data transfer is controlled between the computer and the external device. Figure 6.16 illustrates the concept. GPIB takes care of handshaking with a single, standard method. There is nothing to consider in the design of a system; it either meets GPIB standards and works, or it does not. RS-232 uses both hardware and software handshaking, with variations on each.

Transferring data by telephone requires a method of handshaking that can be encoded for transmission over a standard phone line. The only method that meets this requirement is software handshaking. Indicators to start and stop data transfer are sent as special characters with the other data. The most commonly used software handshake is the *X-ON/X-OFF* (also called the *DC1/DC3*) system. Sending a special character (Control-S) to the device transmitting data causes it to stop. Sending another special character (Control-Q) causes it to start again.

The *ETX/ACK* protocol is another handshaking format. It is a complex control system that eliminates the need for the sending device to constantly monitor the return line for a busy signal. ETX/ACK often is used when data is sent between

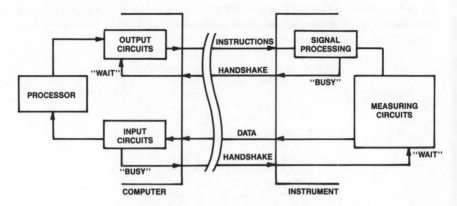

Figure 6.16 Handshaking signals halt the data flow if the device at either end of the cable is busy. Both RS-232 and IEEE-488 use handshaking. (Data from: Gregory D. Carey, "Automated Test Instruments," *Broadcast Engineering* magazine, Intertec Publishing, Overland Park, Kan., November 1989)

mainframe computers. The data is grouped into blocks with the same number of characters. Control characters embedded with the other data mark the end of blocks. The receiving device stores entire blocks of data until it can be processed.

RS-232 systems that do not involve a modem can use either software handshaking or hardware handshaking. With hardware handshaking, two additional wires connect the computer to the external device. One signals the computer to stop sending characters if the remote device is busy; the other serves the same function in the opposite direction. Several pins are used for hardware handshaking. The most common are pins 6 and 20. Others systems, however, might use pin 11 or 19; still others use pins 4 and 5 for different handshaking purposes.

A breakout box is a special tester with lights that show which pins are active when an RS-232 system is connected. This tester often saves time when a device is first connected to a computer. After the connections have been determined, the breakout box is removed, and a cable of the proper configuration is connected in its place.

6.4.4   Software Considerations

Most automated test instrument packages include software that permits the user to customize test instruments and procedures to meet the required task. The software, in effect, writes software. The user enters the codes needed by each automatic instrument, selects the measurements to be performed, and tells the computer where to store the test data. The program puts the final software together after the user answers key configuration questions. The software then looks up the operational codes for each instrument and compiles the software to perform the required tests. Automatic generation of programming greatly reduces the need to have experienced programmers on staff. Once installed in the computer, programming becomes as simple as fitting graphic symbols together on the screen.

6.4.5 Applications

The most common applications for computer-controlled testing are data-gathering, product go/no-go qualification, and troubleshooting. All depend on software to control the instruments. Acquisition of data often can be accomplished with a computer and a single instrument. The computer collects dozens or hundreds of readings until the occurrence of some event. The event may be a preset elapsed time or the occurrence of some condition at a test point, such as exceeding or dropping below a preset voltage. Readings from a variety of test points then are stored in the computer. Under computer direction, test instruments also can run checks that might be difficult or time-consuming to perform manually. The computer may control several test instruments, such as power supplies, signal generators, and frequency counters. The data is stored for later analysis.

6.5 BIBLIOGRAPHY

Carey, Gregory D.: "Automated Test Instruments," *Broadcast Engineering* magazine, Intertec Publishing, Overland Park, Kan., November 1989.

Detwiler, William L.: "Troubleshooting with a Digital Multimeter," *Mobile Radio Technology* magazine, Intertec Publishing, Overland Park, Kan., September 1988.

Gore, George: "Choosing a Hand-Held Test Instrument," *Electronic Servicing & Technology* magazine, Intertec Publishing, Overland Park, Kan., December 1988.

Hamilton, Robert: "Selecting the Right Tool for the Job," *Microservice Management* magazine, Intertec Publishing, Overland Park, Kan., January 1988.

Harju, Rey: "Hands-Free Operation: A New Generation of DMMs," *Microservice Management* magazine, Intertec Publishing, Overland Park, Kan., August 1988.

Ogden, Leonard: "Choosing the Best DMM for Computer Service," *Microservice Management* magazine, Intertec Publishing, Overland Park, Kan., August 1989.

Persson, Conrad: "The New Breed of Test Instruments," *Broadcast Engineering* magazine, Intertec Publishing, Overland Park, Kan., November 1989.

_____: "Test Equipment for Personal Computers," *Electronic Servicing & Technology* magazine, Intertec Publishing, Overland Park, Kan., July 1987.

_____: "IC Tester," *Electronic Servicing & Technology* magazine, Intertec Publishing, Overland Park, Kan., June 1987.

Siner, T. Ann: "Guided Probe Diagnosis: Affordable Automation," *Microservice Management* magazine, Intertec Publishing, Overland Park, Kan., March 1987.

Sokol, Frank: "Specialized Test Equipment," *Microservice Management* magazine, Intertec Publishing, Overland Park, Kan., August 1989.

Stasonis, Robert A.: "Stopping the Profit Drain," *Microservice Management* magazine, Intertec Publishing, Overland Park, Kan., August 1989.

Wickstead, Mike: "Signature Analyzers," *Microservice Management* magazine, Intertec Publishing, Overland Park, Kan., October 1985.

Feeling Comfortable with Logic Analyzers, Hewlett-Packard, Colorado Springs, Colo.

7

OSCILLOSCOPES

7.1 INTRODUCTION

The oscilloscope is one of the most general-purpose of all test instruments. It can be used to measure voltages, examine waveshapes, check phase relationships, examine clock pulses, and perform countless other functions. There has been considerable improvement in oscilloscope technology within the past few years. Now, in addition to familiar specifications such as bandwidth and rise time, and single or multitrace operation, oscilloscopes come in digital as well as analog versions. Some offer logic tracing functions and a hard-copy printout. Table 7.1 lists some basic guidelines on performance characteristics that can be used as a starting point for matching individual requirements with the products available in the marketplace.

7.1.1 Specifications

A number of parameters are used to characterize the performance of an oscilloscope. Key parameters include:

- Bandwidth
- Rise time

Bandwidth. Oscilloscope bandwidth is defined as the frequency at which a sinusoidal signal will be attenuated by a factor of 0.707 (or reduced to 70.7 percent of its maximum value). This is referred to as the –3 dB point. Bandwidth considerations for sine waves are straightforward. For square waves and other complex waveforms, however, bandwidth considerations become substantially more involved.

Square waves are made up of an infinite number of sine waves—the fundamental frequency plus mostly odd harmonics. Fourier analysis shows that a square wave

179

Table 7.1   Minimum Requirements for Oscilloscope Performance for Use with Various Types of Equipment. Data from: Steve Montgomery, "Advances in Digital Oscilloscopes," *Broadcast Engineering* magazine, Intertec Publishing, Overland Park, Kan., November 1989.

Application	Bandwidth	Rise time	Other key features
Video	12.5 MHz	28 ns	TV frame and line triggering
Digital	50 MHz	7 ns	Dual channels, alternate horizontal magnification
Audio	175 kHz	. . .	500 µV sensitivity
Electromechanical	10 MHz	35 ns	Single sweep, HF/LF reject trigger
Robotics	10 MH	35 ns	Single sweep, HF/LF reject trigger, 500 µV sensitivity
Power supplies	175 kHz	. . .	500 µV sensitivity

* Based on 0.14 s sync pulse rise time requirements. At least 50 MHz is needed for IF sections.

consists of the fundamental sine wave plus sine waves that are odd multiples of the fundamental. Figure 7.1 illustrates the mechanisms involved. The fundamental frequency contributes about 81.7 percent of the square wave. The third harmonic contributes about 9.02 percent, and the fifth harmonic about 3.24 percent. Higher harmonics contribute less to the shape of the square wave. As outlined here, approximately 94 percent of the square wave is derived from the fundamental and the third and fifth harmonics. Inaccuracies introduced by the instrument are typically 2 to 3 percent. The user's ability to read a conventional CRT display may contribute another 4 percent error. Thus, at least 6 percent error may be introduced by the instrument and the operator. A 96 percent accurate reproduction of a square wave should, therefore, be sufficient for all but the most critical applications. It follows that oscilloscope bandwidth is important up to and including the fifth harmonic of the desired signal.

Rise Time. The rise time performance required of an analog oscilloscope depends on the degree of accuracy needed in measuring the input signals. A 2 to 3 percent rise time accuracy can be obtained from an instrument that is specified to have approximately five times the rise time of the signal being measured. Rise time can be determined from the bandwidth of the instrument:

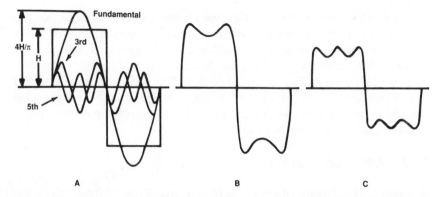

Figure 7.1   The mechanisms involved in square waves: (a) Individual waveforms involved. (b) Waveshape resulting from combining the fundamental and the first odd harmonic (the third harmonic). (c) Waveshape resulting from combining the fundamental and the first two odd harmonics (third and fifth harmonics).

$$T_r = \frac{0.35}{BW}$$

Where:
T_r = instrument rise time
BW = instrument bandwidth

7.1.2   Making Measurements

The oscilloscope is a measurement instrument, displaying voltage waveforms within powered circuits. To conduct certain types of waveform analysis, a separate signal generator is needed to inject test signals. With a conventional analog oscilloscope, actual measurement of the waveform must be performed visually by the operator. The graticule divisions along the trace are counted and multiplied by the time base setting to determine the time interval. The TIME/DIV control determines the value of each increment on the x axis of the display. Typical sweep ranges vary from 0.05 μs to 0.5 s per division. To determine voltage, the operator must estimate the height of a portion of the trace and multiply it by the voltage range selected on the instrument. The VOLT/DIV control determines the value of each increment on the y axis of the display. On a dual-channel oscilloscope, two controls are provided, one for each source. Typical vertical sensitivity ranges vary from 5 mV to 5 V per division.

The need for interpolation introduces inaccuracies in the measurement process. Determination of exact values is especially difficult in the microsecond ranges, where even small increments can make a big difference in the measurement. It also is difficult with a conventional oscilloscope to closely examine specific portions of a waveform. For this purpose, an oscilloscope with two independently adjustable sweep generators is recommended. A small section of the waveform can be selected for closer examination by adjusting the B channel sweep-time and delay-time controls while observing an intensified portion of the waveform. The intensified section then can be expanded across the screen. This type of close-up examination is useful for measuring pulse rise times and viewing details in complex signals, such as video horizontal sync or color burst.

7.1.3   Advanced Features

New oscilloscopes remove the potential inaccuracies of conventional instruments by incorporating advanced features previously found only in waveform analyzers. These features include:

- *Numeric measurement and display.* On-screen readouts provide key measurement parameters for the operator. Readout options include peak waveform voltage, frequency, and sweep and sensitivity settings for the instrument. On-screen readout makes waveform photography more valuable because the photograph displays not only the waveform, but also the instrument settings.
- *Triggering.* A wide selection of trigger modes provides operational flexibility. The trigger mode determines what condition starts the horizontal sweep to display the voltage waveform. The trigger signal can be derived from an external source or from the input channel(s). The trigger usually is continuously adjustable through both positive and negative slopes of the signal. Selectable trigger coupling includes special settings needed in particular applications. A variable hold-off control can be used to obtain a stable display when measuring periodic or complex signals.
- *Control and monitoring.* Microprocessors have been integrated into conventional analog oscilloscopes to reduce the number of mechanical switch contacts, which are prone to failure, and to permit front-panel setting recall capabilities. These features provide the best of both the analog and digital worlds. Auxiliary signal outputs are available to feed other instruments, such as a distortion analyzer or frequency counter.

7.2   DIGITAL OSCILLOSCOPE

The digital storage oscilloscope (DSO) is one of the most exciting developments in oscilloscope technology in recent years. The DSO offers a number of significant advantages beyond the capabilities of analog instruments. It can store in memory the signal being observed, permitting in-depth analysis that was impossible with

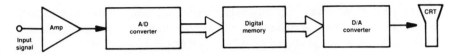

Figure 7.2   Simplified block diagram of a digital storage oscilloscope.

previous technology. Because the waveform resides in memory, the data associated with the waveform can be transferred to a computer for real-time processing, or for processing at a later time.

7.2.1   Operating Principles

Before the introduction of the DSO, the term *storage oscilloscope* referred to an oscilloscope that used a *storage* CRT. This type of instrument stored the display waveform as a trace on the oscilloscope face. The DSO operates on a completely different premise. (Figure 7.2 shows a block diagram.) Instead of being amplified and applied directly to the deflection plates of a CRT, the waveform first is converted into a digital form and stored in memory. To reproduce the waveform on the CRT, the data is sequentially read and converted back into an analog signal for display.

The analog-to-digital (A/D) converter transforms the input analog signal into a sequence of digital bits. The amplitude of the analog signal, which varies continuously in time, is sampled at preset intervals. The analog value of the signal at each sample point is converted into a binary number. This *quantization* process is illustrated in Figure 7.3. The sample rate of a digital oscilloscope must be greater

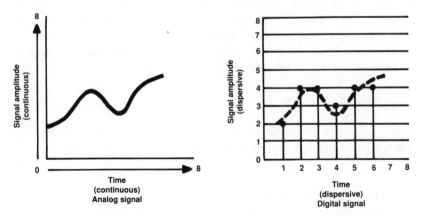

Figure 7.3   The quantization process.

than two times the highest frequency to be measured. The higher the sampling rate relative to the input signal, the greater the measurement accuracy. A sample rate 10 times the input signal should be sufficient for most applications. This rule of thumb applies for single-shot signals, or signals that constantly are changing. The sample rate also may be expressed as the *sampling interval*, or the period of time between samples. The sampling interval is the inverse of the sampling frequency.

Although a DSO is specified by its maximum sampling rate, the actual rate used in acquiring a given waveform usually is dependent on the *time-per-division* setting of the oscilloscope. The *record length* (samples recorded over a given period of time) defines a finite number of sample points available for a given acquisition. The DSO, therefore, must adjust its sampling rate to fill a given record over the period set by the sweep control. To determine the sampling rate for a given sweep speed, the number of displayed points per division is divided into the sweep rate per division. Two additional features can modify the actual sampling rate:

- Use of an external clock for pacing the digitizing rate. With the internal digitizing clock disabled, the digitizer will be paced at a rate defined by the operator.
- Use of a *peak-detection* (or *glitch-capture*) mode. Peak detection allows the digitizer to sample at the full digitizing rate of the DSO, regardless of the time base setting. The minimum and maximum values found between each normal sample interval are retained in memory. These minimum and maximum values are used to reconstruct the waveform display with the help of an algorithm that re-creates a smooth trace along with any captured glitches. Peak detection allows the DSO to capture glitches even at its slowest sweep speed. For higher performance, a technique known as *peak accumulation* (or *envelope mode*) may be used. With this approach, the instrument accumulates and displays the maximum and minimum excursions of a waveform for a given point in time. This builds an envelope of activity that can reveal infrequent noise spikes, long-term amplitude or time drift, and pulse jitter extremes. Figure 7.4 illustrates the advantage of peak accumulation when variations in data pulse width are monitored.

7.2.2   A/D Conversion

The A/D conversion process is critical to the overall accuracy of the oscilloscope. Three types of converters commonly are used:

- Flash
- Successive approximation
- Charge-coupled device (CCD)

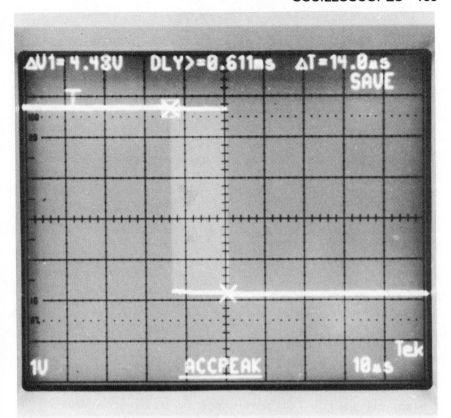

Figure 7.4 Use of the peak accumulation sampling mode to capture variations in pulse width. (Courtesy of Tektronix)

CCD-based instruments currently offer the best performance/cost ratio. The CCD is an analog storage array. The DSO can capture high-speed events in real time using a CCD, then route the captured information to be digitized to low-cost A/D converters.

Conversion resolution is determined by the number of bits into which the analog signal is transformed. The higher the number of bits available to the A/D converter, the more discrete levels the digital signal is able to describe. For example, an 8-bit digitizer has 256 levels, while a 10-bit digitizer has 1024 levels. Although a higher number of bits allows greater discrimination between voltage values, the overall accuracy of the DSO may be limited by other factors, including the accuracy of the analog amplifier(s) feeding the digitizer(s). The accuracy of a DSO normally is

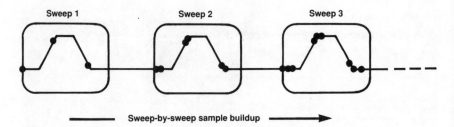

Figure 7.5   Increasing sample density through equivalent-time sampling.

specified consistent with its analog counterpart, typically 2 to 4 percent. The usefulness of a high-conversion digitizer also may be limited by the resolution of the CRT screen.

Although most DSOs provide 8-bit digitizing, there may be significant differences between individual 8-bit instruments. First among these is the *useful storage bandwidth*. This specification is further divided into *single-shot bandwidth* and *equivalent-time bandwidth*.

Achieving adequate samples for a high-frequency waveform places stringent requirements on the sampling rate. High-frequency waveforms require high sampling rates. *Equivalent-time sampling* often is used to provide high-bandwidth capture. This technique relies on sampling a repetitive waveform at a low rate to build up sample density, a concept illustrated in Figure 7.5. When the waveform to be acquired triggers the DSO, several samples are taken over the duration of the waveform. The next repetition of the waveform triggers the instrument again, and more samples are taken at different points on the waveform. Over many repetitions, the number of stored samples can be built up to the equivalent of a high sampling rate.

System operating speed also has an effect on the display update rate. Update performance is critical when measuring waveform or voltage changes. If the display does not track the changes, adjustment of the circuit may be difficult.

In some cases, it may be that high-speed sampling, even with the fastest sweep speed on a DSO, still does not provide a display update as quickly as necessary. In these cases, a conventional analog oscilloscope may provide the best performance. Some DSO instruments offer a switchable analog display path for such applications.

7.2.3   DSO System Design

A detailed block diagram of a DSO is shown in Figure 7.6. The features included in this design are typical of a digital oscilloscope. The major circuit blocks are:

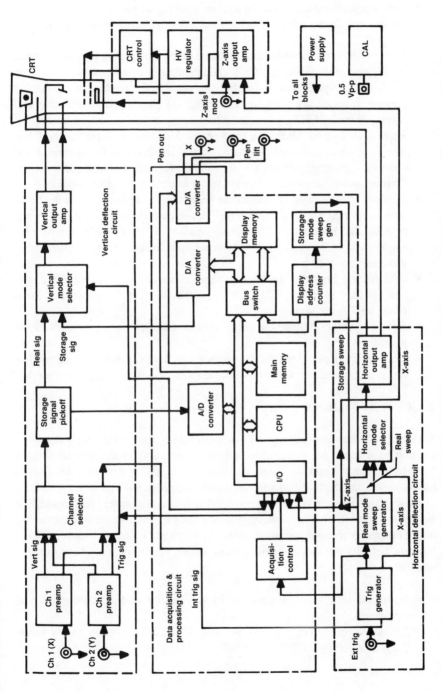

Figure 7.6 Detailed block diagram of a DSO showing the conversion, I/O, memory, and display circuits. (Courtesy of Kikusui)

- *Data acquisition:* Converts the analog input signal into a digital equivalent for storage. Data control signals are routed to a microprocessor through an I/O port.

- *Data processing:* Provides interpolation of the stored data for reproduction of the analog signal on the CRT. The processed data is transferred to the display memory. Except during transfer periods, the content of the display memory is constantly being sent to the digital-to-analog (D/A) converter and displayed on the CRT.

- *Vertical deflection:* Generates the vertical deflection voltages for the CRT. The input signal is impedance-matched and amplified by the channel 1 and channel 2 preamplifiers to a level suitable for driving subsequent stages. Channel-selection switches operate under control of the microprocessor.

- *Horizontal deflection:* Generates the necessary voltages to deflect the beam spot horizontally across the screen. The stage includes trigger and sweep circuits. The horizontal deflection stage also acts as an x axis amplifier for x-y operation. Channel-selection switches operate under control of the microprocessor.

- *CRT driver:* Provides the high voltages necessary to drive the CRT, and controls the z axis of the CRT. The z axis circuit sets the intensity and focus of the CRT beam spot.

- *Power supplies:* Provide the necessary low voltages for other stages of the instrument.

- *Calibration circuit:* Provides a reference signal for calibration of the oscilloscope probe, and for operation of the amplifiers in the noncalibrated mode.

7.2.4   DSO Features

The digital oscilloscope has become a valuable tool in troubleshooting both analog and computer-based products. Advanced components and construction techniques have led to lower costs for these instruments and higher performance. Digital oscilloscopes can capture and analyze transient signals, such as race conditions, clock jitter, glitches, dropouts, and intermittent faults. Automated features reduce testing and troubleshooting costs through the use of recallable instrument setups, direct parameter readout, and unattended monitoring. Digital oscilloscopes have inherent benefits not available from most analog oscilloscopes. These benefits include:

- Increased resolution (determined by the quality of the analog-to-digital converter).
- Memory storage of digitized waveforms. Figure 7.7 shows a complex waveform captured by a DSO.

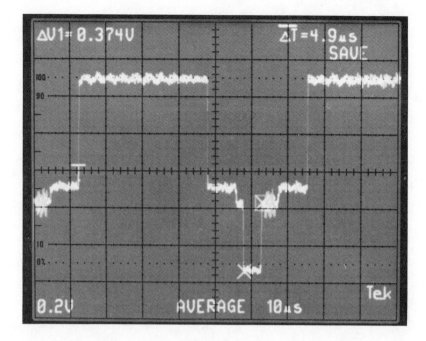

Figure 7.7 A digitized waveform "frozen" on the display to permit measurement and examination of detail. (Courtesy of Tektronix)

- Automatic setup for repetitive signal analysis. For complex multichannel configurations that are used often, front-panel storage/recall can save dozens of manual selections and adjustments. When multiple memory locations are available, multiple front-panel setups can be stored to save even more time.
- Auto-ranging. Many instruments will adjust automatically for optimum sweep, input sensitivity, and triggering. The instrument's microprocessor automatically configures the front panel for optimum display. Such features permit the operator to concentrate on making measurements rather than on adjusting the oscilloscope.
- Instant hard-copy output from printers and plotters.
- Remote programmability via GPIB for automated test applications.
- Trigger flexibility. Single-shot digitizing oscilloscopes capture transient signals and allow the user to view the waveform that *preceded* the trigger point. Figure 7.8 illustrates the use of pre-/post-trigger for waveform analysis.
- Signal analysis. Intelligent oscilloscopes can make key measurements and comparisons. Display capabilities include voltage peak, mean voltage, rms value, rise time, fall time, and frequency. Figure 7.9 shows the voltage measurement option available on one instrument.

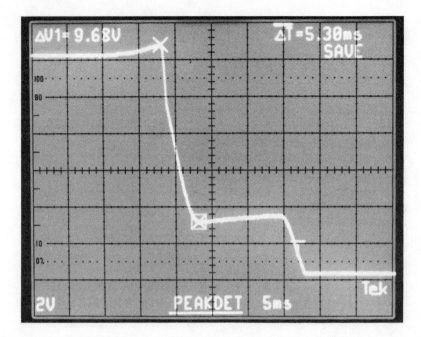

Figure 7.8   Use of pre-/post-trigger function for waveform analysis. (Courtesy of Tektronix)

- Cursor measurement. Advanced oscilloscopes permit the operator to take measurements or to perform comparative operations on data appearing on the display. A measurement cursor consists of a pair of lines or dots that can be moved around the screen as needed. Figure 7.10 shows one such example. Cursors have been placed at different points on the displayed waveform. The instrument automatically determines the phase difference between the measurement points. The cursor readout follows the relative position of each cursor.
- Trace quality. Eye fatigue in viewing low-repetition signals is reduced noticeably with a DSO. For example, a 60 Hz waveform can be difficult to view for extended periods of time on a conventional oscilloscope because the display tends to flicker. A DSO overcomes this problem by writing waveforms to the screen at the same rate, regardless of the input signal. (See Figure 7.11.)

Digital memory storage, rather than CRT storage, produces a number of benefits, including:

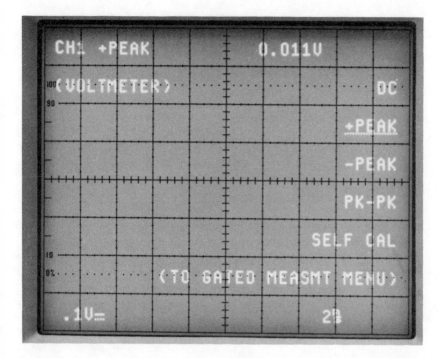

Figure 7.9 Menu of voltage measurement options available from a DSO. (Courtesy of Tektronix)

- Reference memory. A previously acquired waveform can be stored in memory and compared with a sampled waveform. This feature is especially useful for repetitive testing or calibration of a device to a standard waveform pattern. Nonvolatile battery-backed memory permits reference waveforms to be transported to field sites.
- Simple data transfers to a host computer for analysis or archive.
- Local data analysis through the use of a built-in microprocessor.
- Cursors capable of providing a readout of delta and absolute voltage and time.
- No CRT blooming for display of fast transients.
- Full bandwidth capture of long-duration waveforms, thus storing all the signal details. The waveform can be expanded after capture to expose the details of a particular section.

Until recently, digital oscilloscope bandwidth was limited to about 100 MHz. For higher-frequency TTL, ECL, and GaAs circuit measurements, an analog oscilloscope was necessary. But now digital oscilloscope bandwidth rivals that of

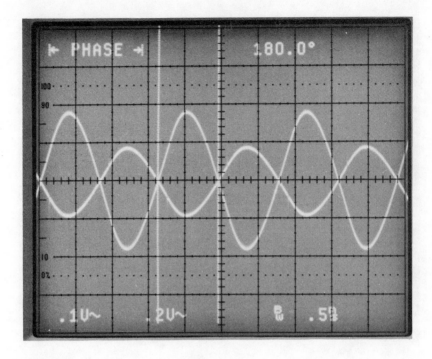

Figure 7.10   Use of cursors to measure the phase difference between two points on a waveform. (Courtesy of Tektronix)

its analog counterpart. Several feature-packed 350 MHz oscilloscopes are available for use with passive probes. Most passive probes are limited to a bandwidth of 300 to 400 MHz; active probes offer extended bandwidth.

Digital oscilloscopes have the potential for superior accuracy. For a high-frequency digital oscilloscope, 3 percent total system accuracy is considered good. Furthermore, digital oscilloscopes can read out amplitude and time directly from data stored in memory. Measurement accuracy is not degraded by operator interpolation of trace positions with respect to a CRT grid, as when using an analog oscilloscope.

7.2.5   Capturing Transient Waveforms

Single-shot digitizing makes it possible to capture and clearly display transient and intermittent signals. Waveforms such as signal glitches, dropouts, logic race conditions, intermittent faults, clock jitter, and power-up sequences can be examined with the help of a digital oscilloscope. With single-shot digitizing, the

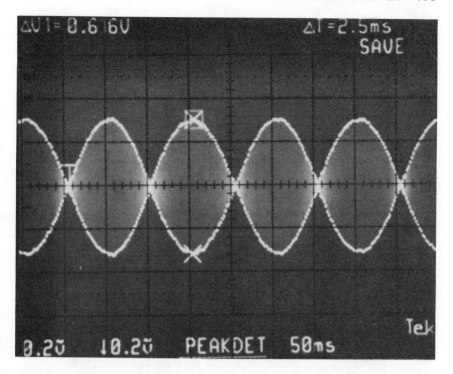

Figure 7.11   The DSO provides a stable, flicker-free display for viewing low-repetition signals. (Courtesy of Tektronix)

waveform is captured the first time it occurs, on the first trigger. It can be displayed immediately or held in memory for analysis at a later date. In contrast, most analog oscilloscopes can capture only repetitive signals. Hundreds of cycles of the signal are needed to construct a representation of the waveshape. Analog oscilloscopes are unable to capture transients. Figure 7.12 illustrates the benefits of digital storage in capturing transient waveforms. An analog oscilloscope often will fail to detect a transient pulse that a DSO can display clearly.

7.2.6   Triggering

Basic triggering modes available on a digital oscilloscope permit the user to select the desired source, its coupling, level, and slope. More advanced digital oscilloscopes contain triggering circuitry similar to that found in a logic analyzer. These powerful features let the user trigger on elusive conditions, such as pulse widths less than or greater than expected, intervals less than or greater than expected, and specified logic conditions. The logic triggering may include digital-pattern, state-

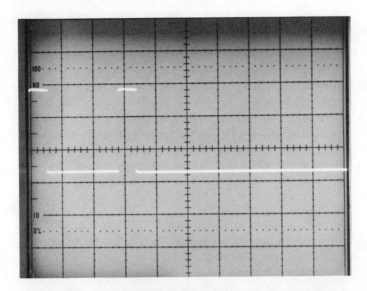

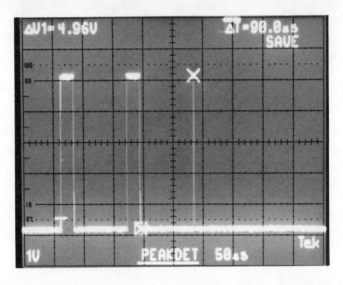

Figure 7.12 The benefits of a DSO in capturing transient signals: (a—top) Analog display of a pulsed signal with transient present, but not visible. (b—bottom) DSO display of the same signal clearly showing the transient. (Courtesy of Tektronix)

qualified, and time-/event-qualified conditions. Many trigger modes are enhanced further by allowing the user to hold off the trigger by a selectable time or number of events. Hold-off is especially useful when the input signal contains bursts of data or follows a repetitive pattern.

Pulse-width triggering lets the operator check quickly for pulses narrower than expected or wider than expected, The pulse-width trigger circuit checks the time from the trigger source transition of a given slope (typically the rising edge) to the next transition of opposite slope (typically the falling edge). The operator can interactively set the pulse-width threshold for the trigger. For example, a glitch may be considered any signal narrower than one-half a clock period. Conditions preceding the trigger may be displayed to show what events led up to the glitch.

Interval triggering lets the operator check quickly for intervals narrower than expected or wider than expected. Typical applications include monitoring for transmission phase changes in the output of a modem or for signal dropouts, such as missing bits on a computer hard disk.

Pattern triggering lets the user trigger on the logic state (high, low, or either) of several inputs. The inputs may be external triggers or the input channels themselves. The trigger may be generated upon either entering or exiting the pattern. Applications include triggering on a particular address select or data bus condition. After the pattern trigger is established, the operator can probe throughout the circuit, taking measurements synchronous with the trigger.

State-qualified triggering enables the oscilloscope to trigger on one source, such as the input signal itself, only after the occurrence of a specified logic pattern. The pattern acts as an *enable* or *disable* for the source.

Advanced Features. Some digital oscilloscopes provide enhanced triggering modes that permit the user to select the level and slope for each input. This flexibility makes it easy to look for odd pulse shapes in the pulse-width trigger mode and for subtle dropouts in the interval trigger mode. It also simplifies testing different logic types (TTL, CMOS, and ECL) and testing analog/digital combinational circuits with dual logic triggering. Additional flexibility is available on multiple-channel oscilloscopes. After a trigger has been sensed, multiple simultaneously sampled inputs permit the user to monitor conditions at several places in the unit under test, with each channel synchronized to the trigger. Additional useful trigger features for monitoring jitter or drift on a repetitive signal include:

- Enveloping
- Extremes
- Waveform delta
- Roof/floor

Each of these functions is related, and in some cases, they describe the same general operating mode. Various oscilloscope manufacturers use different nomenclature to describe proprietary triggering methods. Generally speaking, as the waveshape changes with respect to the trigger, the oscilloscope generates upper and lower traces. For every nth sample point with respect to the trigger, the maximum and minimum values are saved. Thus, any jitter or drift is displayed in the envelope.

Advanced triggering features provide the greatest benefit in conjunction with single-shot sampling. Repetitive sampling oscilloscopes can capture and display only signals that repeat precisely from cycle to cycle. Several cycles of the waveform are required to create a digitally reconstructed representation of the input. If the signal varies from cycle to cycle, such oscilloscopes can be less effective than a standard analog oscilloscope for accurately viewing a waveform.

7.3 BIBLIOGRAPHY

Albright, John R.: "Waveform Analysis with Professional-Grade Oscilloscopes," *Electronic Servicing & Technology* magazine, Intertec Publishing, Overland Park, Kan., April 1989.

Breya, Marge: "New Scopes Make Faster Measurements," *Mobile Radio Technology* magazine, Intertec Publishing, Overland Park, Kan., November 1988.

Harris, Brad: "Understanding DSO Accuracy and Measurement Performance," *Electronic Servicing & Technology* magazine, Intertec Publishing, Overland Park, Kan., April 1989.

_____: "The Digital Storage Oscilloscope: Providing the Competitive Edge," *Electronic Servicing & Technology* magazine, Intertec Publishing, Overland Park, Kan., June 1988.

Hoyer, Mike: "Bandwidth and Rise Time: Two Keys to Selecting the Right Oscilloscope," *Electronic Servicing & Technology* magazine, Intertec Publishing, Overland Park, Kan., April 1990.

Montgomery, Steve: "Advances in Digital Oscilloscopes," *Broadcast Engineering* magazine, Intertec Publishing, Overland Park, Kan., November 1989.

Persson, Conrad: "Oscilloscope: The Eyes of the Technician," *Electronic Servicing & Technology* magazine, Intertec Publishing, Overland Park, Kan., April 1987.

_____: "Oscilloscope Special Report," *Electronic Servicing & Technology* magazine, Intertec Publishing, Overland Park, Kan., April 1990.

Toorens, Hans: "Oscilloscopes: From Looking Glass to High-Tech," *Electronic Servicing & Technology* magazine, Intertec Publishing, Overland Park, Kan., April 1990.

8

TROUBLESHOOTING DIGITAL SYSTEMS

8.1 INTRODUCTION

Repair of microprocessor-based hardware requires a different approach to maintenance than conventional analog circuits. Successful troubleshooting depends on four key elements:

- Understand how the system operates.
- Use the right test equipment.
- Perform troubleshooting steps in the right sequence.
- Examine carefully available data.

It should be noted that the term *microprocessor* may be used in both a generic and in a specific sense. In a generic sense, a microprocessor is any chip-sized device capable of making decisions and sending out instructions based on values entered into it. In a more technical sense, microprocessor refers only to a chip-sized computing device that does not contain its own built-in control program. Instead, the control program is built into a separate read-only memory (ROM). Microprocessors also make use of random-access memory (RAM) as a temporary storage area for data. Less sophisticated microsystems, in contrast, have a built-in control program and little, if any, RAM. Technically speaking, such microsystems are based on microcontrollers, not microprocessors.

8.1.1 Defining Key Terms

An understanding of digital system terminology is essential to any useful discussion of troubleshooting techniques. Definitions of key terms follow:

- Basic input/output system (BIOS): Software instructions contained in ROM that configure and initialize a microcomputer system.
- Direct memory access (DMA): A method of gaining direct access to the computer system memory for data transfer without involving the microprocessor directly.
- Dynamic random-access memory (DRAM): A type of memory device that is faster than static memory and uses less power. However, dynamic memory must be *refreshed* periodically to maintain its data.
- Emulation: A technique using software and hardware in which one device is made to behave like another device.
- Interrupt: A signal that suspends the normal execution of a program and forces the microprocessor to begin executing a special *interrupt handler* routine. Interrupts are used to process keystrokes, critical inputs, or other information to which the microprocessor must respond immediately.
- Kernel: The core of a computer system—the microprocessor, clock, and bus buffers.
- Machine cycle: The shortest amount of time necessary for a microprocessor to complete an operation or a process.
- Memory map: The assignment of various devices to portions of the microprocessor memory space.

8.1.2   Elements of a Microcomputer[1]

A wide variety of microcomputers is available to both consumers and industrial users. Although each manufacturer has its own approach to system design, the basic concepts vary little from one unit to another. Figure 8.1 shows a block diagram of a typical microcomputer. The major elements of the system include:

- Microprocessor
- Clock generator
- Read-only memory
- Random-access memory
- Data/address buses and decoders
- DMA controller
- Input/output interface

---

1   Section 8.1.2 was adapted from: Tom Allen, "Components of Microprocessor-Based Systems," *Electronic Servicing and Technology* magazine, Intertec Publishing, Overland Park, Kan., January 1988.

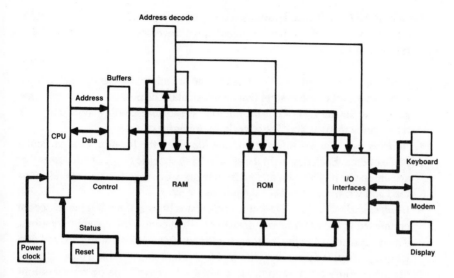

Figure 8.1   Basic elements of a microcomputer system.

Microprocessor. The microprocessor is the central control element of the system. Everything that happens in the circuit is initiated by the microprocessor. The microprocessor executes one operation at a time, usually consisting of data read, data write, or arithmetic functions. In response to instructions contained in the control program, the microprocessor orders data to be placed on the data bus, reads it, performs calculations as required, and writes the resulting data to specific circuit elements over the same bus.

Clock Generator. The system clock determines the rate at which the microprocessor executes instructions. All data movement within the kernel is synchronized with the master clock. A key function of the clock is to make sure that data placed on the bus has time to stabilize before a read or write operation occurs. The clock pulses define *valid data windows* for the various buses. The clock also provides refresh signals for the DRAM chips.

Read-Only Memory. ROM contains built-in instructions for the microprocessor. These instructions are specific to a given type of device, known as the *instruction set.* All microprocessor-based systems require ROM instructions to get them started when they are powered up. This process is referred to as *booting.*

Random-Access Memory. RAM consists of one or more banks of memory devices that serve as a temporary storage area for data. Data contained in RAM is volatile; it is lost when power is removed from the system. Program instructions read from the storage media are stored in and executed from RAM. DRAM chips are used almost exclusively for RAM functions today. Memory density ranges from 64 kbits x 1 to 4 Mbits x 1 or more.

Data and Address Buses. Buses are communications paths that conduct information from one place to another. A typical microprocessor-based system includes three types of buses:

- *Data bus:* Carries information to and from the microprocessor, connecting it with every part of the system that handles data. The number of data bus lines is equal to the number of bits the microprocessor can handle at one time; each bit requires its own line. Data bus lines are bidirectional.
- *Address bus:* Carries information that identifies the location of a particular piece of data. Each device in the system has a specific range of addresses that are unique to it. The data and address buses work together to respond to the read and write commands of the microprocessor.
- *Control and status bus:* The control and status lines of the microprocessor are connected to the other kernel devices to effect control of system operations. These lines allow the microprocessor to specify whether it wants to read data from a particular device, or write data to it. They also provide a means for the device to notify the microprocessor when data is available for transmission.

Both the data and address buses employ buffers located at the outer boundary of the kernel to isolate the buses from other circuits on the system board. Address decoders on the address bus notify each remote device when the address placed on the bus by the microprocessor is within the address range of that particular device. This notification is performed by turning the *chip select* (CS) pin of the device on or off. A device can respond to requests from the microprocessor only when the CS pin is enabled. This prevents the wrong device from responding to commands from the microprocessor.

DMA Controller. The direct memory access controller permits data to move to and from memory without going through the microprocessor. The DMA controller is designed to perform data transfers involving peripheral devices, such as mass storage disks. Like other devices in the microcomputer system, the DMA controller operates only in response to instructions from the microprocessor. When executing instructions, the DMA circuit assumes control over the data, address, and control lines from the microprocessor. This process is know as *cycle stealing*. For one or more clock cycles, the DMA chip takes over the communications buses, and the microprocessor restricts itself to internal functions (such as arithmetic calculations). Because the microprocessor will not relinquish control over the buses for more than a few cycles at a time, the DMA controller and microprocessor may pass control of the buses back and forth several times before the DMA controller completes a given task.

Input/Output Interface. I/O circuitry allows the microprocessor kernel to communicate with peripheral devices, such as the keyboard, monitor, storage media, and communications ports. Unlike the components of the microprocessor kernel, I/O devices are not always governed by the microprocessor clock. Peripherals may operate *asynchronously* with the system clock. The I/O circuitry typically contains buffers that serve to isolate the kernel from the peripherals; they also function as holding points for data on its way to or from the kernel. Data going to the kernel is held by the buffers until the microprocessor is ready to accept it. Data that the microprocessor is sending to a peripheral device is held by the buffers until the device is ready to accept it.

8.2   APPROACH TO TROUBLESHOOTING

The failure modes that may be experienced with a microcomputer system vary widely. Some general rules, however, may be applied to nearly all troubleshooting efforts. First, follow the natural signal flow as it moves from the microprocessor through the system. Begin with the kernel components, and move outward toward the peripheral devices. When a failure node is found, trace backward through the circuit to locate a point at which the input is good, but the output is bad.

This approach may be used to locate failures to the board level or to the component level. Board swapping (or *board float*) is a commonly used repair technique today. Computer systems are designed in a modular fashion to facilitate easy service and upgrade capability. The high cost of skilled technical labor has led many service organizations to avoid component-level troubleshooting unless absolutely necessary. Such detailed work is left to service depots that have sophisticated diagnostic instruments. Still, the cost of board float is significant. It represents money sitting unproductively in the form of unused PWBs. The technician also may replace more than one board in an effort to locate the problem, then simply return all boards replaced to the service depot for analysis. Studies by various organizations have found that as many as 35 percent of all boards returned for rework have been found to be in proper working order. Also, there is no guarantee that the replacement board, especially if it is handled improperly, will perform correctly when installed. The more a PWB is handled and shuffled in inventory, the more likely it is to become damaged before being placed in a system. The use of board swapping vs. repair to the component level is a function of many factors, including:

- *Complexity of the hardware.* As the level of complexity increases, the likeliness of cost-effective component-level repair decreases.
- *Technical documentation available.* Service is always made more difficult by a lack of detailed documentation.
- *Experience level of maintenance personnel.* Solving complex problems usually requires highly skilled technicians.
- *Construction of the hardware.* A system built in a modular fashion is a natural candidate for board swapping. A PWB using socket-mounted devices is a candidate for component-level repair.

- *Types of devices used.* A PWB populated with DIP integrated circuits is much easier to repair than one populated with surface-mounted ICs.
- *Nature of the malfunction.* Some failures can be diagnosed with little more than a DMM and a logic probe. Other failures require complex and expensive test instruments.
- *MTTR considerations.* Hardware used in critical applications where downtime cannot be tolerated is best repaired using the board-swapping technique.

Because of the complexity of computer equipment today, many service departments have established a hierarchical approach to troubleshooting. This technique involves assigning one of three levels to each service problem:

- *First-tier problems:* Obvious failures that usually can be solved without extensive troubleshooting or expensive test equipment. First-tier problems are handled by technology generalists, often less experienced or entry-level personnel. In a sizable maintenance organization, this group is the largest of the three. It handles the majority of service work.
- *Second-tier problems:* Failures that are harder to diagnose than first-tier faults. Test equipment is required to troubleshoot the system. Complex problems often call for sophisticated instruments.
- *Third-tier problems:* The most difficult-to-troubleshoot failures. Complex and expensive test equipment is required, as well as extensive experience on the unit being serviced. In a large maintenance organization, this is the smallest of the three groups.

Such a tiered approach to service results in the most cost-effective use of technical talent. It also facilitates fast turnaround of products to the customer.

8.2.1   Technician Training

The computer business in one in which there is a constant flow of new products. This rapid advancement of technology requires an equally rapid advancement in the training of technical personnel. The situation is made more difficult by a lack of detailed technical information. Sometimes the inadequate information is a matter of oversight on the part of the computer manufacturer; sometimes it is the result of cost-savings efforts. A lack of technical documentation, however, also may be the result of a planned effort on the part of the manufacturer to keep service work in-house.

Many large computer manufacturers offer entry-level maintenance training on their products, but these programs may be restricted to authorized resellers of the product line. Hands-on training is unquestionably the best way to learn how to service a piece of equipment. When a student is allowed to practice a new technique with the supervision of a skilled instructor, the highest level of learning and retention occurs.

8.2.2 Teleservicing

Computer-based test instruments provide the capability to transmit data rapidly from one location to another. Instruments, such as oscilloscopes, are available that can output a waveform or other data to a modem for transmission to a central service facility for analysis. Instead of grappling with a difficult problem all alone, the on-site technician can call on the resources and experience of the service center. Field service thus becomes a team effort that includes the technician in the field and the often more experienced service center engineers and staff. Software programs are available to permit test equipment at a remote location to be configured as required, capture data, and transmit that data to the service center for analysis. Teleservicing also makes possible the creation of reference libraries of key waveforms. These waveforms facilitate troubleshooting, and are useful in documenting the performance characteristics of the equipment being maintained. Over time, such documentation can become a valuable addition to the service record of a piece of equipment.

Service sites are sometimes in less-than-ideal locations—on a mountaintop, at a remote receiver-transmitter site, or on an oil drilling platform in the middle of the ocean. In such cases, teleservicing provides numerous benefits. If the problem involves intermittent failures, the test instrument can be set up in a "babysitting mode" and left to capture a critical signal when it occurs.

8.2.3 Servicing Obsolete Hardware

Maintaining equipment that is no longer manufactured or supported by the original equipment manufacturer (OEM) is a growing problem in computer service. Just because a piece of equipment is "obsolete" does not mean that it is worthless. The marketing life cycle for computer-based hardware is 2 to 4 years. The equipment will, however, continue to provide useful service many years after it is considered obsolete by the OEM. A product typically experiences four major stages in its useful lifetime, as illustrated in Figure 8.2.

The decision to keep or replace a piece of hardware is based primarily on economic factors. New equipment requires an initial outlay of capital. The central question is whether the new gear would have lower life-cycle costs than the older equipment that it would replace. When all factors are considered, the conclusion is often made that the old equipment should be retained, at least until the cost of maintenance becomes prohibitive.

The acquisition of replacement parts in sufficient quantities, and at reasonable prices, is the major challenge in maintaining after-market hardware. Most semi-

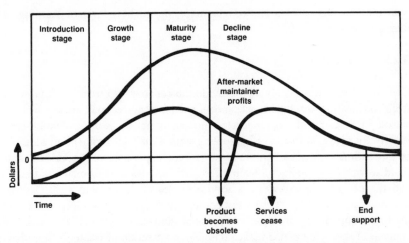

Figure 8.2 The four major cycles of product lifetime. (Data from: Peter Ware, "Servicing Obsolete Equipment," *Microservice Management* magazine, Intertec Publishing, Overland Park, Kan., March 1989)

conductors are generic and can be acquired from one of several sources. ROM and *programmable array logic* (PAL) devices, however, usually must be purchased from the OEM. Typically, mechanical parts also are available only from the OEM. Parts for obsolete equipment also may be obtained by *demanufacturing* (cannibalizing) other units. Buying whole units for demanufacture may be a good solution because assembled equipment is almost always less expensive than the sum of the *field replaceable units* (FRUs).

Perhaps the most difficult aspect of servicing obsolete equipment is the lack of technical documentation. Maintenance and service manuals may no longer be available from the OEM. The service technician, therefore, may have to practice some degree of *reverse engineering* to repair the system. By definition, reverse engineering means working backward from a finished product to develop schematics, parts lists, operating standards, and test procedures. Even if this data is unavailable from the OEM, it may be available from other sources, including:

- *The equipment owner.* Large contract sales often specify that the buyer receive technical documentation as part of the purchase.
- *Parts suppliers to the OEM.* A wealth of technical information is available from semiconductor manufacturers. By ordering application notes from the maker of the microprocessor chip, for example, the technician may gain insight into how the system operates. Most newer PC-based products are built around chip sets (consisting of up to six devices). The chip set manufacturer should be able to supply extensive documentation on how the products operate.

- *Documentation from similar systems.* Product lines tend to share many design concepts. The maintenance manual for a given piece of hardware may not be available, but if documentation for an earlier or later system from the same manufacturer is available, it may contain sufficient information to help the technician to solve the problem at hand.

8.2.4   Shipping Electronic Hardware

It is inevitable that some circuit boards and other computer components will fail during their useful lifetime. Although inoperative, these parts represent a significant investment for the computer user. Proper shipping is an important element in the maintenance chain. PWBs must be protected from bouncing, jostling, and electrostatic discharge (ESD). Careless handling during shipping can magnify what originally may have been a minor problem. In some cases, the board may be destroyed by improper shipping techniques. Protecting sensitive parts from rough handling is easily accomplished with the proper packing materials, such as bubble wrap and antistatic foam sheets. Furthermore, all electronic assemblies should be shipped in antistatic bags to prevent ESD damage.

Large *head disk assemblies* (HDAs) are even more sensitive than PWBs to rough handling. An HDA may be damaged beyond repair if it is subjected to a large shock load from being dropped or hit by another object. The damage may not be readily apparent through physical examination. Only when the unit is installed in a working system will the failure be detected. As a measure of insurance against such problems, some companies use *shock watch* stickers, which show when a package has been subjected to a high *G-force*. A customer who receives a package with the indicator showing that a high-force impact has occurred will know there is a good chance that the part has been damaged during shipping.

8.3   ISOLATING THE DEFECTIVE SUBASSEMBLY

Troubleshooting computer hardware is not as difficult as most technicians might think. With the proper technical documentation, a well-built system should not present significant repair problems. Virtually all computer systems conduct a *power-on self-test* (POST) routine when power is applied. This routine checks the operational condition of each major component of the computer, and initializes system controllers. The initialization and self-test functions of the POST process are interwoven tightly. Functions can be divided into two basic categories:

- *Checking and initializing central system hardware.* A failure during this process may result in a fatal error that will prevent the system from booting. The error may be identified either by a message on the CRT or by a series of beeps from the internal speaker (if the monitor cannot be initialized). If the system detects a fatal error, the ROM monitor will place the processor into a hard loop with the error code in a specified register, and all interrupts disabled. The effect is a processor halt without going into a true halt state. Under this condition, no further checks will be made on the motherboard.

- *Checking and initializing peripheral hardware on the I/O expansion bus.* A failure during this process usually does not result in a fatal error, unless a defective expansion PWB holds a data or address line high or low. The booting process also can be halted by an interrupt or DMA request signal from a peripheral device that will not clear when the *acknowledge signal* is transmitted by the CPU.

8.3.1 POST Routine

Central hardware is checked first during the POST. The system then tests peripheral devices. The POST routine varies from one type of system to another, but computers built with a particular type of microprocessor tend to initialize the same way. A typical POST routine for a 80286-based IBM-compatible AT-type system is shown in Table 8.1. (IBM is a registered trademark of International Business Machines.) A number of checks are made to confirm the operational status of each device or subsystem of the computer. Common routines include the following:

- *CPU test.* Data retention of all internal general-purpose registers is confirmed. Processor flags and conditional jumps are checked for functionality.
- *ROM checksum.* All bytes of the ROM are added to produce one 8-bit sum. The result of the summation should always be zero. During summation, overflow carries are ignored. This process is repeated for each ROM device.
- Parallel peripheral interface (PPI). The PPI is initialized to operate as three output ports. Each port is written with unique data patterns, and the data is verified for accuracy. As a final check, the opposite data pattern is written and verified.
- *Programmable timer.* The timer is initialized as a downcounter. An initial value is programmed into the device, and the processor waits for the counter to go to zero. The processor then reinitializes the counter and waits for it to count up to a given value.
- *DMA.* The direct memory access controller register bank is checked by first writing a value to each port, then writing the opposite data pattern. Verification is performed after each write. After completion of the test, the DMA controller is configured for dynamic RAM refresh.
- *DRAM.* All devices are cleared to zeros at power-up. A value is written into the memory banks, then read out to confirm proper data retention. A value of opposite bit patterns then is written, and data retention confirmed. During the early POST stages, only the first 3 kbytes of DRAM is checked. If the lower

Table 8.1   Typical Power-Up Self-Test Routine for a Personal Computer System

Primary hardware tests
- Central processing unit
- Memory mapper
- Configuration control register
- ROM BIOS checksum
- CMOS RAM
- Peripheral interface
- Programmable interrupt timer (PIT)
- Direct memory access (DMA) controller
- Memory refresh logic
- Base 64-kbyte system memory
- RAM CRT monitor
- Protected-mode memory size
- Integrity of address lines A16 to A23
- Programmable interrupt timer (PIT)
- Memory cache controller

Secondary hardware tests
- CMOS RAM configuration data
- RAM memory above 64 kbytes
- Math coprocessor (if present)
- Keyboard status
- Floppy disk drive status
- Hard disk drive status
- Serial interface
- Parallel interface
- Other expansion bus devices
- Error class check

3 kbytes passes, the processor proceeds to check the remaining DRAM. All memory is cleared to zeros upon completion of the test.

- *Interrupt controller.* The interrupt mask port is checked for data retention by writing two opposite data patterns, and verifying the results. A temporary interrupt vector then is established, and interrupts are masked off. The processor then waits for a given interval to check for spurious interrupts.
- *CRT monitor.* A video reset function is sent to the monitor controller card, which initializes the PWB. A video RAM test then is performed of the screen and attribute memory.
- *Math coprocessor.* Commands are issued to read and write the entire register set from system RAM. All registers and bit patterns are tested.
- *Serial interface.* Opposite data patterns are written to the serial interface controller (SIC), and proper data retention is confirmed.

Table 8.2   CMOS Configuration Data for an AT-Class Personal Computer

ROM BIOS V4.01
Current time is 20:22:13
Current date is 08-23-1990
First floppy drive is 1.2 Mbytes
Second floppy drive is 360 kbytes
First hard drive is type 8
Second hard drive is type 8
Primary display is mono graphics
Enhanced keyboard is installed
Conventional memory size is 640 kbytes
Extended memory size is 1 Mbyte
Number of parallel ports is 2
Number of serial ports is 2

- *Parallel interface.* The operational status of all bidirectional lines are checked.
- *Floppy disk controller (FDC).* The FDC registers are checked for data retention and addressability by writing opposite data patterns to the control registers, and verifying proper data retention.

Error messages typically are displayed on the monitor. Technical documentation for the system usually contains an explanation of the error messages. If the monitor cannot be initialized, a series of beeps may be heard. When a failure is encountered, first check the on-board configuration jumpers and the CMOS RAM configuration data. Both must agree for proper operation. The CMOS configuration routine usually is resident in ROM. A battery-backed CMOS RAM chip is used to store the current configuration. Failure of the battery will result in error messages at POST. Table 8.2 shows a typical CMOS configuration display.

ROM-Based Diagnostic Routines. In addition to POST-related tests, some computer system ROMs include a series of diagnostic routines that can be invoked to check the operational status of individual pieces of hardware. ROM-based diagnostics are preferred to disk-based routines because an operational floppy disk is not required to check the motherboard. Table 8.3 lists the diagnostic routines available in one design. Diagnostic programs that permit examination of status registers provide the maintenance technician with valuable troubleshooting data. The technical service manual for a given system should provide the port addresses for all major registers.

Table 8.3   Test Routines Available in a Common ROM-Based Diagnostic Monitor
Adapted from: Technical Staff, *Technical Reference: ITT XTRA-286 ATW (TM) Business Computer*, ITT Information Systems, San Jose, Calif., 1986.

Diagnostic routines

1. System tests:
   A. 8259 interrupt controller
   B. 8254 timer
   C. ROM checksum
   D. 8042 programmable peripheral interface
   E. 80287 coprocessor
   F. Sound test
2. Memory tests:
   A. Conventional memory (up to 640 kbytes)
   B. Extended memory (above 1 Mbyte)
   C. Display memory map
3. Floppy disk tests:
   A. Format [drive] [type]
   B. Read [drive] [type]
   C. Read/write [drive] [type]
   D. Random read [drive] [type]
   E. Random read/write [drive] [type]
4. Fixed disk tests:
   A. Controller diagnostic test [drive] [type]
   B. Read last cylinder [drive] [type]
   C. Read/write last cylinder [drive] [type]
5. Keyboard tests:
   A. Reset/status return
   B. Check key positions
   C. Interactive key test
6. Video display tests:
   A. Check alignment
   B. Character set
   C. Character attributes
7. ROM monitor tests:
   A. Boot system
   B. Display current configuration
   C. Display contents of I/O port [hex value]
   D. Dump contents of register [address/range]
   E. Go to specified address [address]
   F. Move data from one location to another [address] [address]
   G. Update CMOS RAM configuration

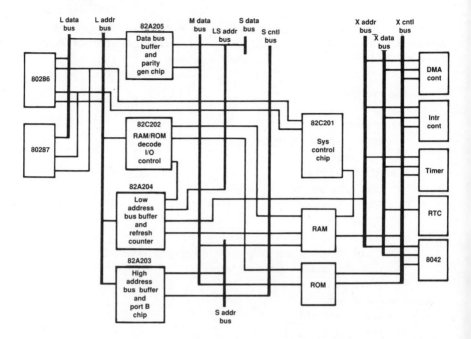

Figure 8.3   Block diagram of an AT-class personal computer based on a chip set design. (Courtesy of Chips and Technologies, Inc. Copyright 1986 Chips and Technologies, Inc. Reprinted with permission. All rights reserved.)

8.3.2   System/CPU Board

VLSI technology has simplified the design of a computer system significantly. Chip sets of four to six devices are used to replace many of the small-scale integration (SSI) devices found in earlier systems. A representative design based on the CS8220 chip set (Chips and Technologies, Inc.) is shown in Figure 8.3. Major elements of the system include:

- *CPU chip (80286).* Manages operation of the entire system and performs mathematical calculations as needed.
- *Math coprocessor chip (80287).* Performs high-level math as required. The coprocessor responds to a unique instruction set.
- *System controller/clock generator VLSI device (82C201).* Generates most of the clock and control signals required for the system; handles decoding and timing of all bus cycles.
- *Memory and I/O decoder VLSI device (82C202).* Manages chip control signals for RAM, ROM, and a portion of the I/O hardware; manages real-time clock and keyboard controller chips.

- *Bus buffer VLSI devices (82C203, 204, and 205).* Provides address latching and control buffering for internal, peripheral, and I/O hardware.
- *RAM and ROM.* Memory and program storage elements.
- *Support hardware controllers.* Includes DMA controllers, interrupt controllers, counter/timer, real time clock, and keyboard controller.

The system motherboard includes three major buses that carry data, address, and control signals. They include:

- CPU bus (LDATA and LADDR)
- Memory and I/O bus expansion (MDATA and LSADDR, SDATA, SADDR, and SCNTL)
- Peripheral bus (XADDR, XDATA, and XCNTL)

Any troubleshooting job is made easier by fully documenting a working system. If the technician understands how the system is supposed to work, repair is simplified. Document the status of the expansion bus pins in a specified mode, such as just after a successful boot. Compare this data with the UUT. The technician can diagnose many computer failures simply by checking the status of the expansion bus. Table 8.4 lists the signals present on the backplane of an AT-class personal computer. System board failures can be divided into two broad categories:

- Catastrophic: A failure that prevents the system from operating.
- Noncatastrophic: A failure that impairs operation of a particular system function.

The first step in the troubleshooting process is to narrow the scope of the work at hand. Before proceeding to troubleshoot the motherboard, confirm that all voltages from the power supply are within acceptable limits. Remove PWBs from the expansion bus one at a time to confirm that a problem really exists on the motherboard. In the case of some boards, such as the video driver, a substitute PWB will be needed.

Never rush into replacing a device on the motherboard. Carefully identify the potential cause and effect of a failure. Proper testing of an SSI device usually is possible with a logic probe. Testing of a VLSI device usually is practical only through substitution. Eliminate socket-mounted devices first as the possible cause of a failure; do not unsolder a device unless absolutely necessary.

Catastrophic Failures. A catastrophic failure will prevent the system from booting. Check the expansion bus for proper signal status. Confirm that all data and address lines are pulsed. Investigate any lines that are stuck high or low. Failures on the expansion bus are commonly the result of ESD-induced damage to line drivers or receivers, or accidental shorting of card edge connectors while power

Table 8.4 I/O Expansion Bus Functional Description for an AT-Class Computer

Pin	I/O	Signal	Description
A1	I	-IOCHCK	I/O check. Indicates that a parity error has been detected on a memory or I/O device on the I/O bus.
A2	I/O	SD7	System data bits SD0-SD7. Provide microprocessor,
A3	I/O	SD6	I/O, and memory data for all 8-bit devices. 16-bit
A4	I/O	SD5	devices on the I/O bus use data lines SD0 through
A5	I/O	SD4	SD15. For 16-bit data transfers to an 8-bit device,
A6	I/O	SD3	data on SD8 through SD15 are gated onto SD0 through
A7	I/O	SD2	SD7.
A8	I/O	SD1	
A9	I/O	SD0	
A10	I	IOCHRDY	I/O channel ready. This signal is disabled by a memory or I/O device to lengthen memory or I/O cycles by a given number of clock periods.
A11	O	AEN	Address latch enable. When this line is low, the microprocessor has control of the system bus. When active (high) the local bus is isolated from the system bus, and the DMA controller manages the data, address, memory, and I/O read/write command lines.
A12	I/O	SA19	Address bus lines A0 through A19. Valid address
A13	I/O	SA18	commands can be generated by the microprocessor,
A14	I/O	SA17	DMA controllers, or other microprocessors on the
A15	I/O	SA16	I/O expansion bus. Addresses are gated onto the bus
A16	I/O	SA15	when BALE is high, and are latched on the falling edge
A17	I/O	SA14	of BALE.
A18	I/O	SA13	
A19	I/O	SA12	
A20	I/O	SA11	
A21	I/O	SA10	
A22	I/O	SA9	
A23	I/O	SA8	
A24	I/O	SA7	
A25	I/O	SA6	
A26	I/O	SA5	
A27	I/O	SA4	
A28	I/O	SA3	
A29	I/O	SA2	
A30	I/O	SA1	
A31	I/O	SA0	
B1	GND	GND	Ground.
B2	O	RESET	Reset. Used to initialize all system hardware.
B3	PWR	+ 5 V	+ 5 V power-supply rail.
B4	I	IRQ9	Interrupt request line 9. A low-to-high transition indicates to the CPU that an I/O device needs service.
B5	PWR	- 5 V	– 5 V power-supply rail.

Table 8.4 (continued)   I/O Expansion Bus Functional Description for an AT-Class
Computer

B6	I	DRQ2	DMA request 2. The DRQ line is brought active (high) and held until the corresponding DMA acknowledge line goes active. DRQ0 has the highest priority; DRQ7 has the lowest priority.
B7	PWR	- 12 V	- 12 V power-supply rail.
B8	I	- OWS	Zero wait state signal. Indicates to the microprocessor that a device on the I/O bus does not require additional wait states.
B9	PWR	+ 12 V	+ 12 V power-supply rail.
B10	GND	Ground	Ground.
B11	O	- SMEMW	System memory write. Signals memory devices to store the data present on the bus. - SMEMW may be driven by a DMA controller or the system microprocessor.
B12	O	- SMEMR	System memory read. Instructs memory devices to drive data onto the data bus. - SMEMR may be driven by a DMA controller or the system microprocessor.
B13	I/O	- IOW	I/O write. Instructs an I/O device to read data on the data bus.
B14	I/O	- IOR	I/O read. Instructs an I/O device to drive its data on to the data bus.
B15	O	- DACK3	DMA acknowledge line 3. Acknowledges DRQ3.
B16	I	DRQ3	DMA request 3.
B17	O	- DACK1	DMA acknowledge line 1. Acknowledges DRQ1.
B18	I	DRQ1	DMA request 1.
B19	I/O	- REF	Memory refresh. Can be driven by any microprocessor on the I/O bus to indicate that a refresh cycle is to occur.
B20	O	SYSCLK	System clock. Buffered clock from the microprocessor used for synchronization of I/O expansion bus operations.
B21	I	IRQ7	Interrupt request 7.
B22	I	IRQ6	Interrupt request 6.
B23	I	IRQ5	Interrupt request 5.
B24	I	IRQ4	Interrupt request 4.
B25	I	IRQ3	Interrupt request 3.
B26	O	- DACK2	DMA acknowledge line 2. Acknowledges DRQ2.
B27	O	T/C	Terminal count. Generates a pulse when any DMA channel terminal count is reached.
B28	O	BALE	Buffered address latch enable. Indicates the presence of a valid address on the system bus from the CPU.
B29	PWR	+ 5 V	+ 5 V power-supply rail.
B30	O	OSC	Oscillator operating at 14.31818 MHz. Used to synchronize video operating modes.

Table 8.4 (continued) I/O Expansion Bus Functional Description for an AT-Class Computer

B31	GND	GROUND	Ground.
C1	I/O	- SBHE	System bus high enable. Indicates that the upper byte of data (SD8-SD15) is active during the current transfer operation.
C2	I/O	LA23	Unlatched high address lines. Used to expand system
C3	I/O	LA22	address capability to 16 Mbytes. These lines may be
C4	I/O	LA21	driven by the CPU, DMA controller, or any
C5	I/O	LA20	microprocessor on the I/O expansion bus. The address
C6	I/O	LA19	lines are valid when BALE is high.
C7	I/O	LA18	
C8	I/O	LA17	
C9	I/O	- MEMR	Memory read. Instructs memory devices to drive data onto the data bus.
C10	I/O	- MEMW	Memory write. Instructs memory devices to store the data present on the data bus.
C11	I/O	SD8	System data bits SD8-SD15. Provides the CPU, DMA
C12	I/O	SD9	controllers, and any other microprocessor on the I/O
C13	I/O	SD10	expansion bus with the capability to conduct 16-bit
C14	I/O	SD11	data transfers.
C15	I/O	SD12	
C16	I/O	SD13	
C17	I/O	SD14	
C18	I/O	SD15	
D1	I	- MEMCS16	Memory chip select 16. Indicates to the system board that the data transfer is a 16-bit operation.
D2	I	- I/OCS16	I/O chip select 16. Indicates to the system board that the data transfer is a 16-bit I/O cycle.
D3	I	IRQ10	Interrupt request 10.
D4	I	IRQ11	Interrupt request 11.
D5	I	IRQ12	Interrupt request 12.
D6	I	IRQ13	Interrupt request 13.
D7	I	IRQ14	Interrupt request 14.
D8	O	- DACK0	DMA acknowledge 0.
D9	I	DRQ0	DMA request 0.
D10	O	-DACK5	DMA acknowledge 5.
D11	I	DRQ5	DMA request 5.
D12	O	- DACK6	DMA acknowledge 6.
D13	I	DRQ6	DMA request 6.
D14	O	- DACK7	DMA acknowledge 7.
D15	I	DRQ7	DMA request 7.
D16	PWR	+ 5 V	+ 5 V power-supply rail.
D17	I	- MASTER	The master signal is used in conjunction with a DRQ line to gain control of the system.
D18	GND	GROUND	Ground.

is applied. Either condition may result in an address or data line that is stuck low. Such a fault usually will prevent the computer from booting. If a line is found to be stuck low (which is usually the case), remove power and make resistance measurements on the lines involved. Values of 1.5 kΩ to ground (or more) are common for a properly operating system.

Line drivers and receivers (transceivers) are usually the first suspects when a bus failure is detected. Transceivers serve to isolate the CPU and other elements of the kernel from the outside world. More often than not, failures are the result of an external influence under the control of the user.

If a bus test reveals that multiple address and/or data lines are stuck high or low, check the address and data lines at the CPU. If the CPU is found to be malfunctioning, carefully check the socket for proper contact with the device. A VLSI socket is delicate, and may be damaged easily. Repair of a damaged socket usually is not possible; replacement may be the only solution.

As noted previously, certain failures will cause the CPU to shut down, preventing data line tracing. In this event, try removing one of the ROM chips from its socket, and then reapplying power. Removal of the ROM will prevent the POST routine from being executed and prevent a programmed system shutdown when a catastrophic failure is detected. If the data and address lines return to a pulsed state, additional troubleshooting will be possible.

Check the master reset bus. Monitor the status of the bus with a logic probe when power is applied. It should hold a given logic level for a brief period after power is applied, and then switch to the alternate state. If a digital storage oscilloscope is available, store the reset waveform, and examine it for glitches. The power supply may provide a "power OK" signal to the motherboard. If so, this signal is required for proper system initialization. Confirm the presence of the "power OK" signal.

Noncatastrophic Failures. The occurrence of a noncatastrophic failure will impair operation of the system, but will not prevent booting. Failures usually are associated with devices outside the kernel. Common problems include interrupt, DMA, and RAM parity errors. Troubleshooting of noncatastrophic failures is aided by the fact that most of the computer is operating. ROM- or disk-based diagnostic routines may be used to troubleshoot the system.

Clock Circuits. A problem in the clock generator can cause intermittent or catastrophic failure of a computer. Confirm that the clock is running at the correct frequency. Use a frequency counter or the frequency function of a waveform analyzer. Check the peak-to-peak amplitude of the clock output. A low-amplitude signal may cause the microprocessor to miss some of the clock pulses. Low output also may cause the system to run slow overall (even though the clock is operating at the proper frequency) or result in erratic operation (some devices may trigger on the low-amplitude clock signals, while others do not). Figure 8.4 shows the usual logic level tolerances for TTL devices. Examine the clock pulse train on an oscilloscope. Look for poorly formed pulses or glitches on the waveform.

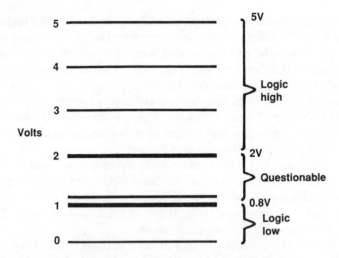

Figure 8.4   Logic level tolerance envelope for TTL-based systems.

Peripheral Interface Problems. The failure of a peripheral interface circuit is usually the result of ESD-induced damage or shorting of one or more output pins. Hardware typically involved includes parallel, serial, display, and keyboard ports. Incorrect cabling during system setup can result in a short circuit that will render the hardware interface inoperative. Depending on the nature of the failure, the POST routine may detect the problem. Use a logic probe to check the output status of the inoperative port. Look for stuck data or control lines. A failed line driver or receiver is often at the root of the problem.

Memory Problems. Most memory-related failures on a motherboard are identified by parity error messages. Never defeat parity error detection. ROM- or disk-based diagnostic routines are available to identify individual defective DRAMs. The failure of an entire bank of devices could point to a shorted address or data pin on one chip, a driver failure, or a short or open circuit on the PWB. Failure of the entire DRAM memory block can be caused by the failure of the refresh circuitry, which may be generated by a timer/counter chip separate from the memory block.

8.3.3   Using an Emulative Test Instrument

Emulative testers are among the most effective troubleshooting tools for microprocessor-based circuitry. As explained in Chapter 6, an emulative tester emulates the function of a key component of the microprocessor kernel—either the microprocessor itself, the ROM chip(s), or DMA controller(s). Emulative testers work in much the same fashion as the components they emulate, sending out read and write commands to various parts of the circuit. Such instruments, therefore, offer the advantage of testing a board from the inside out. This approach is more effective than *backdriving* the circuitry from the edge in because it allows every component on the board to be exercised. Emulative testers also permit the board to run in its native environment during the test process, which allows both static and time-dependent faults to be detected.

Like the DMA controller, a DMA emulator borrows clock cycles from the microprocessor while it performs internal operations. The instrument uses these intervals to perform data read and write functions. The instrument test harness typically clips over the microprocessor to gain access to the needed signals. DMA emulators do not require a circuit design that supports DMA operations. In most cases, DMA functions can be simulated for the purposes of troubleshooting analysis.

Most emulative instruments contain a number of standard test procedures that simplify the troubleshooting process significantly. Although tests differ from one unit to another, generally they fall into these categories:

- Bus tests
- Memory tests
- Input/output tests

In order for an emulative test instrument to function properly, the following conditions must be met:

- The power supply of the UUT must be functioning properly. The emulative tester receives power from the UUT.
- The clock of the UUT must be functioning. The system clock serves to synchronize data transfers on the buses.
- The tester must be able to drive the *DMA request* line. In order to borrow cycles from the microprocessor, the DMA request line must be drivable, not only at power-up, but every time it attempts a DMA access.
- The *DMA acknowledge* line must be operating. After requesting a DMA access, the test instrument waits until receiving a DMA acknowledge signal from the microprocessor before initiating a test.
- The *wait line* must be operative. Long-duration tests require that the instrument temporarily halt operation of the microprocessor. This is accomplished with the wait line.
- The *reset line* must be functional. After the test instrument has issued a wait instruction, a reset signal is needed to resume operation.

Bus Tests. Bus testing verifies the drivability of the address, data, and control lines inside the buffers and address decoders. This process identifies lines that are stuck high, stuck low, or short-circuited together. The integrity of the bus lines always should be checked first. Tests on the remainder of the circuitry will not yield valid results if problems exist on the bus lines.

Memory Tests. Memory testing exercises each of the memory components of the system (RAM, ROM, and peripherals). The operator typically specifies the starting and ending addresses, and the emulative tester performs the required quality assurance checks. ROM devices are checked by determining the checksum and comparing it with a known-good device. RAM devices are checked by reading data into and out of memory. Memory patterns of varying complexity may be used to exercise support circuitry. Data patterns typically used for RAM tests are designed to check each line for both high and low conditions. For example, a hex 55 (10101010) followed by a hex AA (01010101) will confirm that no lines are stuck high or low. Tests for memory-mapped peripherals are identical to RAM tests, except that instead of specifying the RAM address range, the address of the desired peripheral is specified.

Input/Output Tests. I/O checks are used to test peripherals that are accessed through I/O addresses rather than being memory-mapped. I/O tests involve writing data to a specified address range and reading the results. If the data stored at each location corresponds to the data written to the address, the I/O port is functioning properly.

8.3.4 Power-Supply Failures

Power-supply failures usually are obvious, as well as catastrophic to operation of the system. Most switching power supplies used in computer equipment today incorporate protective circuitry to shut down the system if an overvoltage or overcurrent condition is detected. To prevent damage to the computer system board, remove the supply from the chassis before beginning work. The supply usually must be loaded before troubleshooting. Some supplies will not start up unless loaded; others may not operate within normal specifications if no load is applied. Check the technical documentation for the supply, and connect a suitable resistive load. Most specification sheets list minimum and maximum current levels for the various outputs. If a minimum spec is not listed, a load usually is not required. For those specs that list a minimum load, connect a resistor of the proper value and wattage. Remember that the load resistor(s) will run hot. When loaded to specifications, voltages should fall within about 0.2 V of the correct level.

If intermittent disruptions are noted in the computer system, check for noise or ripple on the power-supply output pins. Noise can enter the microprocessor through the supply, causing it to act erratically. Less than 0.1 V of ripple should be observed on an oscilloscope. Check the reference manual for a ripple specification. A defective (open or marginal) filter capacitor or a shorted filter choke may cause 60 Hz ac line ripple. High-frequency noise also may be observed as a result of insufficient filtering of the switching stage.

Most PC-type computer equipment requires +5 V, -5 V, +12 V, and -12 V. A simplified block diagram of a switching supply is shown in Figure 8.5. Major elements of the example system include:

- *Input and energy storage.* Provides input-line conditioning (RFI filtering and current limiting) and ac-to-dc rectification and filtering.
- *Startup and reference.* Supplies unregulated dc to the startup circuit, supplies a +5 V reference for output voltage control, and generates trigger signals to the drive pulse generator circuit (which triggers the power supply into operation).
- *Control and drive.* Manages power pulse-generation during normal operation and during current-limiting conditions, and directs power control pulses to the drive transformer and into the primary inverter circuit.

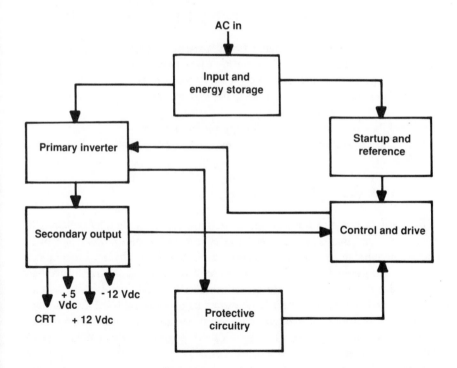

Figure 8.5 Block diagram of a switching power supply for a small computer. (Adapted from: Technical Staff, *Technical Reference: ITT XTRA-XP (TM) Personal Computer,* ITT Information Systems, San Jose, Calif., 1985)

- *Primary inverter.* Provides drive signals for the power switching transistors and supplies current-level status information for the shutdown circuit.
- *Protection.* Detects and limits high current on the primary winding of the primary inverter transformer, directs shutdown in the event of an overvoltage condition on the + 5 V output, and directs shutdown in the event of excessive temperature within the supply chassis.
- *Secondary output.* Provides rectification, filtering, and regulation of the required operating supply voltages.

When troubleshooting a catastrophic power-supply failure, first check the primary line fuse. Consult the technical documentation for the location of internal, PWB-mounted fuses, if used. Check rectifiers and input-side filter capacitors. Both are vulnerable to failure from transient disturbances on the ac line. Check the switching transistors, which are subjected to large switching currents during normal operation. Check the startup circuit for the presence of control pulses.

8.4 TROUBLESHOOTING PERIPHERAL EQUIPMENT

More failures often are experienced with peripheral equipment than with the computer system itself. The peripherals are exposed to the user environment, and usually include moving parts. Together, both conditions result in a high failure rate of some components. In a PC-type computer system, common peripherals include:

- Disk-based memory devices (floppy and hard drives)
- Primary I/O terminal equipment (keyboard and monitor)
- Printer

8.4.1 Disk-Based Memory Devices

In the field, floppy disk drive problems are most easily handled by replacing the drive with a unit that is known good. The usual justification for this repair-by-replacement approach is that floppy drives are inexpensive or that alignment procedures are too time-consuming to do in-house. This approach, however, may lead to excessive repair costs, as Figure 8.6 illustrates. Test instruments and alignment disks are available to simplify the troubleshooting process. Table 8.5 lists the signal interface pinout for a common type of floppy drive. Drives that exhibit intermittent read and/or write errors might be correctable through the use of a cleaning disk. Follow instructions supplied with the product.

Repair of hard drives is usually beyond the scope of most maintenance departments. The two most common failures are head crash and bearing wearout. Both failures require that the disk drive be opened for repair, requiring detailed clean-

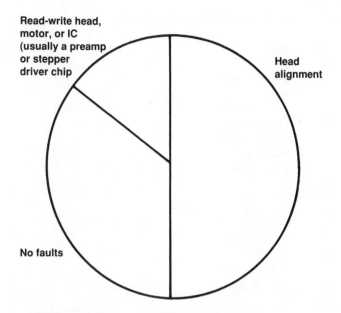

Figure 8.6   Estimation of the failure distribution for floppy disk drives.

room procedures. The purpose of a clean room is to eliminate submicron contamination. Clean-room technology is based on the concept of laminar airflow. Air can be made to flow in unidirectional layers, either horizontally or vertically, when airflow velocities are maintained above 70 ft. per minute. As the air flows from the supply side (usually the ceiling) to the return side (either a perforated floor or return-air register), particulate matter is washed away in a shower of air. The air travels through a return air plenum to a prefilter, and on to a recirculating fan, which forces the air through a high-efficiency particulate air filter. After filtering, the air is pumped back into the clean-room environment. Clean-room design directly relates to the degree or class of cleanliness. Table 8.6 lists clean-room classes as defined by U.S. government standards.

8.4.2   Primary I/O Terminal Equipment

Communication between the computer and the keyboard typically takes place under a serial half-duplex protocol. A single control line is used (the clock line) in conjunction with a bidirectional data line. Transmission consists of a start bit, 8 data bits, a parity bit, and a stop bit. The pinout for a common PC-type keyboard is shown in Table 8.7. A microprocessor within the keyboard scans the key switch matrix and generates separate make and break codes. Characters are stored in a first-in first-out (FIFO) buffer. The buffer is not cleared until all the data is

Table 8.5   Interface Pinout for a Floppy Disk Drive

Pin	I/O	Signal
4	I	Head load
6	I	Drive 3 select
8	O	Index
10	I	Drive select 0
12	I	Drive select 1
14	I	Drive select 2
16	I	Motor on
18	I	Direction select
20	I	Step
22	I	Write data
24	I	Write gate
26	O	Track 00
28	O	Write protect
30	O	Read data
32	I	Side one select
34	O	Diskette change

1, 3, 5,...31, 33 are signal returns

Table 8.6   Clean-Room Classifications as Defined by U.S. Government Standards. Data from: Lawrence E. Wetzel, "Clean Room Considerations," *Microservice Management* magazine, Intertec Publishing, Overland Park, Kan., March 1989.

Class	Max. no. of particles per ft. of 0.5 micron and larger	Max. no. of particles per ft. of 5.0 micron and larger
1	1	0
10	10	0
100	100	0
1,000	1,000	7
10,000	10,000	65
100,000	100,000	700

Table 8.7   Keyboard Pin Assignments for a PC-Type Computer

Pin	Signal
1	Keyboard clock
2	Keyboard data
3	Keyboard reset
4	Ground
5	+ 5 V power

transmitted. Before transmitting, the keyboard checks the status of both the clock and data lines; transmission can begin only if both lines are high. The serial protocol provides two means by which the system can signal the keyboard: *interrupt* and *request to send*. In addition to identifying and transferring scan codes, the keyboard performs auto repeat, key debounce, and self-test functions.

Keyboard failures usually are the result of physical abuse of the switch mechanism or damage to the electronic components as a result of ESD. Some mechanical key switch problems can be repaired; others require replacement of the entire keyboard. Because of the relatively low cost of keyboards, component-level troubleshooting of integrated circuits (unless socket-mounted) is seldom done.

Video monitors are available in a wide variety of types, including monochrome, color graphics adapter (CGA), extended graphics adapter (EGA), and video graphics adapter (VGA). Monochrome monitors make up the majority of units used. A control board interfaces the system expansion bus to the monitor. The board typically includes a CRT controller and a display buffer of 4 kbytes (for monochrome) to 256 kbytes (for VGA) or more of static or dynamic RAM. One or more on-board ROMs contain fonts that translate character codes into dot patterns to be displayed by the monitor. A color/graphics card usually supports a variety of software-selectable modes. One or more VLSI chips generate most of the signals needed to drive the video display. Figure 8.7 shows the pin assignments for a monochrome monitor.

Troubleshooting a video driver card can be a formidable proposition. Most cards contain soldered-in VLSI chips, and detailed documentation may be difficult to locate. Most display driver cards are replaced rather than repaired.

The video monitor itself is the most serviceable of all terminal equipment. A switching power supply typically is used, with conventional flyback-generated CRT anode voltages produced by the horizontal output stage. Video processing and amplification circuitry is usually of conventional design. Step-by-step signal tracing may be used in the video and sweep circuits. Input circuitry converts the digital bit stream into analog voltages of the proper value, which then are amplified by analog stages.

Failure modes for a video monitor are similar to those of a standard television receiver or display. Faults can be grouped as follows:

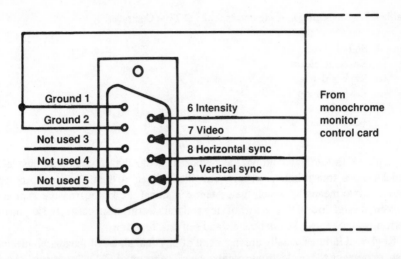

Figure 8.7 Pin assignments for a monochrome video monitor. (Adapted from: Technical Staff, *Technical Reference: ITT XTRA-XP (TM) Personal Computer*, ITT Information Systems, San Jose, Calif., 1985)

- *Power supply.* Failures include: open fuse, shorted or open rectifier or filter capacitor, excessive current demand from a sweep stage, or loss of the horizontal oscillator output (which will remove the CRT anode voltage).
- *Analog video path.* Failures include: defective gun driver transistor, defective intermediate stage amplifier, or failure in the digital interface circuit.
- *Horizontal and video sweep.* Failures include: loss of sweep synchronization input, defective electrolytic interstage coupling or filtering capacitors, defective output stage transistor, or shorted/open yoke winding.
- *CRT.* Failures include: open/shorted filament, low CRT emissions, shorted internal elements, poor connections at the anode or at the neck of the tube, or open/shorted focusing and/or convergence windings.

8.4.3 Printer

The dot-matrix printer is the most common type of printer in use today. It is inexpensive, fast, and reliable. High reliability is a function of the basic design; there are few moving parts. Typically, the electromechanical elements include:

- *Paper feed system.* Includes the platen and paper-feed sprockets.
- *Ribbon mechanism.* Winds the ribbon from one spool to another.

- *Print head.* Travels across the page printing in one or both directions. The print head contains nine or more print wires that make ink impressions on the page by striking the ribbon.

Most printer failures are the result of abuse by operators. Movement of the paper feed mechanism while the print head is engaged can bend the print wires, resulting in flawed reproduction of characters. Some printers include an adjustment for the print head to compensate for paper thickness; improper adjustment can result in poor print quality. If a new print head is required, replacement is relatively straightforward. Many printers include a diagnostic routine that can be engaged independent of the host computer to check operation of the unit.

8.4.4   Cable Considerations

Cabling is a frequent source of failures for computer-based equipment. Connectors are subject to rough handling and abuse that may result in bent or broken pins, or open or shorted internal connections. Many times a defective cable is at the root of what appears to be a computer system problem. Use of excessive lengths of cable also may result in marginal performance. Capacitive loading of high-frequency pulse trains may reduce the amplitude of data signals, and distort the waveform shape. Cable runs for RS-232-C of 50 ft. or less usually present no particular problem. Longer runs, however, may disrupt normal operation. Cables carrying analog or digital data to a high-resolution video monitor usually may stretch 15 ft. or so without noticeable degradation.

In addition to capacitive loading, crosstalk becomes a problem with long cable runs. Because of the low current levels used, the resistance of the cable is, typically, a minor concern. Long cable runs also expose the signal-carrying lines to interference from radio frequency or 60 Hz ac line energy. The likelihood of *electromagnetic interference* (EMI) problems increases as the cable length increases. The use of shielded cable will reduce the effects of EMI.

8.5   BIBLIOGRAPHY

Allen, Tom: "Troubleshooting Microprocessor-Based Circuits: Part 1," *Electronic Servicing & Technology* magazine, Intertec Publishing, Overland Park, Kan., January 1988.

_____: "Troubleshooting Microprocessor-Based Circuits: Part 2," *Electronic Servicing & Technology* magazine, Intertec Publishing, Overland Park, Kan., February 1988.

_____: "Components of Microprocessor-Based Systems," *Electronic Servicing & Technology* magazine, Intertec Publishing, Overland Park, Kan., January 1988.

Bychowski, Phil: "Reverse Engineering," *Microservice Management* magazine, Intertec Publishing, Overland Park, Kan., March 1989.

Carey, Gregory D.: "Isolating Microprocessor-Related Problems," *Electronic Servicing & Technology* magazine, Intertec Publishing, Overland Park, Kan., June 1988.

Clodfelter, Jim: "Troubleshooting the Microprocessor," *Microservice Management* magazine, Intertec Publishing, Overland Park, Kan., November 1985.

Crop, Roland: "A Hierarchical Approach to Servicing," *Microservice Management* magazine, Intertec Publishing, Overland Park, Kan., October 1988.

Evans, Dan: "So You Want to Service Personal Computers," *Electronic Servicing & Technology* magazine, Intertec Publishing, Overland Park, Kan., January 1989.

Greenberg, Bob: "Repairing Microprocessor-Based Equipment," *Sound & Video Contractor* magazine, Intertec Publishing, Overland Park, Kan., February 1988.

Owen, Jeff: "PC/AT Compatible Systems: An Integrated Approach," Chips and Technologies, Inc., Milpitas, Calif., 1986.

Persson, Conrad: "Troubleshooting Dot-Matrix Printer Problems," *Electronic Servicing & Technology* magazine, Intertec Publishing, Overland Park, Kan., January 1989.

Phoenix Technologies, Ltd.: *System BIOS for IBM PC/XT/AT (TM) Computers and Compatibles*, Addison-Wesley Publishing Company, Reading, Mass., 1990.

Smeltzer, Dennis: "Packing and Shipping Equipment Properly," *Microservice Management* magazine, Intertec Publishing, Overland Park, Kan., April 1989.

Ware, Peter: "Servicing Obsolete Equipment," *Microservice Management* magazine, Intertec Publishing, Overland Park, Kan., March 1989.

Wetzel, Lawrence E.: "Clean Room Considerations," *Microservice Management* magazine, Intertec Publishing, Overland Park, Kan., March 1989.

Staff: "Troubleshooting Cable-Related Problems," *Electronic Servicing & Technology* magazine, Intertec Publishing, Overland Park, Kan., November 1988.

Staff: "Teleservicing: A Team Approach to Field Service," *Microservice Management* magazine, Intertec Publishing, Overland Park, Kan., November 1985.

Technical Staff: *Technical Reference: ITT XTRA-XP (TM) Personal Computer*, ITT Information Systems, San Jose, Calif., 1985.

Technical Staff: *Technical Reference: ITT XTRA-286 ATW (TM) Business Computer*, ITT Information Systems, San Jose, Calif., 1986.

9

TROUBLESHOOTING DIGITAL COMMUNICATIONS SYSTEMS

9.1 INTRODUCTION

Getting computer equipment to work as a stand-alone system is one thing; getting computers to talk with each other in a networked system is quite another. In today's business environment, communication is the vital link to success in any enterprise. For many operations, local area networks provide an efficient way to exchange information. LAN technology answers many business problems by providing the means for efficient communication between systems. For maintenance engineers, the introduction of LANs has brought the need for a different approach to service. Because one of the major assets of a network is machine-to-machine communication, service can become complex. The maintenance technician must have an understanding of not only the machines in the network, but the interaction between machines as well.

9.1.1 Communicating Via RS-232-C

RS-232-C is the most common serial communications protocol. It is used to interconnect a wide variety of peripherals with computer equipment. As discussed in Chap. 6 (Sec. 6.4), many variations of RS-232-C currently are used. Table 9.1 lists the interchange circuits of the RS-232-C format by category. A subset of RS-232-C has found wide acceptance in the majority of applications. Common pin and signal assignments on pins 1 to 8, 20, and 22 are found in most systems. The

Table 9.1 RS-232-C Serial Communications Standard Interchange Circuits by Category. Data from: Phillip Buckland, "Solving Data Comm Problems," *Microservice Management* magazine, Intertec Publishing, Overland Park, Kan., September 1988.

Pin no.	Interchange circuit	CCITT equivalent	Description	GND	Data From DCE	Data To DCE	Control From DCE	Control To DCE	Timing From DCE	Timing To DCE
1	AA	101	Protective ground	X						
7	AB	102	Signal ground/common return	X						
2	BA	103	Transmitted data			X				
3	BB	104	Received data		X					
4	CA	105	Request to send					X		
5	CB	106	Clear to send				X			
6	CC	107	Data set ready				X			
20	DC	108.2	Data terminal ready					X		
22	CE	125	Ring indicator				X			
8	CF	109	Received line signal detector				X			
21	CG	110	Signal quality detector				X			
23	CH	111	Data signal rate selector (DTE)					X		
23	CI	112	Data signal rate selector (DCE)				X			
24	DA	113	Transmitter element signal timing (DTE)							X
15	DB	114	Transmitter element signal timing (DCE)						X	
17	DD	115	Receiver signal element timing (DCE)						X	
14	SBA	116	Secondary transmitted data			X				
16	SBB	117	Secondary received data		X					
19	SCA	120	Secondary request to send					X		
13	SCB	121	Secondary clear to send				X			
12	SCF	122	Secondary received line signal detector				X			

RS-232-C standard allows signals to be redefined for specific applications, providing maximum flexibility. This feature, however, also makes system interconnection difficult. Most problems are encountered during setup of a new system or expansion of an existing system. The *breakout box* (BOB) commonly is used to troubleshoot problems associated with RS-232-C. The function of a BOB is to break out the signals present at an interface so they can be opened, monitored, or rerouted. The BOB typically includes:

- A bank of 25 switches for opening any of the 25 RS-232-C lines
- LEDs on both sides of the switch bank for indicating signal status
- A jumper bank for rerouting signal lines
- Gender changers for mating various connectors

Configuring a null-modem is a common application of a BOB. Two possible configurations are shown in Figure 9.1.

9.2   LOCAL AREA NETWORK OPERATION[1]

The *open system interconnections* (OSI) model is the most broadly accepted explanation of LAN transmissions in an open system. The reference model was developed by the International Organization for Standardization (ISO) to define a framework for computer communication. The OSI model breaks the process of data transmission into the following steps:

- Physical layer
- Data-link layer
- Network layer
- Transport layer
- Session layer
- Presentation layer
- Application layer

9.2.1   Physical Layer

Layer 1 of the OSI model is responsible for carrying an electrical current through the computer hardware to perform an exchange of information. The physical layer is defined by the following parameters:

- Bit transmission rate
- Type of transmission medium (twisted pair, coaxial cable, or fiber-optic cable)
- Electrical specifications, including voltage- or current-based, and balanced or unbalanced.
- Type of connectors used (commonly RJ-45 or DB-9)

Many different implementations exist at the physical layer.

Potential Problems. Layer 1 can exhibit error messages as a result of overuse. For example, if a file server is being burdened with requests from workstations, the results may show up in error statistics that reflect the server's inability to handle all incoming requests. An overabundance of *response timeouts* also may be noted in this situation. A response timeout (in this context) is a message sent back to the workstation stating that the waiting period allotted for a response from the file server has passed without action from the server.

---

1 Portions of Section 9.2 were adapted from: Michael W. Dahlgren, "Serving Local Area Networks," *Broadcast Engineering* magazine, Intertec Publishing, Overland Park, Kan., November 1989.

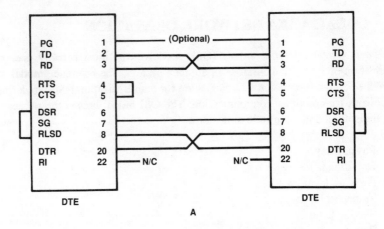

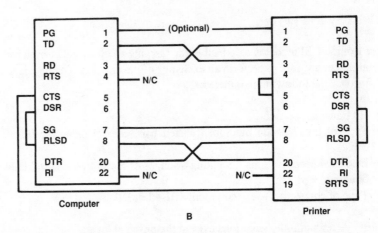

Figure 9.1 Null-modem configurations: (a) Basic interface. (b) Hardware handshaking interface. (Data from: Phillip Buckland, "Solving Data Comm Problems," *Microservice Management* magazine, Intertec Publishing, Overland Park, Kan., September 1988)

Error messages of this sort, which can be gathered by any number of commercially available software diagnostic utilities, may indicate an overburdened file server or a hardware flaw within the system. Intermittent response timeout errors can be caused by a corrupted *network interface card* (NIC) in the server. A steady flow of timeout errors throughout all nodes on the network may indicate the need for another server or bridge. Hardware problems are among the easiest to locate. In simple configurations where something suddenly has gone wrong, the physical layer and the data-link layer usually are the first suspects.

9.2.2 Data-Link Layer

Layer 2 of the OSI model, the data-link layer, describes hardware that enables data transmission (NICs and cabling systems). This layer integrates data packets into messages for transmission and checks them for integrity. Sometimes layer 2 also will send an "arrived safely" or "did not arrive correctly" message back to the transport layer (layer 4), which monitors this communications layer. The data-link layer must define the frame (or package) of bits that is transmitted down the network cable. Incorporated within the frame are several important fields:

- Addresses of source and destination workstations
- Data to be transmitted between workstations
- Error-control information, such as a *cyclic redundancy check (CRC)*, which ensures the integrity of the data

The data-link layer also must define the method by which the network cable is accessed, because only one workstation may transmit at a time on a baseband LAN. The two predominant schemes are:

- Token passing, used with the *ARCnet* and token-ring networks.
- Carrier sense multiple access with collision detection (CSMA/CD), used with *Ethernet* and *starLAN* networks.

At the data-link layer, the true identity of the LAN begins to emerge.

Potential Problems. Because most functions of the data-link layer (in a PC-based system) take place in integrated circuits on NICs, software analysis generally is not required in the event of a failure. As mentioned previously, when something happens on the network, the data-link layer is among the first to suspect. Because of the complexities of linking multiple topologies, cabling systems, and operating systems, the following failure modes may be experienced:

- *RF disturbance.* Transmitters, ac power controllers, and other computers all can generate energy that may interfere with data transmitted on the cable. Radio frequency interference (RFI) is usually the single biggest problem in a broadband network. This problem may manifest itself through excessive checksum errors and/or garbled data.
- *Excessive cable run.* Problems related to the data-link layer may result from long cable runs. Ethernet runs can stretch up to 1000 ft., depending on the cable. A typical token ring system can stretch 600 ft., with the same qualification. The need for additional distance can be accommodated by placing a bridge, gateway, active hub, equalizer, or amplifier on the line.

The data-link layer usually includes some type of routing hardware, such as one or more of the following:

- Active hub
- Passive hub
- Multiple access units (for token ring, starLAN, and ARCnet networks

9.2.3 Network Layer

Layer 3 of the OSI model guarantees the delivery of transmissions as requested by the upper layers of the OSI. The network layer establishes the physical path between the two communicating endpoints through the *communications subnet*, the common name for the physical, data-link, and network layers taken collectively. As such, layer 3 functions (routing, switching, and network congestion control) are critical. From the viewpoint of a single LAN, the network layer is not required. Only one route—the cable—exists. Internetwork connections are a different story, however, because multiple routes are possible. The *internet protocol* (IP) and *internet packet exchange* (IPX) are two examples of layer 3 protocols.

Potential Problems. The network layer confirms that signals get to their designated targets, then translates logical addresses into physical addresses. The physical address determines where the incoming transmission is stored. Lost data errors usually can be traced back to the network layer, in most cases incriminating the network operating system. The network layer is also responsible for statistical tracking and communications with other environments, including gateways. Layer 3 decides which route is the best to take, given the needs of the transmission. If *router tables* are being corrupted or excessive time is required to route from one network to another, an operating system error on the network layer may be involved.

9.2.4 Transport Layer

Layer 4, the transport layer, acts as an interface between the bottom three and the upper three layers, ensuring that the proper connections are maintained. It does the same work as the network layer, only on a local level. The network operating system driver performs transport-layer tasks.

Potential Problems. Connection flaws between computers on a network sometimes can be attributed to the shell driver. The transport layer may be able to save transmissions that were en route in the case of a system crash, or to reroute a transmission to its destination in case of primary route failure. The transport layer also monitors transmissions, checking to make sure that packets arriving at the destination node are consistent with the *build specifications* given to the sending node in layer 2. The data-link layer in the sending node builds a series of packets according to specifications sent down from higher levels, then transmits the packets to a *destination node*. The transport layer monitors these packets to ensure they arrive according to specifications indicated in the original build order. If they do not, the transport layer calls for a retransmission. Some operating systems refer to this technique as a *sequenced packet exchange* (SPX) transmission, meaning that the operating system guarantees delivery of the packet.

9.2.5 Session Layer

Layer 5 is responsible for turning communications on and off between communicating parties. Unlike other levels, the session layer may receive instructions from the application layer through the network basic input/output operation system (netBIOS), skipping the layer directly above it. The netBIOS protocol allows applications to "talk" across the network. The session layer establishes the session, or logical connection, between communicating host processors. Name-to-address translation is another important function; most communicating processors are known by a common name, rather than a numerical address.

Potential Problems. Multiple-vendor problems often can crop up in the session layer. Failures relating to gateway access usually fall into layer 5 for the OSI model, and often are related to compatibility issues.

9.2.6 Presentation Layer

Layer 6 translates application-layer commands into syntax understood throughout the network. It also translates incoming transmissions for layer 7. The presentation layer masks other devices and software functions, allowing a workstation to emulate a 3270 terminal through an emulation card and software. Reverse video, blinking cursors, and graphics also fall into the domain of the presentation layer. Layer 6 software controls printers and plotters, and may handle encryption and special file formatting. Data compression, encryption, and ASCII translations are examples of presentation-layer functions.

Potential Problems. Failures in the presentation layer often are the result of products that are not compatible with the operating system, an interface card, a resident protocol, or another application. *Terminate-and-stay-resident* (TSR) programs are particularly troublesome.

9.2.7 Application Layer

At the top of the 7-layer stack is the application layer. It is responsible for providing protocols that facilitate user applications. Print spooling, file sharing, and E-mail are components of the application layer, which translates local application requests into network application requests. Layer 7 provides the first layer of communications into other open systems on the network.

Potential Problems. Failures at the application layer usually center on software quality and compatibility issues. The program for a complex network may include latent faults that will manifest themselves only when a given set of conditions are present. The compatibility of the network software with other programs, particularly TSR utilities, is another source of potential problems.

9.2.8 Wide Area Networks

Various options are available for linking intelligent devices beyond the traditional local serial interface. The evolution of wide area network technology has permitted efficient 2-way transmission of data between distant computer systems. High-speed facilities are now cost-effective and available from the telephone company (telco) central office to the customer premises. LANs have proliferated and are being integrated with WANs through bridges and gateways. Interconnections via fiber-optic cable are common.

9.3 NETWORK INSTALLATION AND SERVICE

Fiber optics is the medium of choice for computer communications networks. Fiber systems offer numerous benefits, including immunity to EMI, small physical size, and zero capacitive loading. The cost of fiber interface hardware, however, is high as of this writing, relative to copper. Fiber installation poses several distinct challenges:

- Fiber is difficult to splice, requiring the use of special equipment that aligns the glass filament and ensures that the ends of the fiber meet perfectly so there is no deflection.
- Network drops require an optic splitter. A drop cable cannot be tapped in, as with coax.
- Equipment standards relating to fiber have yet to be resolved formally by industry.

9.3.1 Acceptance Testing

Acceptance testing of equipment during the installation process is basic to establishing any type of network. Testing involves verifying the performance of network components in the operating environment according to the requirements of the user. Acceptance testing may occur during installation, during troubleshooting, or after a repair or modification. Successful acceptance testing requires test equipment that can simulate the operating environment. For terminal testing, the network must be simulated.

It is desirable that information generated on one system be capable of being processed or displayed on another, even if manufactured by different companies. The process of ensuring compatibility between systems is referred to as *interoperability testing*. Interoperability testing during product development increases the likelihood of interoperability with complementary equipment but does not guarantee it; there are simply too many devices and systems to test. Plus, the standards themselves may change. Field interoperability testing should occur during network installation, while troubleshooting, and in verifying performance after repair or modification.

9.3.2   Test Instruments

Data communication test instruments can be classified into four general categories:

- *Interface tester:* A powered or nonpowered breakout box device.
- *Equipment tester:* An instrument that monitors the status of the physical interface between two network components, such as a terminal and a modem, or a PBX and a channel service unit (CSU). Equipment testers interact with a specific network component, such as a printer, terminal, or front-end processor to check its functionality and performance. Such instruments can range from simple testers with preprogrammed messages to complex PC-based emulators.
- *Link tester:* An instrument that checks the transmission integrity of the channel between two network components. *Bit error rate test (BERT)* sets are used for checking digital channels. *Transmission impairment measurement* (TIM) sets are used for checking analog lines that carry digital data.
- *Protocol analyzer:* An instrument that monitors data on the network. A protocol analyzer also can simulate network components when substitution troubleshooting is required.

Instrument criteria for interoperability testing are similar to those for acceptance testing. The ability of the gear to faithfully capture every detail of the exchange of information between interoperating systems is a primary consideration. Its ability to simulate the target environment (terminal equipment, PBX, or switch) to the component being tested is another key consideration.

Physical, functional, electrical, and procedural problems related to the interface between two devices constitute a large percentage of OSI network problems. Therefore, field service equipment should be able to monitor and analyze all parameters of layers 1 to 3. A breakout box, TIM set, transmission test set, and BERT set are useful tools for making functional and parametric checks of the physical interface and transmission media.

Data Line Monitor. A data line monitor is used to capture transmitted or received information for analysis. This approach represents an improvement over

BERT because it not only identifies transmission errors, but also shows the reaction of the system to those errors. Data line monitoring also pinpoints protocol errors generated by *data terminal equipment* (DTE) devices. Most monitors provide some means of indicating the presence of transmission errors. The types of errors encountered will vary according to the format and protocol used. Asynchronous data is subject to framing errors, while parity errors may occur in both asynchronous and synchronous data. Bit-oriented protocols are vulnerable to *frame check sequence* (FCS) errors, aborts, and short frames. Character-oriented protocols are subject to *block check character* (BCC) and timeout errors. When the data line monitor indicates the presence of errors, the instrument may be moved to different points along the communications path to determine where the errors are originating.

Some data line monitors include a *trapping* feature, by which the instrument searches the incoming data stream for a user-specified condition (the *trigger*). The trigger signal may be an error condition, a signal transition, or receipt of a message signifying that a problem exists. When the trigger is detected, the monitor stores the data that was received before and/or after the occurrence. The data then may be examined for error conditions.

Integrated Services Digital Network (ISDN). Systems built around the ISDN protocol present special instrumentation requirements. With ISDN, two or more physical channels are time-division multiplexed onto a single interface. Therefore, the instrument must be capable of extracting (demultiplexing) the channel to be measured. It also is necessary to compare the *signaling channel* (D channel) with one of the data channels (B channel). Measurement of *pulse code modulation* (PCM) interfaces involves a similar consideration—the ability to extract data from a multiplexed data stream. For applications requiring simulation, this function commonly is referred to as "drop and insert."

PC-Based Test Instruments. An add-on board and the appropriate software permit an off-the-shelf computer to perform sophisticated protocol analysis. PC-based instruments provide a number of features, including:

- Trigger flexibility
- Mixed-mode (transmit/receive) monitoring
- Data communication equipment (DCE) and DTE emulation
- BERT functions
- Clock signal outputs
- User-defined messages of up to 512 bytes
- Remote operation capability with appropriate control software
- Time stamps for all recorded events

PC-based systems can display data in real time or store it directly on disk for examination at a later date. Graphics capabilities provide a more understandable display of the data present on the network. Figure 9.2 shows the display screen of a protocol analyzer operating on a PC platform.

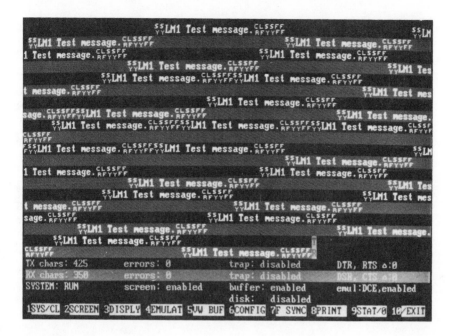

Figure 9.2   Display page from a PC-based protocol analyzer. (Courtesy of Progressive Computing)

9.4   TROUBLESHOOTING NETWORK SYSTEMS

Experience has shown that hardware failures account for as much as 80 percent of all LAN difficulties; software accounts for the remaining 20 percent. In most cases, hardware failures are easier to isolate and repair; software failures can prove elusive. Most data communications problems can be located with fairly simple test equipment, such as a breakout box, cable tester, BERT instrument, or TIM set. Sometimes, however, a difficult problem emerges that defies common attempts at troubleshooting. Such problems are best solved by simulation testing or data line monitoring.

9.4.1   Hardware Failures

Hardware failures occur at the physical and data-link layers of the OSI model. A broken cable, loose or detached cable connector, defective NIC, or faulty repeater between cable segments can bring a LAN down. Isolating these types of failures generally is straightforward, if the fault can be isolated to a given portion of the network. If the entire network is down, a fault in the backbone cabling system or the network server should be suspected. If only one workstation is unable to communicate, there are four possible hardware causes:

- *The workstation itself.* Many network failures are really workstation failures in disguise. Determine whether the workstation can function in a stand-alone (nonnetworked) manner. If it cannot, the problem is a PC failure, not a LAN failure.
- *Connection between the workstation and the backbone cable.* The problem may be as simple as a BNC connector that was removed when the workstation was moved a few feet, or a drop cable to a wall outlet that has pulled loose.
- *The NIC at the workstation.* Several possible problems can occur, such as a poor connection between the NIC and the bus connector. Clean the gold finger contacts on the NIC, and firmly reseat it into the connector. This procedure often will solve the problem.
- *Conflict between the workstation options and other add-on cards inside the workstation.* Conflicts may exist in the RAM buffer address, I/O base address, or interrupt request line selection. If none of these attempts are effective, swap the NIC with a known-good spare. Test both transmit and receive functions of the board.

Cable-related problems vary from simple open or short circuits that can be diagnosed with a DMM to unique failure modes that require a *time domain reflectometer* (TDR) to troubleshoot. A TDR instrument transmits a pulse from one end of the cable and monitors the line for a response. A cable short or open circuit or an impedance mismatch, caused by a crimp, kink, or poor connection, will reflect the transmitted signal. The TDR measures the time difference between the original pulse and any reflection. The instrument then calculates the distance to the mismatch, based on the applicable cable constants (nominal velocity of propagation and other parameters).

Most interfaces between the LAN workstation and backbone cable are proprietary to a given network architecture. Most NICs, however, come equipped with diagnostic software that allows in-depth analysis of the board while it is still plugged into the PC. In typical tests, the technician:

- Reads the contents of the address ROM on the NIC.
- Performs a loop-around test from the transmit to the receive circuit.
- Performs internal diagnostics of the protocol handler ICs located on the NIC.

External interfaces between LAN workstations and printers, plotters, and modems can be tested and diagnosed individually.

Many mysterious network and PC problems can be related to a lack of clean ac power to devices in the system. Surge suppressors, isolation transformers, power conditioners, and voltage regulators commonly are used to guarantee the quality and reliability of the power feed. Power-line test equipment is required to determine the need for supplemental power conditioning.

9.4.2   Software Failures

Most software failures involve parameters associated with the network operating system (NOS), such as log-in sequences, passwords, or file read/write permission modes. If a problem is experienced, first determine whether any parameters at the network server have been altered. Examine the network administration screens. Compare the current documentation with previous documentation for that server. If problems persist, go to the workstation that is experiencing difficulty and attempt to log in, thus eliminating any password or procedure mistakes. If these basic steps are not successful in correcting the problem, it will be necessary to examine the data going to and from that workstation with a protocol analyzer.

Software analysis begins by dissecting each field of the frame that is defined by the data-link layer. Next, the protocols used at the network through application layers are examined. A protocol analyzer commonly is used for solving difficult software problems.

9.4.3   T-1 Data System Failure Modes

The T-1 service is a digital communications path over telco facilities with a bandwidth of 1.544 Mb/s. A T-1 circuit might be routed through several carrier networks before reaching its final destination. BERT instruments commonly are used to identify T-1 network problems. The G.821 performance analysis is the most powerful way to evaluate a T-1 line. The G.821 standard defines error-rate performance categories. Figure 9.3 illustrates the performance categories with respect to total test time. A description of each performance category is listed in Table 9.2. The division includes:

- *Unavailable time:* Periods of catastrophic circuit failure, such as inadvertent line and/or repeater disconnections.
- *Unacceptable time:* Periods of failure caused by error bursts resulting from lightning strikes, line switching, or frame/clock slips.
- *Degraded time:* Periods of failure, usually long-term in nature, resulting from microwave radio fading, atmospheric effects on satellite links, degraded copper cable, or fiber-optic line driver problems.
- *Acceptable time:* Periods when the error rate is less than 10^{-6}, and when voice signal quality is considered good.

G.821 performance analysis is applied in a preventive maintenance program or when the circuit is experiencing transmission difficulty.

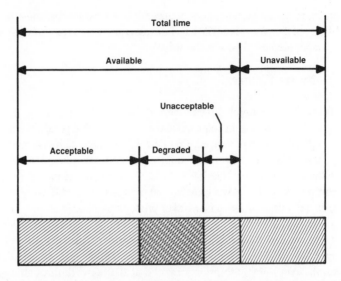

Figure 9.3 Performance classifications for a T-1 system. (Data from: Phil Ardire, "Bit Error Rate Testing," *Microservice Management* magazine, Intertec Publishing, Overland Park, Kan., February 1990)

Error Types. Errors on a T-1 network can be classified as either format or logic errors. Format errors include deviations from the specified signal format, bipolar violations, excess 0s, and loss of frame synchronization. Logic errors consist of any change in the original signal, including the actual data bits, framing bits, and CRC characters. Format errors and framing bit/CRC logic errors can be detected and measured. Data bit logic errors, however, may not be detectable and measured unless a known pattern is being sent.

The measurement of a network's true data bit error rate requires the T-1 link to be taken out of service, and a known pattern transmitted to a remote point in the system. Standard patterns exist for such tests, known in the industry as *quasirandom signal sources* (QRSS). These patterns include:

- *QRSS 2E15-1:* A quasirandom 15-bit test signal source.
- *QRSS 2E20-1:* A quasirandom 20-bit signal.
- *QRSS 2E23-1:* A quasirandom 23-bit signal.

Telco-supplied T-1 circuits (tariffed) include a digital service unit (DSU) that typically provides a monitor jack for test purposes. Use this port to conduct all tests. Hazardous voltages may exist on the T-1 line itself. Some equipment manufacturers provide 15-pin D-type connectors to facilitate bit error rate testing. These test points

Table 9.2   Performance Classifications of the G.821 T-1 Communications Standard. Data from: Phil Ardire, "Bit Error Rate Testing," *Microservice Management* magazine, Intertec Publishing, Overland Park, Kan., February 1990.

Performance Category	Description
Unavailable	The T-1 link is considered unavailable when the error rate has exceeded 10E-3 for at least 10 consecutive seconds. These 10 seconds are considered part of the unavailable time. The T-1 link remains unavailable until the error rate has remained below 10E-3 for at least 10 seconds. These 10 seconds are considered part of the available time.
Unacceptable	Unacceptable performance are consecutive periods of time less than 10 seconds in duration that have an error rate exceeding 10E-3.
Degraded	Degraded performance are cumulative periods of time that have an error rate less than 10E-3 but greater than 10E-6.
Acceptable	Acceptable performance are cumulative periods of time that have an error rate below 10E-3.
Available	Available performance is the total time less the time that the link is unavailable.

enable the user to send BERT patterns in a loopback configuration to the equipment. This feature aids quick isolation of faulty components on the T-1 line.

9.5   BIBLIOGRAPHY

Ardire, Phil: "Bit Error Rate Testing," *Microservice Management* magazine, Intertec Publishing, Overland Park, Kan., February 1990.

Blog, Thomas: "Solving Data Comm Problems with a PC," *Microservice Management* magazine, Intertec Publishing, Overland Park, Kan., July 1987.

Buckland, Phillip: "Solving Data Comm Problems," *Microservice Management* magazine, Intertec Publishing, Overland Park, Kan., September 1988.

Dahlgren, Michael W.: "Servicing Local Area Networks," *Broadcast Engineering* magazine, Intertec Publishing, Overland Park, Kan., November 1989.

Gorman, Ron: "Transmission Line Testing," *Microservice Management* magazine, Intertec Publishing, Overland Park, Kan., November 1986.

Miller, Mark A.: "Servicing Local Area Networks," *Microservice Management* magazine, Intertec Publishing, Overland Park, Kan., February 1990.

_____: *LAN Troubleshooting Handbook*, M&T Books, Redwood City, Calif., 1990.

Nystrom, Peter: "Power Conditioning Devices," *Microservice Management* magazine, Intertec Publishing, Overland Park, Kan., February 1987.

Waller, Mark: *PC Power Protection*, Howard W. Sams, Indianapolis, Ind., 1989.

Wilkin, Donald: "Trends in Data Communications and Test Equipment," *Microservice Management* magazine, Intertec Publishing, Overland Park, Kan.

International Organization for Standardization: "Information Processing Systems—Open Systems Interconnection—Basic Reference Model," ISO 7498, 1984.

10

RF SYSTEM
MAINTENANCE

10.1 INTRODUCTION

Radio frequency (RF) equipment is unfamiliar to many persons entering the electronics industry. Colleges do not routinely teach RF principles, instead favoring digital technology. Unlike other types of products, however, RF equipment often requires preventive maintenance to achieve its expected reliability. Maintaining RF gear is a predictable, necessary expense that facilities must include in their operating budgets. Tubes (if used in the system) will have to be replaced from time to time, no matter what the engineer does; components fail every now and then; and time must be allocated for cleaning and adjustments. Anticipating these possibilities can help prevent unnecessary downtime.

Although the reason generally given for minimum RF maintenance is a lack of time and/or money, the cost of such a policy can be deceptively high. Problems that could be solved for a few dollars may, if left unattended, result in considerable damage to the system and a large repair bill. A standby system often can be a lifesaver, but its usefulness sometimes is overrated. The best standby RF system is a *main system* in good working order.

10.2 ROUTINE MAINTENANCE

Most RF system failures can be prevented through regular cleaning, inspection, and close observation. The history of the unit also is important in a thorough maintenance program so that trends can be identified and analyzed.

Parameter	Typical value	Measured value
RF power output	18.3 kW	_____
Plate current	2.8 A	_____
Plate voltage	7.55 kV	_____
Screen current	380 mA	_____
Screen voltage	650 V	_____
PA grid current	110 mA	_____
PA bias voltage	490 V	_____
PA filament voltage	6 V	_____
Left driver cathode current	142 mA	_____
Right driver cathode current	142 mA	_____
Driver screen voltage	275 V	_____
Driver screen current	35 mA	_____
Driver grid current	1 mA	_____
Driver plate voltage	1.85 kV	_____
28-V power supply	27 V	_____
Reflected power	15 W	_____
Transmission-line pressure	3.9 psi	_____
Tank pressure	1500 psi	_____
Transmitter hours	5412	_____
Exciter AFC	Center scale	_____

Figure 10.1 Example of a transmitter operating log that should be filled out regularly by maintenance personnel.

10.2.1 The Maintenance Log

The control-system front panel can tell a great deal about what is going on inside an RF generator. Develop a maintenance log, and record all front-panel meter readings, as well as the positions of critical tuning controls, on a regular basis. (See Figure 10.1.) This information provides a history of the system and can be a valuable tool in noting problems at an early stage. The most obvious application of such logging is to spot failing power tubes, but other changes occurring in components can be identified as well.

Creating a history of the line and tank pressure for a pressurized transmission line can help to identify developing line or antenna problems. After the regulator is set for the desired line pressure, record the tank and line readings each week, and chart the data. If possible, make the observations at the same time of day each week. Ambient temperature can have a significant effect on line pressure; note any temperature extremes in the transmission-line log when the pressure is recorded. The transmission-line pressure usually will change slightly between carrier-on and carrier-off conditions (depending on the power level). The presence of RF can heat the inner conductor of the line, causing the pressure to increase. After the gradual loss of tank pressure has been charted for a few months, a pattern should become obvious. Investigate any deviation from the normal amount of tank pressure loss over a given period.

Whenever a problem occurs with the RF system, make a complete entry describing the failure in the maintenance log. Include a description of all maintenance activities required to return the system to operational condition. All entries should be complete and clear. Include the following data:

- Description of the nature of the malfunction, including all observable symptoms and performance characteristics
- Description of the actions taken to return the system to a serviceable condition
- Complete list of the components replaced or repaired, including the device schematic number and part number
- Total system downtime as a result of the failure
- Name of the engineer who made the repairs

The importance of regular, accurate logging can best be emphasized through the following examples:

Case Study No. 1. Improper neutralization is detected on an AM broadcast transmitter IPA (intermediate power amplifier), shown in Figure 10.2. The neutralization adjustment is made by moving taps on a coil, and none have been changed. The history of the transmitter (as recorded in the maintenance record) reveals, however, that the PA grid tuning adjustment has, over the past two years, been moving slowly into the higher readings. An examination of the schematic diagram leads the technician to conclude that C-601 is the problem.

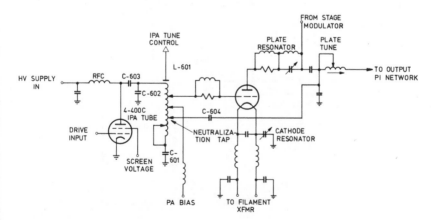

Figure 10.2   AM transmitter IPA/PA stage exhibiting neutralization problems. A history of IPA returning (through adjustment of L-601) helped determine that loss of neutralization was the result of C-601 changing in value. (Courtesy of *Broadcast Engineering* magazine, Intertec Publishing, Overland Park, Kan.)

The tuning change of the stage was so gradual that it was not thought significant until an examination of the transmitter history revealed that continual retuning in one direction only was necessary to achieve maximum PA grid drive. Without a record of the history of the unit, time probably would have been wasted in substituting expensive capacitors in the circuit, one at a time. Worse yet, the engineer might have changed the tap on coil L-601 to achieve neutralization, further masking the real cause of the problem.

Case Study No. 2. A UHF broadcast transmitter is found to exhibit decreasing klystron body-current. The typical reading with average picture content is 50 mA, but over a 4-week period, the reading dropped to 30 mA. No other parameters show deviation from normal. Yet, the decrease in the reading indicates an alternate path (besides the normal body-current circuitry) by which electrons return to the beam power supply. A schematic diagram of the system is shown in Figure 10.3. Several factors could cause the body-current variation, including water leakage into the

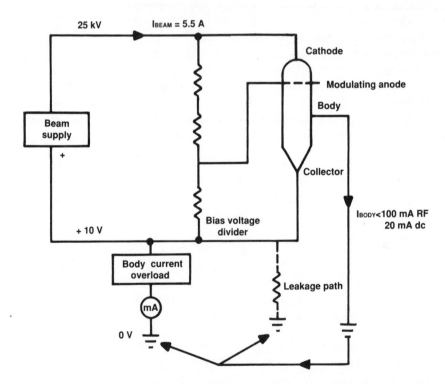

Figure 10.3   Simplified high-voltage schematic of a klystron amplifier showing the parallel leakage path that can cause a reduction in protection sensitivity of the body-current circuit. (Courtesy of *Broadcast Engineering* magazine, Intertec Publishing, Overland Park, Kan.)

body-to-collector insulation of the klystron. In time, this water can corrode the klystron envelope, possibly leading to vacuum loss and klystron failure.

Water leakage also can cause partial bypassing of the body-current circuitry, an important protection system in the transmitter. It is essential that the circuit functions normally at all times and at full sensitivity to detect change when a fault condition occurs. Regular logging of transmitter parameters ensures that developing problems are caught early.

10.2.2 Preventive Maintenance Routine

Thorough inspection of the RF system on a regular basis is the key to minimizing equipment downtime. Component problems often can be spotted at an early stage through regular inspection of the system. Remember to discharge all capacitors in the circuit with a grounding stick before touching any component in the high-voltage sections of the system. Confirm that all primary power has been removed before any maintenance work begins.

Special precautions must be taken with systems that receive ac power from two independent feeds. Typically, one ac line provides 208 V 3-phase service for the high-voltage section of the system, and a separate ac line provides 120 V power for low-voltage circuits. Older transmitters or high-power transmitters often utilize this arrangement. Be sure that all ac is removed before any maintenance work begins.

Consider the following preventive maintenance procedures:

Resistors and Capacitors
- Inspect resistors and RF capacitors for signs of overheating (see Figure 10.4).
- Inspect electrolytic or oil-filled capacitors for signs of leakage (see Figure 10.5).
- Inspect feedthrough capacitors and other high-voltage components for signs of arcing.

Transmitting capacitors—mica vacuum and *doorknob* types—should never run hot. They may run warm, but that is usually the result of thermal radiation from nearby components (such as power tubes) in the circuit. An overheated transmitting capacitor is often a sign of incorrect tuning. Vacuum capacitors present special requirements for the maintenance technician. Care in handling is essential to maximum service life. Because the vacuum capacitor is evacuated to a higher degree than most vacuum tubes, it is particularly susceptible to shock and rough handling. Provide adequate protection to vacuum capacitors whenever maintenance is performed. The most vulnerable parts of the capacitor are the glass-to-metal seals on each end of the unit. Exercise particular care during removal or installation.

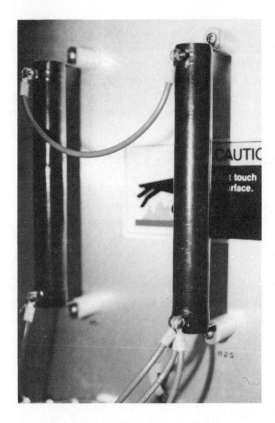

Figure 10.4   Check resistors, particularly high-power units, regularly for signs of premature wear caused by excessive heating.

The current ratings of vacuum capacitors are limited by the glass-to-metal seal temperature and the temperature of the solder used to secure the capacitor plates. Seal temperature is increased by poor connecting clip pressure, excessive ambient temperatures, corrosion of the end caps and/or connecting clip, excessive dust and dirt accumulation, or excessive currents. Dust accumulation on sharp points in high-voltage circuitry near the vacuum capacitor can cause arcs or corona that actually may burn a hole through the glass envelope.

Power-Supply Components
- Inspect the mechanical operation of circuit breakers. Confirm that they provide a definite *snap* to the off position (remove all ac power for this test), and that they reseat firmly when restored. Replace any circuit breaker that is difficult to reset.

Figure 10.5   Inspect high-voltage capacitors for signs of leakage around the case feed-through terminals.

- Inspect power transformers and reactors for signs of overheating or arcing (see Figure 10.6).
- Inspect oil-filled transformers for signs of leakage.
- Inspect transformers for dirt build-up, loose mounting brackets and rivets, and loose terminal connections.
- Inspect high-voltage rectifiers and transient-suppression devices for overheating and mechanical problems (see Figure 10.7).

Power transformers and reactors normally run hot to the touch. Check both the transformer frame and the individual windings. On a 3-phase transformer, each winding should produce about the same amount of heat. If one winding is found to run hotter than the other two, further investigation is warranted. Dust, dirt or

Figure 10.6   Inspect power transformers just after shutdown for indications of overheating, or leakage in the case of oil-filled transformers.

moisture between terminals of a high-voltage transformer may cause flashover failures. Insulating compound or oil around the base of a transformer indicates overheating or leakage.

Coils and RF Transformers
- Inspect coils and RF transformers for indications of overheating (see Figure 10.8).
- Inspect connection points for arcing or loose terminals.

Coils and RF transformers operating in a well-tuned system rarely will heat appreciably. If discoloration is noticed on several loops of a coil, consult the factory service department to find out whether the condition is normal. Pay particular attention to variable tap inductors, often found in AM broadcast transmitters and phasers. Closely inspect the roller element and coil loops for overheating or signs of arcing.

Relay Mechanisms
- Inspect relay contacts, including high-voltage or high-power RF relays, for signs of pitting or discoloration.

Figure 10.7   Check the heating of individual rectifiers in a stack assembly. All devices should generate approximately the same amount of heat.

- Inspect the mechanical linkage to confirm proper operation. The contactor arm (if used) should move freely, without undue mechanical resistance.
- Inspect vacuum contactors for free operation of the mechanical linkage (if appropriate) and for indications of excessive dissipation at the contact points and metal-to-glass (or metal-to-ceramic) seals.

Unless problems are experienced with an enclosed relay, do not attempt to clean it. More harm than good can be done by disassembling properly working components for detailed inspection.

Figure 10.8 Clean RF coils and inductors as often as needed to keep contaminants from building up on the device loops.

Connection Points
- Inspect connections and terminals that are subject to vibration. Tightness of connections is critical to the proper operation of high-voltage and RF circuits (see Figure 10.9).
- Inspect barrier strip and printed circuit board contacts for proper termination.

Although it is important that all connections are tight, be careful not to over-tighten. The connection points on some components, such as doorknob capacitors, can be damaged by excessive force. There is no section of an RF system where it

Figure 10.9   Check the tightness of connectors on an occasional basis, but do not stress the connection points.

is more important to keep connections tight than in the power amplifier stage. Loose connections can result in arcing between components and conductors that can lead to system failure. The cavity access door is a part of the outer conductor of the coaxial transmission-line circuit in FM and TV transmitters, and in many RF generators operating at VHF and above. High-potential RF circulating currents flow along the inner surface of the door, which must be fastened securely to prevent arcing.

10.2.3   Cleaning the System

Cleaning is a large part of a proper maintenance routine. Generally, a shop vacuum and clean brush are the main requirements. Use isopropyl alcohol and a soft cloth for cleaning insulators on high-voltage components (see Figure 10.10). Cleaning affords the opportunity to inspect each component in the system and observe any changes.

Regular maintenance of insulators is important to the proper operation of RF final amplifier stages because of the high voltages usually present. Pay particular attention to the insulators used in the PA tube socket (see Figure 10.11). Because

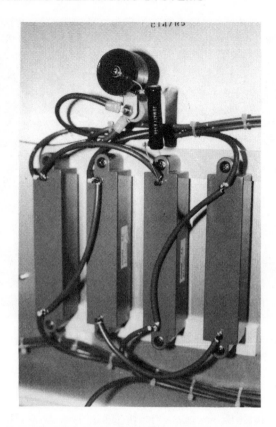

Figure 10.10   Keep all high-voltage components, such as this rectifier bank, free of dust and contamination that might cause short-circuit paths.

the supply of cooling air is passed through the socket, airborne contaminants may be deposited on various sections of the assembly. These can create a high-voltage arc path across the socket insulators. Perform any cleaning work around the PA socket with extreme care. Do not use compressed air to clean out a power tube socket. Blowing compressed air into the PA or IPA stage of a transmitter merely moves the dirt from places where it *can* be seen to places where it *cannot*. Use a vacuum instead. When cleaning the socket assembly, do not disturb any components in the PA circuit (see Figure 10.12). Visually check the tube anode to see whether dirt is clogging any of the heat-radiating fins.

   Cleaning also is important to proper cooling of solid-state components in the transmitter. A layer of dust and dirt can create a thermal insulator effect and prevent proper heat exchange from a device into the cabinet (see Figure 10.13).

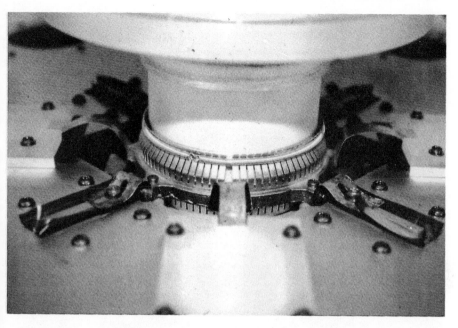

Figure 10.11 Carefully inspect the PA tube socket assembly. Do not remove the PA tube unless necessary.

10.2.4 Klystron Devices

Klystrons are expensive to buy and expensive to operate. Compared with tetrodes, they require larger auxiliary components (such as power supplies and heat exchangers), and are physically larger. Yet, they are stable, provide high gain, and may be driven easily by solid-state circuitry. Klystrons are relatively simple to cool and are capable of long life with a minimum of maintenance. Two different types of klystrons are in use today:

- Integral cavity klystrons, in which the resonant cavities are built into the body, as shown in Figure 10.14
- External cavity klystrons, in which the cavities are clamped mechanically onto the body and are outside the vacuum envelope of the device, as shown in Figure 10.15.

This difference in construction requires different maintenance procedures. The klystron body (the RF *interaction region* of the integral cavity device), is cooled by the same liquid that is fed to the collector. Required maintenance involves

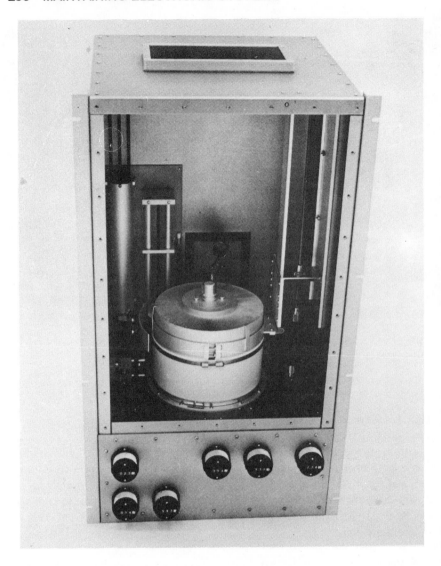

Figure 10.12   Clean the PA cavity assembly to prevent an accumulation of dust and dirt. Check hardware for tightness. (Courtesy of Varian Associates)

checking for leaks and adequate coolant flow. Although the cavities of the external cavity unit are air-cooled, the body may be water- or air-cooled. Uncorrected leaks in a water-cooled body can lead to cavity and tuning mechanism damage. Look inside the magnet frame with a flashlight once a week. Correct leaks immediately, and clean away coolant residues.

Figure 10.13   Clean power semiconductor assemblies to ensure efficient dissipation of heat from the devices.

The air-cooled body requires only sufficient airflow. The proper supply of air can be monitored through one or two adhesive temperature labels and close visual inspection. Watch for discoloration of metallic surfaces. The external cavities need a clean supply of cooling air. Dust accumulation inside the cavities will cause RF arcing. Check air-supply filters regularly. Some cavities have a mesh filter at the inlet flange. Inspect this point as required.

It is possible to make a visual inspection of the cavities of an external-cavity device by removing the loading loops and/or air loops. This procedure is recommended only when unusual behavior is experienced, and not as part of routine maintenance. Generally, there is no need to remove a klystron from its magnet frame and cavities during routine maintenance.

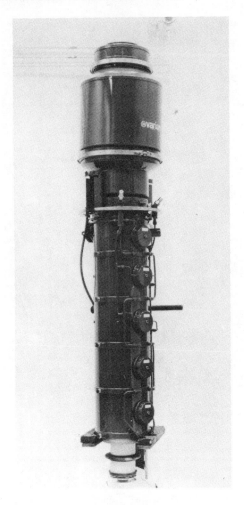

Figure 10.14   Integral cavity klystron. (Courtesy of Varian Associates)

10.3   POWER GRID TUBES

The power tubes used in RF generators and transmitters are perhaps the most important and least understood components in the system. The best way to gain an understanding of the capabilities of a PA tube is to secure a copy of the tube manufacturer's data sheet for each type of device. They are available from either the tube or transmitter manufacturer. The primary value of the data sheets to the end-user is the listing of maximum permissible values. These give the maintenance

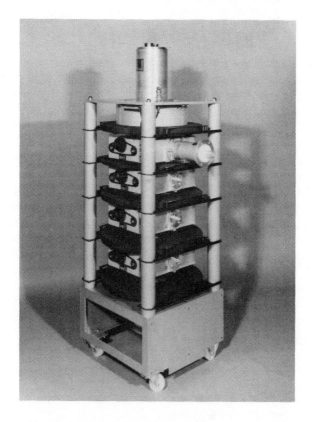

Figure 10.15   External cavity klystron. (Courtesy of EEV)

engineer a clear rundown of the maximum voltages and currents that the tube can withstand under normal operation. Note these values, and avoid them.

An examination of the data sheet will show that a number of operating conditions are possible, depending upon the class of service required by the application. As long as the maximum ratings of the tube are not exceeded, a wide choice of operating parameters, including plate voltage and current, screen voltage, and RF grid drive, are possible. When studying the characteristic curves of each tube, remember that they represent the performance of a *typical* device. All electronic products have some tolerance among devices of a single type. Operation of a given device in a particular system may be different than that specified on the data sheet or in the transmitter instruction manual. This effect is more pronounced at VHF and above.

10.3.1 Tube Dissipation

Proper cooling of the tube envelope and seals is a critical parameter for long tube life. Deteriorating effects that result in shortened tube life and reduced performance increase as the temperature increases. Excessive dissipation is perhaps the single greatest cause of catastrophic failure in a power tube. PA tubes used in broadcast, industrial, and research applications can be cooled using one of three methods: forced-air, liquid, and vapor-phase cooling. In radio and VHF-TV transmitters, forced-air cooling is, by far, the most common method used. Forced-air systems are simple to construct and easy to maintain.

The critical points of almost every PA tube type are the metal-to-ceramic junctions or seals. At temperatures below 250°C, these seals remain secure, but above that temperature, the bonding in the seal may begin to disintegrate. Warping of grid structures also may occur at temperatures above the maximum operating level of the tube. The result of prolonged overheating is shortened tube life or catastrophic failure. Several precautions usually are taken to prevent damage to tube seals under normal operating conditions. Air directors or sections of tubing may be used to provide spot-cooling to critical surface areas of the device. Airflow sensors prevent operation of the system in the event of a cooling-system failure.

Tubes that operate in the VHF and UHF bands are inherently subject to greater heating action than devices operated at lower frequencies (such as AM service). This effect is the result of larger RF charging currents into the tube capacitances, dielectric losses, and the tendency of electrons to bombard parts of the tube structure other than the grid and plate in high-frequency applications. Greater cooling is required at higher frequencies.

The technical data sheet for a given power tube will specify cooling requirements. The end-user normally is not concerned with this information; it is the domain of the transmitter manufacturer. The end-user, however, is responsible for proper maintenance of the cooling system.

10.3.2 Air-Handling System

All modern air-cooled PA tubes use an air-system socket and matching chimney for cooling. Never operate a PA stage unless the air-handling system provided by the manufacturer is complete and in place. For example, the chimney for a PA tube often can be removed for inspection of other components in the circuit. Operation without the chimney, however, may reduce airflow through the tube significantly and result in overdissipation of the device. It also is possible that operation without the proper chimney could damage other components in the circuit because of excessive radiated heat. Normally, the tube socket is mounted in a pressurized compartment so that cooling air passes through the socket and then is guided to the

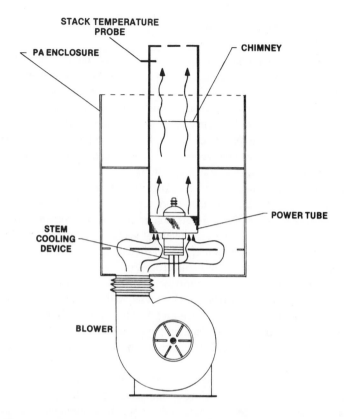

Figure 10.16 Airflow system for an air-cooled power tube. (Courtesy of *Broadcast Engineering* magazine, Intertec Publishing, Overland Park, Kan.)

anode cooling fins, as illustrated in Figure 10.16. Do not defeat any portion of the air-handling system provided by the manufacturer.

Cooling of the socket assembly is important for proper cooling of the tube base, and for cooling of the contact rings of the tube itself. The contact fingers used in the *collet* assembly of a socket typically are made of beryllium copper. If subjected to temperatures above 150°C for an extended period of time, the beryllium copper will lose its temper (springy characteristic) and will no longer make good contact with the base rings of the device. In extreme cases, this type of socket problem can lead to arcing, which can burn through the metal portion of the tube base ring. Such an occurrence ultimately can lead to catastrophic failure of the device due to a loss of the vacuum envelope. Other failure modes for a tube socket include arcing between the collet and tube ring that can weld together a part of the socket and tube. The result is failure of both the tube and the socket.

10.3.3 Ambient Temperature

The temperature of the intake air supply is a parameter that is usually under the control of the maintenance engineer. The preferred cooling-air temperature is no higher than 75°F, and no lower than the room dew point. The air temperature should not vary because of an oversized air-conditioning system or because of the operation of other pieces of equipment at the transmission facility. Monitoring the PA exhaust stack temperature is an effective method of evaluating overall RF system performance. This can be accomplished easily, and it also provides valuable data on the cooling system and final stage tuning.

Another convenient method for checking the efficiency of the transmitter cooling system over a period of time involves documenting the back pressure that exists within the PA cavity. This measurement is made with a *manometer*, a simple device that is available from most heating, ventilation, and air-conditioning (HVAC) suppliers. The connection of a simplified manometer to a transmitter PA input compartment is illustrated in Figure 10.17.

When using the manometer, be careful that the water in the device is not allowed to backflow into the PA compartment. Do not leave the manometer connected to the PA compartment when the transmitter is on the air. Make the necessary measurement of PA compartment back pressure, then disconnect the device. Seal the connection point with a subminiature plumbing cap or other appropriate hardware.

By charting the manometer readings, it is possible to accurately measure the performance of the transmitter cooling system over time. Changes resulting from the build-up of small dust particles (*microdust*) may be too gradual to be detected except through back-pressure charting. Be certain to take the manometer readings during periods of calm weather. Strong winds may result in erroneous readings because of pressure or vacuum conditions at the transmitter air intake or exhaust ports.

Deviations from the typical back-pressure value, either higher or lower, could signal a problem with the air-handling system. Decreased PA input compartment back pressure could indicate a problem with the blower motor or a build-up of dust and dirt on the blades of the blower assembly. Increased back pressure, on the other hand, could indicate dirty PA tube anode cooling fins or a build-up of dirt on the PA exhaust ducting. Either condition is cause for concern. A system suffering from reduced air pressure into the PA compartment must be serviced as soon as possible. Failure to restore the cooling system to proper operation may lead to premature failure of the PA tube or other components in the input or output compartments. Cooling problems do not improve—they always get worse.

If the PA compartment air-interlock switch fails to close reliably, it may be an early indication of impending cooling-system trouble. This situation could be caused by normal mechanical wear or vibration of the switch assembly, or it may

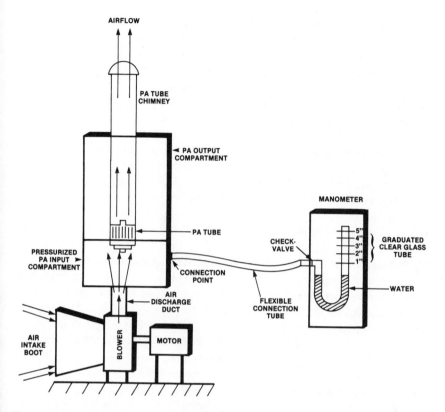

Figure 10.17   A manometer device used for measuring back pressure in the PA compartment of a transmitter. (Courtesy of *Broadcast Engineering* magazine, Intertec Publishing, Overland Park, Kan.)

signal that the PA compartment air pressure has dropped. In such a case, documentation of manometer readings will show whether the trouble is caused by a failure of the air pressure switch or a decrease in the output of the air-handling system.

10.3.4   Thermal Cycling

Most power grid tube manufacturers recommend a warm-up period between application of *filament-on* and *plate-on* commands. Most RF equipment manufacturers specify a warm-up period of about 5 minutes. The minimum warm-up time is 2 minutes. Some RF generators include a time delay relay to prevent the application of a plate-on command until a predetermined warm-up cycle is completed. Do not defeat these protective circuits. They are designed to extend PA tube life. Most manufacturers also specify a recommended cool-down period between the application of *plate-off* and *filament-off* commands. This cool-down, generally about 10 minutes, is designed to prevent excessive temperatures on the PA tube surfaces when the cooling air is shut off. Large vacuum tubes contain a significant mass of metal, which stores heat effectively. Unless cooling air is maintained at the base of the tube and through the anode cooling fins, excessive temperature rise can occur. Again, the result can be shortened tube life, or even catastrophic failure because of seal cracks caused by thermal stress.

Most tube manufacturers recommend that cooling air continue to be directed toward the tube base and anode cooling fins after filament voltage has been removed to further cool the device. Unfortunately, however, not all control circuits are configured to permit this mode of operation.

10.3.5   Tube Changing Procedure

Plug-in power tubes must be seated firmly in their sockets, and the connections to the anodes of the tubes must be tight. Once it is in place, do not remove a tube assembly for routine inspection unless it is malfunctioning. Whenever a tube is removed from its socket, carefully inspect the fingerstock for signs of overheating or arcing. Keep the socket assembly clean and all connections tight. If any part of a PA tube socket is found to be damaged, replace the defective portion immediately. In many cases, the damaged fingerstock ring can be ordered and replaced. In other cases, however, the entire socket must be replaced. This type of work is a major undertaking, requiring an experienced maintenance engineer.

10.4   EXTENDING VACUUM TUBE LIFE

RF power tubes probably are the most expensive replacement parts that a transmitter or RF generator will need on a regular basis. With the cost of new and rebuilt tubes continually rising, maintenance engineers should do everything possible to extend tube life.

10.4.1   Conditioning a Power Tube

Whenever a new tube is installed in a transmitter, inspect the device for cracks or loose connections (in the case of tubes that do not socket-mount). Also check for interelectrode short circuits with an ohmmeter. Tubes must be seated firmly in their sockets to allow a good, low-resistance contact between the fingerstock and contact rings. After a new tube, or one that has been on the shelf for some time, is installed in the transmitter, run it with *filaments only* for at least 30 minutes, after which plate voltage may be applied. Next, slowly bring up the drive (modulation), in the case of an amplitude modulated transmitter. Residual gas inside the tube may cause an interelectrode arc (usually indicated by the transmitter as a plate overload) unless it is burned off in such a warm-up procedure.

Keep an accurate record of performance for each tube. Shorter-than-normal tube life could point to a problem in the RF amplifier itself. The average life that may be expected from a power grid tube is a function of many parameters, including:

- Filament voltage
- Ambient operating temperature
- RF power output
- Operating frequency
- Operating efficiency

The best estimate of life expectancy for a given system at a particular location comes from on-site experience. As a general rule of thumb, however, at least 12 months of service can be expected from most power tubes. Possible causes of short tube life include:

- Improper transmitter tuning
- Inaccurate panel meters or external wattmeter, resulting in more demand from the tube than actually is required
- Poor filament voltage regulation
- Insufficient cooling-system airflow
- Improper stage neutralization

10.4.2 Filament Voltage

A *true-reading* rms voltmeter is necessary for accurate measurement of filament voltage. Make the measurement directly from the tube socket connections. Secure the voltmeter test leads to the socket terminals, and carefully route the cables outside the cabinet. Switch off the plate power-supply circuit breaker. Close all interlocks, and apply a *filament on* command. Do not apply the high voltage during filament voltage tests. Serious equipment damage and/or injury to the maintenance engineer may result.

A true-reading rms meter, instead of the more common *average-responding* rms meter, is suggested because the true-reading meter can measure a voltage accurately despite an input waveform that is not a pure sine wave. Some filament voltage regulators use silicon-controlled rectifiers (SCRs) to regulate the output voltage. Do not put too much faith in the front-panel filament voltage meter. It is seldom a true-reading rms device; most are average-responding meters.

Long tube life requires filament voltage regulation. Many RF systems have regulators built into the filament supply. Older units without such circuits often can be modified to provide a well-regulated supply through the addition of a ferroreson-ant transformer or motor-driven auto-transformer to the ac supply input. A tube whose filament voltage is allowed to vary along with the primary line voltage will not achieve the life expectancy possible with a tightly regulated supply. This problem is particularly acute at mountaintop installations, where utility regulation generally is poor.

To extend tube life, some broadcast engineers leave the filaments on at all times, not shutting down at sign-off. If the sign-off period is 3 hours or less, this practice can be beneficial. Filament voltage regulation is a must in such situations because the primary line voltages may vary substantially from the carrier-on to carrier-off value. Do not leave voltage on the filaments of a klystron for a period of more than 2 hours if no beam voltage is applied. The net rate of evaporation of emissive material from the cathode surface of a klystron is greater without beam voltage. Subsequent condensation of the material on gun components may lead to voltage holdoff problems and an increase in body current.

10.4.3 Filament Voltage Management

Accurate management of the filament voltage of a thoriated tungsten power tube can extend the useful life of the device considerably, sometimes to twice the normal life expectancy. The following procedure is recommended:

- Operate the filament at its full-rated voltage for the first 200 hours following installation.
- Following the burn-in period, reduce the filament voltage by 0.1 V per step until power output begins to fall (for frequency-modulated systems) or until modulating waveform distortion begins to increase (for amplitude-modulated systems).
- When the *emissions floor* has been reached, raise the filament voltage 0.2 V.

Long-term operation at this voltage may result in a substantial extension in the useful life of the tube, as illustrated in Figure 10.18.

Do not operate the tube with a filament voltage that is at or below 90 percent of its rated value. At regular intervals—about every 3 months—check the filament voltage, and increase it if power output begins to fall or distortion begins to rise. Never increase filament voltage to more than 105 percent of rated voltage. Some tube manufacturers place the minimum operating point at 94 percent. Others recommend that the tube be set for 100 percent filament voltage and left there. The choice of which approach to follow is left to the user.

When it becomes necessary to boost filament voltage to more than 103 percent, order a new tube. If the old device is replaced while it still has some life remaining, the facility will have a standby tube that will perform well as a spare.

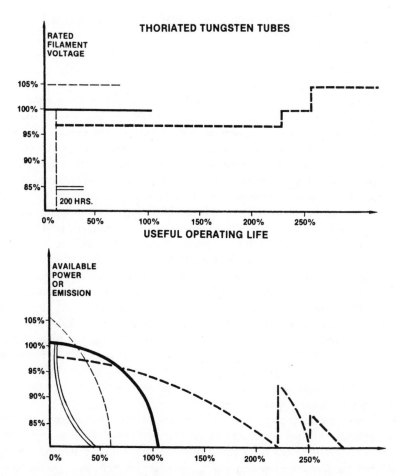

Figure 10.18 The effects of filament voltage management on the useful life of a thoriated tungsten filament power tube. Note the dramatic increase in emission hours when filament voltage management is practiced. (Courtesy of *Broadcast Engineering* magazine, Intertec Publishing, Overland Park, Kan.)

Check the filament current when the tube is installed, and at annual intervals thereafter, to ensure that the filament draws the desired current. A tube may fail early in life because of an open filament bar that would have been discovered during the warranty period if a current check had been made upon installation.

For one week of each year of tube operation, run the filament at full-rated voltage. This will operate the *getter* and clean the tube of gas.

Filament voltage is an equally important factor in achieving long life in a klystron. The voltages recommended by the manufacturer must be set and checked on a regular basis. Measure the voltage at the filament terminals, and calibrate the front-panel meter as needed.

10.4.4   PA Stage Tuning

There are probably as many ways to tune the PA stage of an RF generator or transmitter as there are types of systems. Experience is the best teacher when it comes to adjusting for peak efficiency and performance. Compromises often must be made among various operating parameters. Some engineers follow the tuning procedures contained in the transmitter instruction manual to the letter. Others never open the manual, preferring to tune according to their own methods. Whatever procedure is used, document the operating parameters and steps for future reference. Do not rely on memory for a listing of the typical operating limits and tuning procedures for the system. Write down the information, and post it at the facility. The manufacturer's service department can be an excellent source for information about tuning a particular unit. Many times, the factory can provide pointers on how to simplify the tuning process, or what interaction of adjustments may be expected. Make notes of the information learned from such conversations.

Table 10.1 shows a typical tuning procedure for an FM transmitter. The actual steps vary, of course, from transmitter to transmitter. However, when the tuning characteristics of a given unit are documented in a detailed manner, future repair work is simplified. This record can be of great value to an engineer who is fortunate enough to have a reliable system that does not require regular service. Many of the tuning tips learned during the last service session may be forgotten by the time maintenance work must be performed again.

When to Tune. Tuning can be affected by any number of changes in the PA stage. Replacing the final tube in an AM transmitter or low- to medium-frequency RF generator usually does not significantly alter stage tuning. It is advisable, however, to run through a touch-up tuning procedure just to be sure. On the other hand, replacing a tube in an FM or TV transmitter or high-frequency RF generator can alter stage tuning significantly. At high frequencies, normal tolerances and variations in tube construction result in changes in element capacitance and inductance. Likewise, replacing a component in the PA stage may cause tuning changes because of normal device tolerances.

Stability is one of the primary objectives of transmitter tuning. Avoid tuning positions that do not provide stable operation. Adjust for broad peaks or dips, as required. Tune so that the system is stable from a cold start-up to normal operating temperature. Readings should not vary measurably after the first minute of operation.

Adjust tuning not only for peak efficiency, but also for peak performance. These two elements of transmitter operations, unfortunately, do not always coincide. Tradeoffs sometimes must be made to ensure proper operation of the system. For

Table 10.1 A Sample Documented Transmitter Tuning Procedure

PA tuning adjustments
Unload the transmitter (switch the loading control to *lower*) to produce a PA screen current of 400 mA to 600 mA.

Peak the PA screen current with the plate-tuning control.

Maintain screen current at or below 600 mA by adjusting the loading control (switch it to *raise*).

Position the plate-tuning control in the center of travel by moving the coarse-tune shorting plane up or down as needed.

If the screen current peak is reached near the raise end of plate tune travel, raise the shorting plane slightly.

If the peak is reached near the lower end of travel, lower the plane slightly.

After the screen current has been peaked, adjust the loading control for maximum power output and minimum synchronous AM.

Peak the driver screen current with C-37.

The driver screen peak should coincide with PA screen peak and PA grid peak.

Driver screen peak also should coincide with a dip in the left and right driver cathode currents.

example, FM or TV aural transmitter loading can be critical to wide system bandwidth and low synchronous AM. Loading beyond the point required for peak efficiency often must be used to broaden cavity bandwidth. Heavy loading lowers the PA plate impedance and cavity Q. A low Q also reduces RF circulating currents in the cavity.

10.4.5 Vacuum Tube Life

Failures in semiconductor components result primarily from deterioration of the device caused by exposure to environmental fluctuations and voltage extremes. The vacuum tube, on the other hand, suffers wearout because of a predictable chemical reaction. Life expectancy is one of the most important factors to be considered in the use of vacuum tubes. In general, manufacturers specify maximum operating parameters for power grid tubes so that operation within the ratings will provide for a minimum useful life of 1000 hours.

The cathode is the heart of any power tube. The device is said to *wear out* when filament emissions are inadequate for full power output or acceptable distortion levels. In the case of a thoriated tungsten filament tube, three primary factors determine the number of hours a device will operate before reaching this condition:

- The rate of evaporation of thorium from the cathode
- The quality of the tube vacuum
- The operating temperature of the filament

In the preparation of thoriated tungsten, 1 to 2 percent of thorium oxide (thoria) is added to the tungsten powder before it is sintered and drawn into wire form. After being mounted in the tube, the filament usually is *carburized* by being heated to a temperature of about 2000°K in a low-pressure atmosphere of hydrocarbon gas or vapor until its resistance increases by 10 to 25 percent. This process allows reduction of the thoria to metallic thorium. The life of the filament as an emitter is increased because the rate of evaporation of thorium from the carburized surface is several times less than from a surface of pure tungsten.

Despite the improved performance obtained by carburization of thoriated-tungsten filaments, they are susceptible to deactivation by the action of positive ions. Although the deactivation process is negligible for anode voltages below a critical value, a trace of residual gas pressure too small to affect the emission from a pure tungsten filament can cause rapid deactivation of a thoriated-tungsten filament. This restriction places stringent requirements on vacuum-processing the tube.

These factors, taken together, determine the wearout rate of the tube. Catastrophic failures because of interelectrode short-circuits or failure of the vacuum envelope are considered abnormal and usually are the result of some external influence. Catastrophic failures that are not the result of the operating environment usually are caused by a defect in the manufacturing process. Such failures generally occur early in the life of the component.

The design of the equipment can have a substantial impact on the life expectancy of a vacuum tube. Protection circuitry must remove applied voltages rapidly to prevent damage to the tube in the event of a failure external to the device. The filament turn-on circuit also can have an effect on PA tube life expectancy. The surge current of the filament circuit must be maintained at a low level to prevent thermal cycling problems. This consideration is particularly important in medium- and high-power PA tubes. When the heater voltage is applied to a cage-type cathode, the tungsten wires expand immediately because of their low thermal inertia. However, the cathode support, which is made of massive parts (relative to the tungsten wires) expands more slowly. The resulting differential expansion can cause permanent damage to the cathode wires. It also can cause a modification of the tube operating characteristics and, occasionally, arcs between the cathode and the control grid.

10.4.6   Examining Tube Performance

Examination of a power tube after it has been removed from a transmitter or other type of RF generator can tell a great deal about how well the transmitter-tube combination is working. Compare the appearance of a new power tube, shown in Figure 10.19, with a component at the end of its useful life. If a power tube fails prematurely, examine the device to determine whether an abnormal operating condition exists within the transmitter. Consider the following examples:

- *Figure 10.20:* Two 4CX15,000A power tubes with differing anode heat-dissipation patterns. Tube (a) experienced excessive heating because of a lack of PA compartment cooling air or excessive dissipation because of poor tuning. Tube (b) shows a normal thermal pattern for a silver-plated 4CX15,000A. Nickel-plated tubes do not show signs of heating because of the high heat resistance of nickel.
- *Figure 10.21:* Base heating patterns on two 4CX15,000A tubes. Tube (a) shows evidence of excessive heating because of high filament voltage or lack of cooling air directed toward the base of the device. Tube (b) shows a typical heating pattern with normal filament voltage.
- *Figure 10.22:* A 4CX5,000A with burning on the screen-to-anode ceramic. Exterior arcing of this type generally indicates a socketing problem, or another condition external to the tube.

Figure 10.19   A new, unused 4CX15,000A tube. Compare the appearance of this device with the tubes that follow. (Courtesy of Varian/Eimac)

Figure 10.20 Anode dissipation patterns on two 4CX15,000A tubes: Tube (a), on the left, shows excessive heating. Tube (b), on the right, shows normal wear. (Courtesy of Econco Broadcast Service)

- *Figure 10.23:* The stem portion of a 4CX15,000A that had gone down to air while the filament was on. Note the deposits of tungsten oxide that formed when the filament burned up. The grids are burned and melted because of the ionization arcs that subsequently occurred. A failure of this type will trip overload breakers in the transmitter. It is indistinguishable from a short-circuited tube in operation.

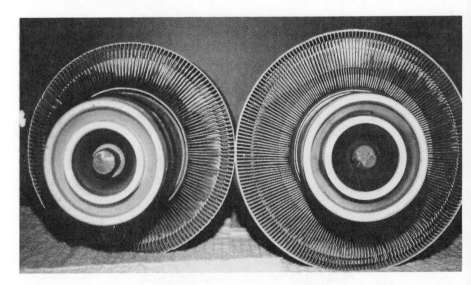

Figure 10.21 Base heating patterns on two 4CX15,000A tubes: Tube (a), on the left, shows excessive heating. Tube (b), on the right, shows normal wear. (Courtesy of Econco Broadcast Service)

Figure 10.22 A 4CX5,000A tube that appears to have suffered socketing problems. (Courtesy of Econco Broadcast Service)

Figure 10.23 The interior elements of a 4CX15,000A tube that had gone to air while the filament was lit. (Courtesy of Econco Broadcast Service)

Figure 10.24 A 4CX10,000D tube showing signs of external arcing. (Courtesy of Econco Broadcast Service)

- *Figure 10.24:* A 4CX10,000D that experienced arcing typical of a bent fingerstock, or exterior arcing caused by components other than the tube.

10.5 BIBLIOGRAPHY

Gray, T. S.: *Applied Electronics*, Massachusetts Institute of Technology, 1954.

High Power Transmitting Tubes for Broadcasting and Research, Phillips technical publication, Eindhoven, Netherlands, 1988.

Power Grid Tubes for Radio Broadcasting, Thomson-CSF publication no. DTE-115, Thomson-CSF, Dover, N.J., 1986.

The Care and Feeding of Power Grid Tubes, Varian Eimac, San Carlos, Calif., 1984.

11

PREVENTING RF
SYSTEM FAILURES

11.1 INTRODUCTION

The reliability and operating costs over the lifetime of an RF system can be affected significantly by the effectiveness of the preventive maintenance program designed and implemented by the engineering staff. When it comes to a *critical-system unit* such as a broadcast transmitter or other RF generator that must operate on a daily basis, maintenance can have a major impact—either positive or negative—on downtime and bottom-line profitability of the facility. The sections of a transmitter most vulnerable to failure are those exposed to the outside world: the ac-to-dc power supplies and RF output stage. These circuits are subject to high-energy surges from lightning and other sources.

The reliability of a communications system may be compromised by an *enabling event phenomenon*. An enabling event phenomenon is one that, although it does not cause a failure by itself, sets up (or enables) a second event that can lead to failure of the communications system. This phenomenon is insidious because the enabling event often is not self-revealing. Examples include:

- A warning system that has failed or been disabled for maintenance.
- One or more controls set incorrectly so that false readouts are provided for operations personnel.
- Redundant hardware that is out of service for maintenance.
- Remote metering that is out of calibration.

11.1.1 Common Mode Failures

A common mode failure is one that can lead to the failure of all paths in a redundant configuration. In the design of redundant systems, therefore, it is important to

275

identify and eliminate sources of common mode failures, or to increase their reliability to at least an order of magnitude above the reliability of the redundant system. Common mode failure points in a transmission system include the following:

- Switching circuits that activate standby or redundant hardware.
- Sensors that detect a hardware failure.
- Indicators that alert personnel to a hardware failure.
- Software that is common to all paths in a redundant system.

The concept of software reliability in control and monitoring has limited meaning in that a good program always will run, and copies of the program always will run. On the other hand, a program with one or more errors always will fail, and so will copies of the program fail, given the same input data. The reliability of software, unlike hardware, cannot be improved through redundancy if the software in the parallel path is identical to that in the primary path.

11.1.2 Modifications and Updates

If problems are experienced with a system, examine what can be done to prevent the failure from recurring. Installing various protection devices or consulting the factory for updates to the hardware may be useful in preventing a repeat of the problem. If the transmitter is several years old, the factory service department can detail any changes that may have been made in the unit to provide more reliable operation. Many of these modifications are minor and can be incorporated into older models with little cost or effort. Modifications might include the following:

- Changing a variable capacitor in a critical tuning stage to a vacuum variable for more stability.
- Installing additional filtering in the high-voltage power supply to improve AM noise performance.
- Replacing older-technology transistorized circuit boards with newer IC and power semiconductor PWBs to improve reliability and performance.
- Improving the overload protection circuitry through the addition of solid-state logic circuits.
- Adding transient-protection devices at critical states of the transmitter.

11.1.3 Spare Parts

The spare parts inventory is a key aspect of any successful equipment maintenance program. Having adequate replacement components on hand is important not only to correct equipment failures, but to identify those failures as well. Many parts, particularly in the high-voltage power supply and RF chain, are difficult to test under static conditions. The only reliable way to test the component may be to substitute one of known quality. If the system returns to normal operation, then the substituted component is defective. Substitution also is a valuable tool in troubleshooting intermittent failures caused by component breakdown under peak power conditions.

11.2 TRANSMISSION-LINE/ANTENNA PROBLEMS

The *voltage standing wave ratio* (VSWR) of an antenna and its transmission line is a vital parameter that has a considerable effect on the performance and reliability of a transmission system. VSWR is a measure of the amount of power reflected back to the transmitter because of an antenna and/or transmission-line mismatch. Figure 11.1 provides a chart for VSWR calculation. A mismatched or defective transmission system will result in a high degree of reflected power, or a higher VSWR.

The amount of reflected power that a given system can accept is a function of the application. For example, it is common practice in FM broadcasting to maintain a VSWR of 1.1:1 as the maximum level within the transmission channel that can be tolerated without degrading the quality of the on-air signal. For television broadcasting, a VSWR into the antenna feeder of more than 1.04:1 will start to degrade picture quality, particularly on systems that use a long transmission line. Reflections down the line from a mismatch at the antenna disrupt the performance of the transmitter output stage. The reflections also cause multipath distortion *within* the transmission line itself. When power is reflected back to the transmitter, it causes the RF output stage to look into a mismatched load with unpredictable phase and impedance characteristics. Because of the reflective nature of VSWR on a transmission system, the longer the transmission line (assuming the reflection is originating at the antenna), the more severe the problem may be for a given VSWR. A longer line means that reflected power seen at the RF output stage has greater time (phase) delays, increasing the reactive nature of the load.

VSWR is affected not only by the rating of the antenna and transmission line as individual units, but also by the combination of the two as a system. The worst-case system VSWR is equal to the antenna VSWR multiplied by the transmission-line VSWR. For example, if an antenna with a VSWR of 1.05:1 is connected to a line with a VSWR of 1.05:1, the resulting worst-case system VSWR would be 1.1025:1. Given the right set of conditions, an interesting phenomenon may occur: The VSWR of the antenna cancels the transmission-line VSWR, resulting in a perfect 1:1 match. The determining factors for this condition are the point of origin of the antenna VSWR, the length of transmission line, and the observation point.

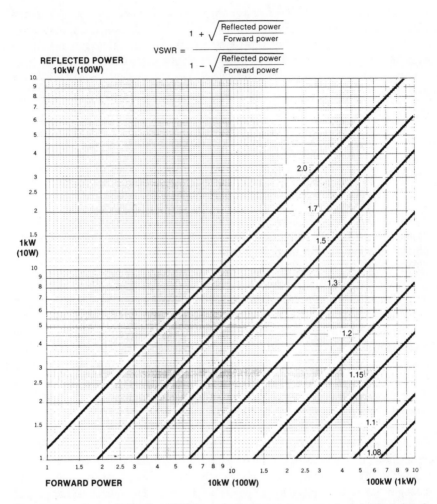

Figure 11.1 Transmission-line VSWR graph. For low-power operation, use the values in parentheses. (Courtesy of *Broadcast Engineering* magazine, Intertec Publishing, Overland Park, Kan.)

11.2.1 Effects of Modulation

The VSWR of a transmission system is a function of frequency, and it changes with carrier modulation. This change may be large or small, but it will occur to some extent. The cause can be traced to the frequency-dependence of the VSWR of the antenna (and to a lesser extent, of the transmission line). The effects of frequency

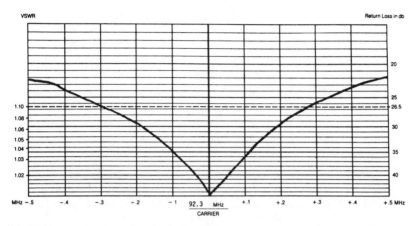

Figure 11.2   The measured performance of a single-channel FM antenna (tuned to 92.3 MHz). The antenna provides a VSWR of below 1.1:1 over a frequency range of nearly 300 kHz. (Courtesy of Jampro Antenna Company)

on VSWR are shown in Figure 11.2. Although the plot of VSWR-vs.-frequency for a common FM antenna is good, notice that with no modulation the system VSWR is one figure. VSWR measurements are different with *positive modulation* (carrier plus modulation) and *negative modulation* (carrier minus modulation).

VSWR is complicated further because power reflected back to the transmitter from the antenna may not come from a single point, but from a number of different points. One reflection might be caused by the antenna-matching unit, another by various flanges in the line, and a third by a damaged part of the antenna system. Because these reflection points are different lengths from the transmitter PA plate, a variety of standing waves may be generated along the line, varying with the modulating frequency.

Energy reflected back to the transmitter from the antenna is not lost. A small percentage of the energy is turned into heat, but most of it is radiated by the antenna, delayed in time by length of the transmission line.

11.2.2   Maintenance Considerations

To maintain low VSWR, regularly service the transmission-line and antenna system. Use the following guidelines:

- Inspect the antenna elements, interconnecting cables, impedance transform-ers and support braces at least once each year. Falling ice can damage FM, TV, and communications antenna elements if proper precautions are not taken. Icing on the elements of an antenna will degrade the antenna VSWR because ice lowers the frequency of the electrical resonance of the antenna. Two methods commonly are used to prevent a buildup of ice on high-power transmitting antennas: electrical deicers and *radomes*.

- Check AM antennas regularly for structural integrity. Because the tower itself is the radiator, bond together each section of the structure for good electrical contact.
- Clean base insulators and guy insulators (if used) as often as required.
- Keep lightning ball gaps or other protective devices clean and adjusted properly.
- Inspect the transmission line for signs of damage. Check supporting hardware, and investigate any indication of abnormal heating of the line.
- Keep a detailed record of VSWR in the facility maintenance log, and investigate any increase above the norm.
- Check the RF system test load regularly. Inspect the coolant filters and flow rate, as well as the resistance of the load element.

11.2.3   UHF Transmission Systems

The RF transmission system of most high-power UHF stations is externally diplexed after the final RF amplifiers. Either coaxial or waveguide-type diplexers may be used. Maintenance of the combining sections is largely a matter of careful observation and record-keeping:

- Monitor the reject loads on diplexers and power combiners regularly to ensure adequate cooling.
- Check the temperature of the transmission line and components, particularly coaxial elements. Keep in mind that coax of the same size and carrying the same RF power runs warmer in UHF systems than in VHF systems. This phenomenon is caused by the reduced penetration depth (the *skin effect*) of UHF signal currents. Hot spots in the transmission line can be caused by poor contact areas or by high VSWR. If they are the result of a VSWR condition, the hot spots will be repeated every 1/2-wavelength toward the transmitter.
- Monitor the reverse power/VSWR meters closely. Some daily variation is not unusual, in small amounts. Greater variations that are cyclical in nature are an indication of a long-line problem, most likely at the antenna. The fact that transmission lines usually are long at UHF (a taller tower allows greater coverage) and the wavelength is small leads to large phase changes of a mismatch at the antenna. Mismatches inside the building do not cause the same cyclical variation. If reverse power is observed to vary significantly, run the system with the test load to see whether the problem disappears. If the variations are not present with a test load, arrange for an RF sweep of the line.

A change in klystron output power is another effect of VSWR variation on a UHF transmitter. The output coupler transforms the line characteristic impedance upward to approximately match the beam impedance. This provides maximum power transfer from the cavity. Large VSWR phase variations associated with long lines change the impedance that the output coupler sees. This causes the output power to vary, sometimes more significantly than the reverse power metering indicates. *Ghosting* on the output waveform is a common indication of antenna VSWR problems in a long-line television system. If the input signal is clean, and the output has a ghost, arrange for an RF sweep.

11.3   HIGH-VOLTAGE POWER-SUPPLY PROBLEMS

The high-voltage plate supply is the first line of defense between external ac line disturbances and the power amplifier stage(s). Next to the output circuit itself, the plate supply is the second section of a transmitter most vulnerable to damage because of outside influences.

11.3.1   Power-Supply Maintenance

Figure 11.3 shows a high-reliability power supply of the type common in transmission equipment. Many transmitters use simpler designs, without some of the protection devices shown, but the principles of preventive maintenance are the same:

- Thoroughly examine every component in the high-voltage power supply. Look for signs of leakage on the main filter capacitor (C2).
- Check all current-carrying meter/overload shunt resistors (R1 to R3) for signs of overheating.
- Carefully examine the wiring throughout the power supply for loose connections.
- Examine the condition of the filter capacitor series resistors (R4 and R5), if used, for indications of overheating. Excessive current through these resistors could point to a pending failure in the associated filter capacitor.
- Examine the condition of the bleeder resistors (R6 to R8). A failure in one of the bleeder resistors could result in a potentially dangerous situation for maintenance personnel by leaving the main power-supply filter capacitor (C2) charged after the removal of ac input power.
- Examine the plate voltage meter multiplier assembly (A1) for signs of resistor overheating. Replace any discolored resistors with the factory-specified type.

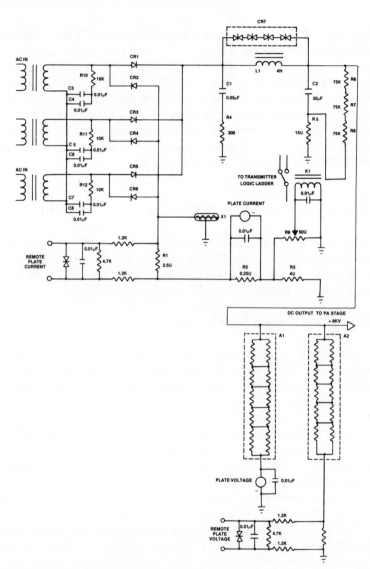

Figure 11.3 A common high-voltage transmitter power-supply circuit design. (Courtesy of *Broadcast Engineering* magazine, Intertec Publishing, Overland Park, Kan.)

When changing components in the transmitter high-voltage power supply, be certain to use parts that meet with the approval of the manufacturer. Do not settle for a close match of a replacement part. Use the exact replacement part. This ensures that the component will work as intended and will fit into the space provided in the cabinet.

11.3.2   Power-Supply Metering

Proper metering is an effective way to prevent failures in transmission equipment. Accurate readings of plate voltage and current are fundamental to RF system maintenance. Check each meter for proper mechanical and electrical operation. Replace any meter that sticks or will not zero.

With most transmitter plate current meters, accuracy of the reading can be verified by measuring the voltage drop across the shunt element (R2 of the previous figure) and using Ohm's law to determine the actual current in the circuit. Be certain to take into consideration the effects of the meter coil itself. Contact the transmitter manufacturer for suggestions on how best to confirm the accuracy of the plate current meter.

The plate voltage meter can be checked for accuracy by using a high-voltage probe and a high-accuracy external voltmeter. Be extremely careful when making such a measurement. Follow, to the letter, instructions regarding use of the high-voltage probe. Do not defeat transmitter interlocks to make this measurement. Instead, fashion a secure connection to the point of measurement, and route the meter cables carefully out of the transmitter. Never use common test leads to measure a voltage of more than 600 V. Standard test lead insulation for most meters is not rated for use above 600 V.

11.3.3   Overload Sensor

The plate supply overload sensor in most transmitters is arranged as shown in Figure 11.3. An adjustable resistor—either a fixed resistor with a movable tap or a potentiometer—is used to set the sensitivity of the plate overload relay. Check potentiometer-type adjustments periodically. Fixed-resistor-type adjustments rarely require additional attention. Most manufacturers have a chart or mathematical formula that may be used to determine the proper setting of the adjustment resistor (R9) by measuring the voltage across the overload relay coil (K1) and observing the operating plate current value. Clean the overload relay contacts periodically to ensure proper operation. If a relay exhibits mechanical problems, replace it.

Transmitter control logic for a high-power UHF system usually is configured for two states of operation:

1. An *operational level,* which requires all the "life-support" systems to be present before the high voltage (HV) command is enabled.
2. An *overload level,* which removes HV when one or more fault conditions occur.

Inspect the logic ladder for correct operation at least once a month. At longer intervals, perhaps annually, check the speed of the trip circuits. (A storage oscilloscope is useful for this measurement.) Most klystrons require an HV removal time of less than 100 ms from the occurrence of an overload. If the trip time is longer, damage may result to the klystron. Pay particular attention to the body-current overload circuits. Occasionally check the body current without applied drive to ensure that the dc value is stable. A relatively small increase in dc body current can lead to overheating problems.

The RF arc detectors in a UHF transmitter also require periodic monitoring. External cavity klystrons generally have one detector in both the third and fourth cavities. Integral devices use one detector at the output window. A number of factors can cause RF arcing, including:

- Overdriving the klystron
- Mistuning the cavities
- Poor cavity fit (external type only)
- Undercoupling of the output
- High VSWR

Regardless of the cause, arcing can destroy the vacuum seal, if drive and/or HV are not removed quickly. A lamp is included with each arc detector photocell for test purposes. If the lamp fails, a flashlight can provide sufficient light to trigger the cell until a replacement can be obtained.

11.3.4   Transient Disturbances

Every electronic installation requires a steady supply of clean power to function properly. Recent advances in technology have made the question of ac power quality even more important, as microcomputers are integrated into transmission equipment. Different types and makes of RF generators offer varying degrees of transient overvoltage protection. Given the experience of the computer industry, it is hard to overprotect electronic equipment from ac line disturbances.

Figure 11.3 shows surge suppression at two points in the power-supply circuit. C1 and R4 make up an R/C snubber network that is effective in shunting high-energy, fast-rise-time spikes that may appear at the output of the rectifier assembly (CR1 to CR6). Similar R/C snubber networks (R10 to R12 and C3 to C8) are placed across the secondary windings of each section of the three-phase power transformer. Any signs of resistor overheating or capacitor failure are an indication of excessive transient activity on the ac power line. Transient disturbances should be suppressed before the ac input point of the transmitter.

Assembly CR7 is a surge-suppression device that should be given careful attention during each maintenance session. CR7 is typically a selenium thyrector assembly that is essentially inactive until the voltage across the device exceeds a predetermined level. At the *trip point*, the device will break over into a conducting state, shunting the transient overvoltage. CR7 is placed in parallel with L1 to prevent damage to other components in the transmitter in the event of a loss of RF excitation to the final stage. A sudden drop in excitation will cause the stored energy of L1 to be discharged into the power supply and PA circuits in the form of a high-potential pulse. Such a transient can damage or destroy filter, feedthrough, or bypass capacitors; damage wiring; or cause PA tube arcing. CR7 prevents these problems by dissipating the stored energy in L1 as heat. Investigate discoloration or other outward signs of damage to CR7. They may indicate a problem in the exciter or IPA stage of the transmitter. Immediately replace CR7 if it appears to have been stressed.

Check spark-gap surge-suppressor X1 periodically for signs of overheating. X1 is designed to prevent damage to circuit wiring in the event that one of the meter/overload shunt resistors (R1 to R3) opens. Because the spark-gap device is nearly impossible to test accurately in the field and is relatively inexpensive, it is an advisable precautionary measure to replace the component every few years.

11.3.5   Single Phasing

Any transmitter using a three-phase ac power supply is subject to the problem of *single-phasing*, the loss of one of the three legs from the primary ac power distribution source. Single phasing is usually a utility company problem, caused by a downed line or a blown pole-mounted fuse. The loss of one leg of a three-phase line results in a particularly dangerous situation for three-phase motors, which will overheat and sometimes fail. AM transmitters utilizing *pulse width modulation* (PWM) systems also are vulnerable to single-phasing faults. PWM AM transmitters can suffer catastrophic failure of the plate power-supply transformer as a result of the voltage regulation characteristics of the modulation system. The PWM circuit will attempt to maintain carrier and sideband power through the remaining leg of the three-phase supply. This forces the active transformer section and its associated rectifier stack to carry as much as three times the normal load.

Figure 11.4 shows a simple protection scheme that has been used to protect transmission equipment from damage caused by single phasing. At first glance, the system looks as if it would handle the job easily, but operational problems can result. The loss of one leg of a three-phase line rarely results in zero (or near-zero) voltages in the legs associated with the problem line. Instead, a combination of leakage currents caused by *regeneration* of the missing legs in inductive loads and the system load distribution usually results in voltages of some sort on the fault legs of the three-phase supply. It is possible, for example, to have phase-to-phase voltages of 220 V, 185 V, and 95 V on the legs of a three-phase, 208 V ac line experiencing a single-phasing problem. These voltages often change, depending upon the equipment turned on at the transmitter site.

Integrated circuit technology has provided a cost-effective solution to this common design problem in medium- and high-power RF equipment. Phase-loss protection modules, available from several manufacturers, provide a contact clo-

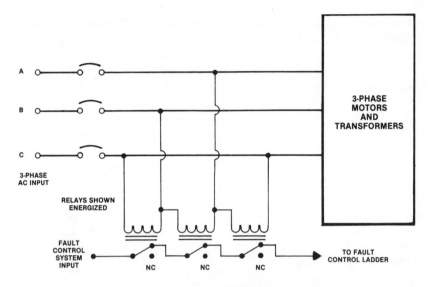

Figure 11.4   A protection circuit using relays for utility company ac phase-loss protection. (Courtesy of *Broadcast Engineering* magazine, Intertec Publishing, Overland Park, Kan.)

sure when voltages of proper magnitude and phase are present on the monitored line. The relay contacts can be wired into the logic control ladder of the transmitter to prevent the application of primary ac power during a single-phasing condition. Figure 11.5 shows the recommended connection method. Note that the input to the phase monitor module is taken from the final set of three-phase blower motor fuses. In this way, any failure inside the transmitter that might result in a single-phasing condition is taken into account. Because three-phase motors are particularly sensitive to single-phasing faults, the relay interlock is tied into the filament circuit logic ladder. For AM transmitters utilizing PWM schemes, the input of the phase-loss protector is connected to the load side of the plate circuit breaker. The phase-loss protector shown in the figure includes a sensitivity adjustment for various nominal line voltages. The unit is small and relatively inexpensive. If the transmitter does not have such a protection device, consider installing one. Contact the factory service department for recommended connection methods.

11.4   TEMPERATURE CONTROL

The environment in which the transmitter is operated is a key factor in determining system reliability. Proper temperature control must be provided for the transmitter to prevent thermal fatigue in semiconductor components and shortened life in vacuum tubes. Problems can be avoided if preventive maintenance work is performed on a regular basis.

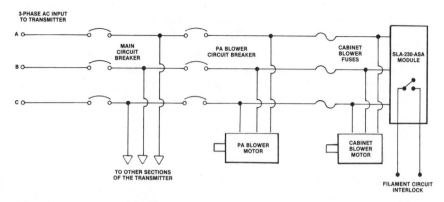

Figure 11.5   A high-performance single-phasing protection circuit using a phase-loss module as the sensor. (Courtesy of *Broadcast Engineering* magazine, Intertec Publishing, Overland Park, Kan.)

11.4.1   Cooling-System Maintenance

Each RF transmission system is unique, and each requires an individual assessment of cooling-system needs. Still, a number of preventive maintenance tasks apply to nearly all systems:

- Keep all fans and blowers clear of dirt, dust, and other foreign material that might restrict airflow. Check the fan blades and blower impellers for any imbalance conditions that could result in undue bearing wear or damage. Inspect belts for proper tension, wear, and alignment.
- Clean the blower motor regularly. Motors usually are cooled by the passage of air over the component. If the ambient air temperature is excessive or the airflow is restricted, the lubricant gradually will be vaporized from the motor bearings, and bearing failure will occur. If dirty air passes over the motor, the accumulation of dust and dirt must be blown out of the device before the debris impairs cooling.
- Follow the manufacturer's recommendations for frequency and type of lubrication. Bearings and other moving parts normally require some lubrication. Carefully follow any special instructions on operation or maintenance of the cooling equipment.
- Inspect motor-mounting bolts periodically. Even well-balanced equipment experiences some vibration. This can cause bolts to loosen over time.
- Inspect air filters weekly, and replace or clean them as necessary. Replacement filters should meet original specifications.

- Clean dampers and all ducting to avoid airflow restrictions. Lubricate movable and mechanical linkages in dampers and other devices as recommended.
- Check actuating solenoids and electromechanical components for proper operation. Movement of air throughout the transmitter causes static electrical charges to develop. Static charges can result in a buildup of dust and dirt in ductwork, dampers, and other components of the system. Filters should remove the dust before it gets into the system, but no filter traps every dust particle.
- Check thermal sensors and temperature system control devices for proper operation.

11.4.2   Air-Cooling-System Design

Transmitter cooling-system performance is not necessarily related to airflow volume. The cooling capability of air is a function of its mass, not its volume. The designer must determine an appropriate airflow rate within the equipment and establish the resulting resistance to air movement. A specified static pressure that should be present within the ducting of the transmitter can be a measure of airflow. For any given combination of ducting, filters, heat sinks, RFI honeycomb shielding, tubes, tube sockets, and other elements in the transmitter, a specified system resistance to airflow can be determined. It is important to realize that any changes in the position or number of restricting elements within the system will change the system resistance and, therefore, the effectiveness of the cooling. The altitude of operation also is a consideration in cooling-system design. As altitude increases, the density (and cooling capability) of air decreases. A calculated increase in airflow is required to maintain the cooling effectiveness that the system was designed to achieve.

Transmitter-room cooling requirements vary considerably from one location to another, but some general guidelines on cooling apply to all installations:

- A transmitter with a power output greater than 1 kW must have its exhaust ducted to the outside whenever the outside temperature is greater than 50°F.
- Transmitter buildings must be equipped with refrigerated air-conditioning units when the outside temperature is greater than 80°F. The exact amount of cooling capacity needed is subject to a variety of factors, such as actual transmitter efficiency, thermal insulation of the building itself, and size of the transmitter room.
- Radio transmitters up to and including 5 kW usually can be cooled (if the exhaust is efficiently ducted outside) by a 10,000 Btu air conditioner. A 10 kW installation will require a minimum of 17,500 Btu's of air-conditioning, and 20 kW plants need at least 25,000 Btu's of air-conditioning. For larger radio installations or television systems, consult an air-conditioning expert.

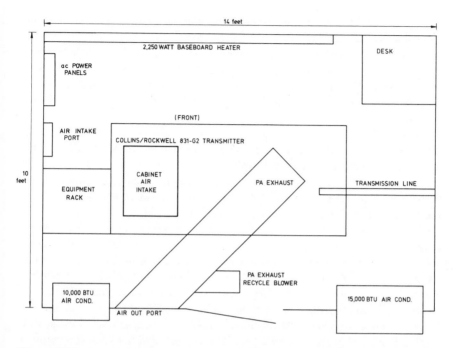

Figure 11.6 A typical heating and cooling arrangement for a 20 kW FM transmitter installation. Ducting of PA exhaust air should be arranged so that it offers minimum resistance to airflow. (Courtesy of *Broadcast Engineering* magazine, Intertec Publishing, Overland Park, Kan.)

Figure 11.6 shows a typical 20 kW FM transmitter plant. The building is oriented so that the cooling activity of the blowers is aided by normal wind currents during the summer months. Air brought in from the outside for cooling is well-filtered in a hooded air-intake assembly that holds several filter panels. The building includes two air conditioners, one 15,000 Btu and the other 10,000 Btu. The thermostat for the smaller unit is set for slightly greater sensitivity than the larger air conditioner, allowing small temperature increases to be handled more economically.

It is important to keep the transmitter room warm during the winter, as well as cool during the summer. Install heaters and PA exhaust recycling blowers as needed. A transmitter that runs 24 hours a day normally will not need additional heating equipment, but facilities that sign off for several hours during the night should be equipped with electric room heaters (baseboard types, for example) to keep the room temperature above 50°F. PA exhaust recycling can be accomplished by using a thermostat, relay logic circuit, and solenoid-operated register or electric blower. If the room temperature is maintained at between 60°F and 70°F, tube and component life will be improved substantially.

Layout considerations. The layout of a transmitter room HVAC (heating, ventilation and air-conditioning) system can have a significant impact on the life of the PA tube(s) and the ultimate reliability of the transmitter. Air-intake and output ports must be designed with care to avoid airflow restrictions and back-pressure problems. This process, however, is not as easy as it may seem. The science of airflow is complex and generally requires the advice of a qualified HVAC consultant.

To help illustrate the importance of proper cooling system design and the real-world problems that some facilities have experienced, consider the following examples taken from actual case histories:

Case 1: A fully automatic building ventilation system (Figure 11.7) was installed to maintain room temperature at 20°C during the fall, winter, and spring. During the summer, however, ambient room temperature would increase to as much as 60°C. A field survey showed that the only building exhaust route was through the transmitter. Therefore, air entering the room was heated by test equipment, people, solar radiation on the building, and radiation from the transmitter itself as it made its way to the transmitter. The problem was solved through the addition of a 3000 ft³/min exhaust fan. The 1 HP fan lowered room temperature by 20°C.

Case 2: A simple remote installation was constructed with a heat-recirculating feature for the winter (Figure 11.8). Outside supply air was drawn by the transmitter cooling system blowers through a bank of air filters, and hot air was exhausted through the roof. A small blower and damper were installed near the roof exit point. The damper allowed hot exhaust air to blow back into the room through a tee duct during winter months. For summer operation, the roof damper was switched open

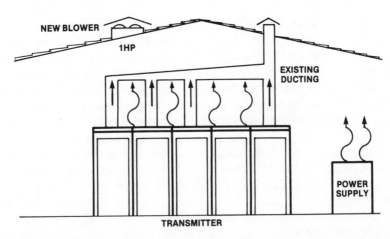

Figure 11.7   Case study in which excessive summertime heating was eliminated through the addition of a 1 HP exhaust blower to the building. (Courtesy of Harris Corporation)

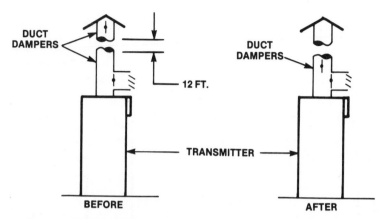

Figure 11.8 Case study in which excessive back-pressure to the PA cavity occurred during winter periods, when the rooftop damper was closed. The problem was eliminated by repositioning the damper as shown. (Courtesy of Harris Corporation)

and the room damper closed. For winter operation, the arrangement was reversed. The facility, however, experienced short tube life during winter operation, even though the ambient room temperature during winter was not excessive.

The solution was to move the roof damper 12 ft down to just above the tee. This eliminated the stagnant "air cushion" above the bottom heating duct damper and significantly improved airflow in the region. Cavity back-pressure, therefore, was reduced. With this relatively simple modification, the problem of short tube life disappeared.

Case 3: An inconsistency regarding test data was discovered within a transmitter manufacturer's plant. Units tested in the engineering lab typically ran cooler than those at the manufacturing test facility. Figure 11.9 shows the test station difference, a 4 ft exhaust stack that was used in the engineering lab. The addition of the stack increased airflow by up to 20 percent because of reduced air turbulence at the output port, resulting in a 20°C decrease in tube temperature.

These examples point out how easily a cooling problem can be caused during HVAC system design. All power delivered to the transmitter either is converted to RF energy and sent to the antenna or becomes heated air. Proper design of a cooling system, therefore, is a part of transmitter installation that should not be taken lightly.

11.4.3 Air Filters

After the transmitter is clean, keeping it that way for long periods of time may require improving the air-filtering system. Most filters are inadequate to keep out very small dirt particles (microdust), which can become a serious problem if the

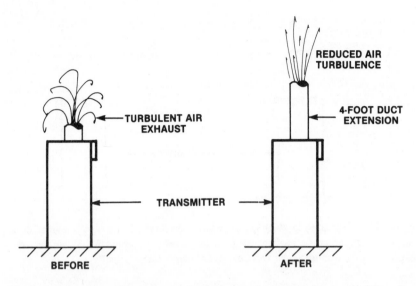

Figure 11.9 Case study in which air turbulence at the exhaust duct resulted in reduced airflow through the PA compartment. The problem was eliminated by adding a 4 ft extension to the output duct. (Courtesy of Harris Corporation)

environment is unusually dirty. Microdust also can become a problem in a relatively clean environment after a number of years of operation. In addition to providing a well-filtered air-intake port for the transmitter building, consider placing an extra air filter in front of the normal transmitter filter assembly. A computer-grade filter panel can be secured to the air-intake port to provide greater protection. With the extra filter in place, it generally is necessary only to replace or clean the outer filter panel. The transmitter's integral filter assembly will stay clean, eliminating the work and problems associated with pulling the filter out while the transmitter is operating. Be certain that the supplemental filtering does not restrict airflow into the transmitter.

11.5 KLYSTRON WATER-COOLING SYSTEMS

The cooling system is vital to any transmitter. In a UHF unit, the cooling system may dissipate as much as 70 percent of the input ac power in the form of waste heat in the klystron collector. For vapor-phase-cooled klystrons, pure (distilled or demineralized) water must be used. Because the collector is only several volts above ground potential, it is not necessary to use deionized water. The collector and its water jacket act like a distillery. Any impurities in the water eventually will find their way into the water jacket and cause corrosion of the collector. It is essential to use high-purity water with low conductivity, less than 10 mS/cm (millisiemens per centimeter), and to replace the water in the cooling jacket as needed. Efficient heat transfer from the collector surface into the water is essential for long klystron life. Oil, grease, soldering flux residue, and pipe sealant containing silicone compounds must be excluded from the cooling system. This applies to both vapor- and liquid-conduction cooling systems, although it is usually more critical in the vapor-phase type.

Water quality is essential to proper operation of a liquid-cooled klystron. In general, greater flows and greater pressures are inherent in liquid-cooled vs. vapor-phase systems, and when a leak occurs, large quantities of coolant may be lost before the problem is discovered. Inspect the condition of gaskets, seals and fittings regularly. Most liquid-cooled klystrons use a distilled water and ethylene glycol mixture. Do not exceed a 50:50 mix by volume. The heat transfer of the mixture is lower than that of pure water, requiring the flow to be increased, typically by 20 to 25 percent. Greater coolant flow means higher pressure and calls for close observation of the cooling system after adding the glycol. Allow the system to heat and cool several times. Then check all plumbing fittings for tightness.

The action of heat and air on ethylene glycol causes the formation of acidic products. The acidity of the coolant can be checked with litmus paper. Buffers can and should be added with the glycol mixture. Buffers are alkaline salts that neutralize acid forms and prevent corrosion. Because they are ionizable chemical salts, the buffers cause the conductivity of the coolant to increase. Measure the collector-to-ground resistance periodically. Coolant conductivity is acceptable if the resistance caused by the coolant is greater than 20 times the resistance of the body-metering circuitry.

11.5.1 Scale and Corrosion

The high-power levels of present-day microwave tubes require that careful attention be given to the design and operation of cooling systems. In some cases, inadequate or improper cooling because of scale or corrosion may be the limiting factor in tube life. Scale is formed as a deposit upon the wetted surface of the tube cooling system. It results from the conversion of a coolant-soluble salt into an insoluble compound because of a chemical reaction in the coolant. Corrosion is the result of chemical reaction products on some portion of the wetted surface.

The liquid-cooled klystron generally is wetted with two cooling paths, one for the collector, the other for the tube body. Other elements of the system, which usually are liquid-cooled by the same heat-exchange system, include the electromagnet focusing coils, RF dummy load, and certain microwave transmission components between the tube and antenna. In some klystrons, as much as 2 kW of heat per square inch must be transmitted through the collector wall and dissipated into the coolant. Frequently, the heat flux in the collector is not distributed uniformly. At high values of heat dissipation, a small amount of scale can cause a large rise in temperature in one portion of the collector, thereby greatly increasing the possibility of premature tube failure.

11.5.2 Coolants

Distilled water is chemically stable and has high heat-transfer capability. It is, therefore, preferred over antifreeze mixtures as a tube coolant. If protection against low temperatures is required, draining the system or using electric water heaters during nonoperating periods will prevent the problems caused by antifreeze. If a freezing-point depressant is necessary, a closed cooling system — complete with a purification loop — can be used with an uninhibited solution of ethylene glycol and water as the coolant. However, because antifreeze mixtures have lower cooling capabilities than water, tube ratings established for water may not apply if an antifreeze is used.

Continuous purification of the coolant is desirable whether water or antifreeze is used. The cooling system usually includes the components needed to accomplish this function, such as a purification loop. The purification loop processes a small amount of coolant from the main recirculating loop, removing soluble salts by ion-exchange, dissolved oxygen and carbon dioxide, small particulate matter, and other contaminants. If a mixture of ethylene glycol and water is used, organic breakdown products from the glycol must be removed. Packaged purification systems suitable for this purpose are available.

Tubing, fittings, pumps, and other material that will be in direct contact with the coolant must be selected to minimize galvanic action. It is best to use only metals and alloys at the "noble" end of the electromotive force (EMF) series, such as copper, nickel, bronze, and Monel. The EMF differences between these metals/alloys and the copper collector are small. Metals, such as steel, cast iron, galvanized iron, aluminum, and magnesium should not be used in direct contact with the coolant; brass should be used sparingly.

In addition to the deterioration of the piping, manifolding, and radiator because of electrolysis between dissimilar metals, oxidation of the copper material of the tube can cause corrosion in the cooling system. This normally happens in areas of high heat transfer to the coolant, primarily in the collector. The corrosion rate is related directly to the amount of dissolved oxygen in the coolant, and the temperature of the copper collector core. Corrosion-free operation can be attained only by holding the dissolved oxygen level to 1.25 parts per million or less. Measurement of the oxygen level must be made frequently. One alternative to the relatively expensive design and maintenance problems associated with tight control of oxygen in the system is to institute flushing procedures for the tube. This can be done on a routine maintenance schedule.

11.5.3   Cooling System Maintenance

All contaminants, such as oils, grease, and particulate matter, must be removed from the system because they might deposit on the heat-exchange surfaces inside the tube and reduce the heat-transfer capability of the device. Degrease the cooling loop with a solvent or detergent, followed by a number of clean water flushings. Do not add soluble oil inhibitors or stop-leak compounds to the coolant; they may cause foaming.

New transmitter water lines frequently contain contaminants. When the water lines are installed, these contaminants must be flushed and cleaned from the system before the klystron and magnet are connected. Before cleaning the transmitter closed circulating water system, disconnect tube and magnet. The following general procedures are recommended:

- Add jumper hoses between the input and output of the klystron, and the electromagnet water lines.
- Disconnect or bypass the pump motor.
- Fill the system with hot tap water, if available. Open a drain in the transmitter cabinet and flush for 15 minutes or until clean.
- Flush the water lines between the tank and pump separately with hot tap water until clean.
- Reconnect all water lines, and fill the system with hot tap water, if available, and one (1) cup of nonsudsing detergent. Trisodium phosphate is recommended.
- Operate the water system with hot tap water for 30 minutes.
- Drain and flush the system with hot tap water for 30 minutes.
- Remove and clean the filter element.
- Refill the water system with tap water at ambient temperature.
- Operate the water system. Maintain the water level while draining and flushing the system until no detergent, foam, or foreign objects or particles are visible in the drained or filtered element.
- Drain and refill the system with distilled water when the tube and transmitter water lines are both clean.
- Remove, clean, and replace the filter before using.

Tubes with contaminated water lines, corrosion, scale, or blocked passages must be flushed. Magnet water lines may be flushed in the same manner. The following procedure is recommended:

- Remove the input water fitting (Hansen type), and add a straight pipe extension to the tube.
- Attach a hose to the fitting, using a hose-clamp. This is the drain line.

- On some tubes, the normal body-cooling output line is fed to the base of the Vapotron boiler. Remove the hose at base of Vapotron boiler. Do not damage this fitting; it must be reused.
- Attach a 2- to 3-in extension pipe with fittings for the small hose at one end and a garden hose at the other. Secure with clamps.
- Connect a garden hose to the tap water faucet (preferably hot).
- Backflush the klystron cooling passages for 10 to 15 minutes at full pressure until clean.
- Reconnect the input and output water lines to klystron.

If scale is present on the Vapotron collector, use a solution of trisodium phosphate for the first cleaning. Clean the transmitter water system before performing this step. Use the following procedure:

- Connect the klystron to the transmitter water lines, and fill the system with tap water and one cup of nonsudsing detergent.
- Operate the water system for 15 minutes, making certain that the water level is sufficient to cover the collector core completely.
- Drain and flush the system for 30 minutes, or until no detergent foam is present.
- Remove and clean the filter element.
- Fill the system with distilled water.
- Drain the system water and refill with distilled water.

Magnet and klystron water lines may be cleaned in the same manner. If there is heavy scaling on the Vapotron collector and/or blocked water passages, they may be cleaned by using a stronger cleaning agent. The following procedure is recommended:

- Clean the transmitter water system first.
- Connect the tube to the transmitter water system.
- Fill the system with hot tap water, and add 2 gallons of cleaning solution for every 50 gallons of water.
- Operate the water system for 15 minutes or until scales have been removed and the collector has a clean copper color.
- Confirm that water covers the top of the collector during cleaning.
- Drain and flush the system with tap water.
- Remove and clean (or replace) the filter element.
- Refill the system with distilled water, and flush for 30 minutes to 1 hour.
- Check for detergent foam and pH factor at end of flushing.
- Continue flushing with distilled water until foam is gone and the pH factor is within the specified range.
- Refill the system with clean, distilled water.

A thorough flushing with distilled water will remove harmful chlorine traces from the tap water.

Two remaining items must be clean to achieve efficient operation: the sight glass and the floats in the water-flow indicators. The water-flow indicators may become contaminated during use. Contaminants collect on the sight glass and on the float, making flow-reading difficult. If excessive contamination is present on the glass and float, sticking or erroneous readings may result. Detergent and cleaning solutions may not remove all of this contamination. Remove and clean the flow meter if necessary. Use a brush to clean the glass surface.

Cooling System Checklist: In general, the following procedures will provide long tube life:

- Keep the coolant temperature constant and as low as ambient weather conditions and other system requirements will allow.
- Use clean, distilled water for original flushing, final filling, and makeup.
- Use ethylene glycol only. Do not use automobile radiator antifreeze.
- Monitor the condition of the ion-exchange cartridge. Rapid exhaustion of the cartridge may indicate a source of contamination, electrolysis, or the use of inhibited glycol. Rapid exhaustion may also indicate that the purification loop lacks sufficient capacity for the bulk coolant being processed.
- Keep the main loop and branch filters clean through routine inspection.
- Keep the system free of dissolved oxygen. Flush the tube when the collector differential pressure increases by 25 percent above the original value at the equivalent flow rate.
- Follow the recommendations of the purification loop manufacturer when replacing filter membranes and cartridges.
- The foregoing cleaning and flushing procedure is not meant to supersede the recommendations of the transmitter manufacturer.

11.5.4   Checking Water Purity

Water purity is a major factor in the life expectancy and operating efficiency of vapor-cooled tubes. If impurities are present, foaming may occur. This will inhibit heat transfer, thereby lowering the cooling efficiency of the system. A typical klystron cooling system with a purification loop is shown in Figure 11.10. Impurities that may cause foaming include the following:

- Cleaning-compound residue
- Detergents
- Joint-sealing compounds
- Oily rust preventatives in pumps and other components
- Valve-stem packing
- Impurities in tap water

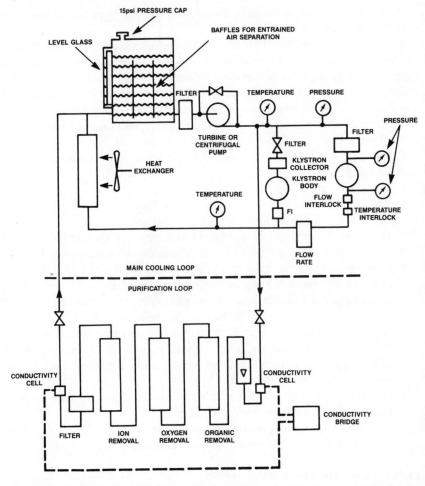

Figure 11.10 Klystron cooling system with purification loop. (Source: Colin Erridge, "Servicing Your Klystrons, Part 3," *Broadcast Engineering* magazine, Intertec Publishing, Overland Park, Kan., December 1990)

The following equipment is required to test water purity:

- A 1/2-in x 4-in glass test tube with a rubber stopper
- A 1-pint glass or polypropylene bottle with cap

Perform the following test after each water change, system cleaning, or modification.

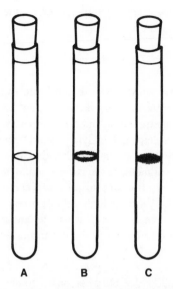

Figure 11.11   Testing a cooling system for foaming: (a) No foam-producing impurities. (b) Foam-producing impurities at a temporarily acceptable level. (c) Excessive foam-producing impurities. (Source: Colin Erridge, "Servicing Your Klystrons, Part 3," *Broadcast Engineering* magazine, Intertec Publishing, Overland Park, Kan., December 1990)

- Fill the cooling system with water, and circulate until thoroughly mixed (approximately 30 minutes).
- Drain a sample of water into the bottle, and cool to room temperature. If the water sample stands for more than 1 hour, slowly invert the capped bottle about 10 times. Avoid shaking the bottle because this will create air bubbles. (When the water is static, foaming impurities tend to collect at the surface. This step mixes the sample without generating foam.)
- Using the sample water, rinse the test tube and stopper three times.
- Half-fill the test tube with the sample water.
- Shake the test tube vigorously for 15 seconds.
- Let the sample stand for 15 seconds.
- Observe the amount of foam remaining on top of the water, and compare with the drawings in Figure 11.11.

A completely foam-free water surface and test-tube wall (shown in a), indicates no foam-producing impurities. If the water surface and test-tube wall are partly covered with foam, but a circle of clear water appears in the center (shown in b), the impurity level is temporarily acceptable. Conduct a second test in approximately one week. If the foam layer completely bridges the inside of the test tube (shown in c), flush and clean the system.

11.5.5   Power-Up/Down Procedures

Powering up and shutting down klystrons requires care, especially in a remote-control operation. Improper procedures can destroy the device. The order in which the components of a system are turned on and off is important.

At shutdown, the high voltage to the klystron must be removed before any other voltages or signals. Verify this condition by checking for a positive reading of zero on the transmitter anode voltmeter or the corresponding remote-control telemetry channel. If this reading does not drop to zero, proceed no further until the high voltage is removed manually via the high-voltage off switch, or by mechanically tripping the high-voltage breaker. Only when the anode meter shows that the voltage is zero should power be cut to the electromagnets, pumps, and control circuits of the transmitter.

An orderly shutdown is not always possible. Even at facilities equipped with a backup generator, interruptions of utility ac power to the transmitter sufficient to cause shutdown will occur from time to time. In such cases, klystron damage can occur because of one or more of the following events:

- A mechanical and/or electrical malfunction of the main circuit breaker causes it to remain closed when power is lost. When power returns, high voltage is applied to the klystron before the magnet supply can provide adequate beam focusing. This allows full beam power to be dissipated in the klystron body, thereby melting the drift tube.
- The time constant of one subsystem power-supply filter may differ significantly from another. Varying time constants will cause each power supply to exhibit a different discharge time when ac power is removed. If the focusing electromagnet supply has no filter capacitance, it will discharge almost instantly, while the klystron beam filter circuitry allows the high voltage to ramp off more slowly. As before, this defocuses the beam, dissipating the remaining energy into the drift tube.
- A single-phasing fault may cause the 3-phase magnets and protection circuits to become inoperative or unreliable. If the ac voltage-sensing relay on the main breaker is misadjusted or not working, the high voltage will stay on, but the focusing and protection circuits will shut down. Once again, beam defocus will occur, and the tube body will melt.

There are a number of ways to prevent failures resulting from the foregoing events.

- Test the main circuit breaker on a regular basis. Because most power outages are a seasonal phenomenon, inspect and service the breaker just before the prime outage period. If a particularly high number of outages are experienced, more frequent breaker service is recommended.
- Check the discharge times of the filtering components in the beam and magnet power supplies. Capacitance may be added to the magnet supply if its time constant is less than that of the beam supply. Use computer-grade electrolytic capacitors with adequate voltage ratings. Keep in mind, however, that added capacitance on the focusing magnet supply may also increase the ramp-up time upon application of power.
- Install a 3-phase vacuum relay on the ac power mains, between the main breaker and the high-voltage supply. This will prevent the instantaneous return of high voltage described in the first breaker failure scenario. Configure the relay to prevent reclosure before the focusing magnet field has fully built up. If it is not already provided, install a loss-of-leg sensor to prevent the single-phasing problem from occurring.
- Install an undercurrent-sensing interlock on the focusing magnet supply. This will prevent the premature restoration of high voltage to the klystron before the focusing magnets have returned to full field strength. It will also shut down the high voltage if the magnet supply current falls below a preset level.

11.5.6   Caring for Spare Tubes

Before placing a spare klystron into storage, drain and then blow dry all water lines. Any water — even distilled water — that remains in cooling passages over extended periods can cause corrosion. If the klystron is stored in a place where it is exposed to freezing temperatures, physical damage could also result from the expansion of residual water.

When draining an integral-cavity klystron, place it in a horizontal position, with the tuners facing downward. Keep the device well covered while in storage to prevent any accumulation of dirt or moisture.

Burn-in new tubes for 200 to 300 hours as soon as possible after receipt. Once in service, klystrons can be rotated periodically between visual and aural service (for television applications), but any handling of the tubes should be kept to a minimum. Most manufacturers recommend keeping dedicated spare tubes, using the following procedure: When a klystron fails, a spare is taken from storage and put into service. When the new replacement tube arrives, it is installed in the transmitter as soon as possible, returning the spare to storage. Occasional operation of the spare klystron will disperse any vacuum contaminants that may have accumulated while the tube was in storage.

Check the vacuum integrity of spare klystrons every 90 days or so. A gas-checking procedure may be used in which the electron gun of the klystron is used as a triode ionization gauge. This test may be performed with the klystron mounted in its electromagnet, or while it is still in the shipping container. In the latter case, remove any gun-clamping hardware before proceeding. The equipment required for this test includes:

- A digital multimeter of at least 11 MΩ input resistance.
- One 45 V battery.
- Three 67.5 V batteries (connected in series to provide 202.5 V dc).
- A 110 MΩ 0.50 W resistor.
- A dc milliammeter with 0-5 mA or 0-10 mA range. (A single digital multimeter fitting these specifications may be used in lieu of both meters, because they are not required simultaneously.)

To avoid the possibility of contaminated results from ground loops, ripple current, and electromagnetic interference (EMI), use batteries instead of ac supplies. Current drain on the batteries will be low. A single set should serve for many tests.

Set up the test rig as shown in Figure 11.12, keeping all connecting leads as short as possible. Use an outboard supply or the transmitter supply for the klystron heater. In the latter case, disconnect the negative high-voltage lead from the klystron cathode.

Begin your observations after all connections have been made, but *before* the application of heater voltage. Then, verify that the DMM reading is near zero and is stable. Be aware that body motion near the tube or test rig may cause some

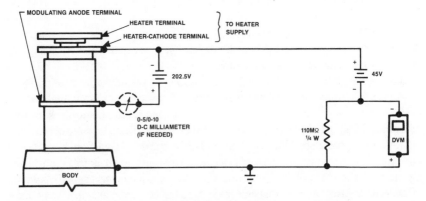

Figure 11.12 Gas-checking procedure for a klystron. (Source: Colin Erridge, "Servicing Your Klystrons, Part 5," *Broadcast Engineering* magazine, Intertec Publishing, Overland Park, Kan., February 1991)

fluctuation. If there is a significant meter reading, check and clean all connections, wipe exposed ceramics with alcohol, and look for any leakage paths in the test setup. Once the meter reading has been reduced to a small and steady value, note any remaining leakage current for subtraction from later measurements. Then apply the heater voltage, allowing 5 to 15 minutes for cathode warmup.

During the warmup time, observe that the DMM reading typically will rise, come to a peak value, fall off a bit, then stabilize. A steady reading of less than 0.1 V on the DMM indicates a gas current below 0.01 mA, denoting an excellent vacuum within the klystron under test. Vacuum integrity is considered satisfactory with DMM readings of less than 5 V, corresponding to gas currents up to 0.5 mA. Measurements of 5 V or more indicate that the tube should be put into service for a few days to reduce gas current to below 0.5 mA. After the tube has been reconditioned in this manner, return the device to storage. Keep written records of all tube checks and reconditioning.

In some cases, no reading will be seen on the DMM. To verify that this condition is the result of an extremely good vacuum, and not a fault in the test setup, connect the dc milliammeter in series with the 202.5 V battery, as shown in the previous figure. For this test, the DMM, the resistor, and the 45 V battery may be disconnected (keep the heater supply up and running). A good vacuum is verified by a milliammeter reading between 1 mA and 4 mA. While performing this test, never energize the klystron heater for more than 30 minutes without auxiliary forced-air cooling.

Use extreme caution when connecting and disconnecting the heater leads. Although low voltages are used to drive the heater, high currents are present. Use cable of sufficient capacity to handle the full rated current.

11.5.7   Electron Gun Leakage

Klystrons that have been in operation for some time may be susceptible to high-voltage breakdown faults when they are moved from visual to aural operation (in television service), or from continuous to pulsed service. With either of these changes, the range of voltage potential that the tube encounters is greater than what it has been conditioned to accept. This situation can cause internal arcing. A buildup of electrical leakage between elements is a normal part of tube aging, but the process can be reduced or eliminated by high-voltage reconditioning. This technique is recommended whenever an older klystron is about to be rotated to a different service. *High-potting*, the application of a high dc potential across the klystron elements, can be used to check electron gun leakage. It also may be used, in many cases, to recondition tube elements. The only equipment needed is a continuously variable dc supply with a range of 0-30 kV. The output of the power supply must be current-limited to a 5 mA maximum. This instrument is often referred to as a *high-potter*.

To perform the operation, turn off the klystron heater, and give the cathode an hour or two to completely cool off. Then connect the negative terminal of the high-potter to the heater-cathode terminal of the klystron, and the positive ground terminal to the modulating anode. Observe the leakage current drawn from the supply, and slowly increase the voltage at the supply output. Stop increasing the voltage when the leakage current reads approximately 2 mA. Wait for it to drop or "burn off." Then continue to increase the voltage, holding each time the 2 mA level is reached. Repeat the process until the voltage of the supply is equal to the normal operating voltage of the klystron. At this voltage, continue the burn-off process until a leakage current of 200 to 400 μA is achieved. This is the normal leakage level in a good tube.

Next, shut down the external power supply, and connect the *negative* terminal of the supply to the modulating anode of the klystron. Move the positive ground lead to the tube body. Repeat the process to check and reduce leakage between the tube elements. If the klystron full-voltage leakage current cannot be reduced to the 200-400 μA range, it still should operate normally, but may be more susceptible to internal arcing.

Because of the high voltages involved, extreme caution must be exercised when conducting the foregoing tests. It also should be noted that subtle differences exist in various klystron designs. Consult the tube manufacturer before conducting these tests.

11.6   BIBLIOGRAPHY

Erridge, Colin: "Servicing Your Klystrons," *Broadcast Engineering* magazine, Intertec Publishing, Overland Park, Kan., October 1990.

_____: "Cleaning and Flushing Klystron Water and Vapor Cooling Systems," *Broadcast Engineering* magazine, Intertec Publishing, Overland Park, Kan., November 1990.

_____: "Servicing Your Klystrons, Part 3," *Broadcast Engineering* magazine, Intertec Publishing, Overland Park, Kan., December 1990.

_____: "Servicing Your Klystrons, Part 4," *Broadcast Engineering* magazine, Intertec Publishing, Overland Park, Kan., January 1991.

_____: "Servicing Your Klystrons, Part 5," *Broadcast Engineering* magazine, Intertec Publishing, Overland Park, Kan., February 1991.

Svet, Frank A.: "Factors Affecting On-Air Reliability of Solid State Transmitters," *Proceedings* of the SBE Broadcast Engineering Conference, Society of Broadcast Engineers, Indianapolis, Ind., October 1989.

12

TROUBLESHOOTING TRANSMISSION EQUIPMENT

12.1 INTRODUCTION

Problems will occur from time to time with any piece of equipment. The best way to prepare for a transmitter failure is to know the equipment well. Study the transmitter design and layout. Know the schematic diagram and what each component does. Examine the history of the transmitter by reviewing old maintenance logs to see what components have failed in the past.

12.1.1 Troubleshooting Procedure

When a problem occurs, the first task is to keep the transmitter on the air. If a standby transmitter is available, the solution is obvious. If the facility does not have a standby, quick thinking will be needed to minimize downtime and keep the unit running until repairs can be made. Because most transmitters have sufficient protective devices, it usually is impossible to operate them if they have serious problems. If the transmitter will not stay on the air on normal power, try lowering the power output to determine whether the trip-offs are eliminated. Failing this, many transmitters have driver outputs that can be connected to the antenna on a temporary basis, thereby bypassing the final amplifiers—provided, of course, the failure is in one of the PA stages. Do not allow the transmitter to operate at any power level if the meter readings are out of tolerance. Serious system damage may result.

When presented with a problem, proceed in an orderly manner to track it down. Stop and think about what's happening. Examine the last set of transmitter readings, and make a complete list of meter readings in the failure mode. Note which overload lamps are lit, and what other indicators are in an alarm state. Once this information is assembled, the cause of the failure often can be identified. Invest at least 15 to 30 minutes in looking over the available data and the schematic diagram; it can save hours of trial-and-error troubleshooting.

When checking inside the unit, look for changes in the physical appearance of components in the problem area. An overheated resistor or leaky capacitor may be the cause of the problem, or may point to the cause. Devices never fail without a reason. Try to piece together the sequence of events that led to the problem. Then, the cause of the failure—not just the obvious symptoms—will be corrected. In a circuit using direct-coupled transistors, a failure in one device often causes a failure in another, so check all semiconductors associated with one found to be defective.

In high-power transmitters, look for signs of arcing in the RF compartments. Loose connections and clamps can cause failures that are hard to locate. Never rush through a troubleshooting job. A thorough knowledge of the theory of operation and history of the transmitter is a great aid in locating problems in the RF sections. Do not overlook the possibility of tube failure when troubleshooting a transmitter. Tubes can fail in unusual ways; substitution may be the only practical test for power tubes used in modern transmitters.

Study the control ladder of the transmitter to identify interlock or fail-safe system problems. Most newer transmitters have troubleshooting aids built in to help locate problems in the control ladder. Older transmitters, however, often require a moderate amount of investigation before repairs can be accomplished.

The "Quick Fix." There is no such thing as a quick fix when it comes to transmission equipment. Think out the problem, and allow ample time for repair. It makes little sense to rush through a repair job to get the system back on the air simply to suffer another failure as soon as the technician walks out the door. Careful analysis of the cause and effects of the failure will ensure that the source of the trouble is uncovered and the problem solved. On the other hand, an attempt to achieve a quick fix probably will result in the treatment of symptoms, not causes. If temporary repairs must be made to return the transmitter to a serviceable condition, make them, then finish the job as soon as the needed replacement parts are available.

12.1.2 Factory Service Assistance

Factory service engineers are available to aid in troubleshooting transmission equipment, but such services have their limits. No factory engineer can fix a transmitter over the phone. The factory can suggest areas of the system to investigate and relate the solutions to similar failure modes, but the facility engineer is the person who does the repair work. If the engineer knows the equipment and has done a good job of analyzing the problem, the factory can help. When calling the factory service department, have the following basic information on hand:

- The type of transmitter and the exact failure mode. The service department will need to know the what the meter readings were before and after the problem occurred, and whether any unusual circumstances preceded the failure. For example, it would be important for the factory to know that the failure occurred after a brief power outage, or during an ice storm.
- A list of what already has been done in an effort to correct the problem. All too often, the factory is called *before* any repair efforts are made. The service engineer will need to know what happens when the high voltage is applied, and what overloads may occur.
- A copy of the transmitter diagram and component layout drawings. A thorough knowledge of the transmitter design and construction allows the maintenance engineer to converse intelligently with the factory service representative.

12.2 PLATE OVERLOAD FAULT

Of all the problems that can occur in a transmitter, probably the most familiar (and most feared) is the plate supply overload. Occasional plate trip-offs—one or two a month—are not generally cause for concern. Most of these occurrences can be attributed to power-line transients. More frequent trip-offs require a closer inspection of the transmission system. For the purposes of this discussion, assume that the plate supply overload occurs frequently enough to make continued operation of the transmitter difficult.

12.2.1 Troubleshooting Procedure

The first step in any transmitter troubleshooting procedure is to switch the system to *local control* so that the maintenance technician, not an off-site operator, has control over the unit. This is important for safety reasons. Next, switch off the transmitter automatic recycle circuit. During troubleshooting, the transmitter should not be allowed to cycle through an overload any more times than absolutely necessary. Such action only increases the possibility of additional component damage. Use a logical, methodical approach to finding the problem. The following procedure is recommended:

- Determine the fault condition. When the maintenance engineer arrives at the transmitter site, the unit probably will be down. The carrier will be off, but the filaments still will be lit. Check all multimeter readings on the transmitter and exciter. If they indicate a problem in a low-voltage stage, troubleshoot that failure before bringing the high voltage up.

- If all low-voltage systems are operating normally, switch the filaments off and make a quick visual check inside the transmitter cabinet. Determine whether there is any obvious problem. Pay particular attention to the condition of power transformers and high-voltage capacitors. Check for signs of arcing in the PA compartment. Look on the floor of the transmitter and in the RF compartments for any pieces of components lying around. Sniff inside the cabinet for hints of smoke. Check the circuit breakers and fuses for indications of failure.

- After running through these preliminary steps, restart the filaments. Then bring up the high voltage. Watch the front-panel meters to see how they react. Observe what happens, and listen for any sound of arcing. If the transmitter will come up, quickly run through the PA and IPA meter readings. Check the VSWR meter for excessive reflected power.

If problems persist, determine whether the plate supply overload is RF- or dc-based. With the plate off, switch the exciter off. Bring up the high voltage (plate supply). If the overload problem remains, the failure is based in the dc high- voltage power supply. If the problem disappears, the failure is centered in the transmitter RF chain. Proper bias must be present on all vacuum tube stages of the RF system when this test is performed. The PA tube bias supply, usually switched on with the filaments, generally can be read from the front panel of the transmitter. Confirm proper bias before applying high voltage with no excitation. It is also important that the exciter is switched off while the high voltage is off. Removing excitation from a transmitter while it is on the air can result in a large transient overvoltage, which can cause arcing or component damage.

If the overload is based in the high-voltage dc power supply, shut down the transmitter and check the schematic diagram for the location in the circuit of the plate overload sensor relay (or comparator circuit). This will indicate within what limits component checking will be required. The plate overload sensor usually is found in one of two locations: the PA cathode dc return or the high-voltage power supply negative connection to ground. Transmitters using a cathode overload sensor generally have a separate high-voltage dc overload sensor in the plate power supply.

A sensor in the cathode circuit will reduce substantially the area of component checking required. A plate overload with no excitation in such an arrangement would almost certainly indicate a PA tube failure, because of either an inter-electrode short circuit or a loss of vacuum. Do not operate the transmitter when the PA tube is out of its socket. This is not an acceptable method of determining whether a problem exists with the PA tube. Substitute a spare tube instead. Operating a transmitter with the PA tube removed can result in damage to other tubes in the transmitter when the filaments are on, as well as damage to the driver tubes and driver output/PA input circuit components when the high voltage is on.

If circuit analysis indicates a problem in the high-voltage power supply itself, use an ohmmeter to check for short circuits. Remove all power from the transmitter, and discharge all filter capacitors before beginning any troubleshooting work inside the unit. When checking for short circuits with an ohmmeter, take into account the effects that bleeder resistors and high-voltage meter multiplier assemblies can have on resistance readings. Most access panels on transmitters use an interlock system that will remove the high voltage and ground the high-voltage supplies when a panel is removed. For the purposes of ohmmeter tests, these interlocks may have to be defeated temporarily. Never defeat the interlocks unless all ac power has been removed from the transmitter and all filter capacitors have been discharged using the grounding stick supplied with the transmitter.

Following the preliminary ohmmeter tests, check the following components in the dc plate supply:

- Oil-filled capacitors for signs of overheating or leakage.
- Feedthrough capacitors for signs of arcing or other damage.
- The dc plate blocking capacitor for indications of insulation breakdown or arcing.
- All transformers and chokes for signs of overheating or winding failure.
- Transient-suppression devices for indications of overheating or failure.
- Bleeder resistors for signs of overheating.
- Any surge-limiting resistors placed in series with filter capacitors in the power supply for indications of overheating or failure. A series resistor that shows signs of overheating can be an indication that the associated filter capacitor has failed.

If the plate overload trip-off occurs only at elevated voltage levels, ohmmeter checks will not reveal the cause of the problem. It may be necessary, therefore, to troubleshoot the problem using the process of elimination.

12.2.2 Process of Elimination

Troubleshooting through the process of elimination involves isolating various portions of the circuit, one section at a time, until the defective component is found. Special precautions are required before performing such work:

- Never touch anything inside the transmitter without first removing all ac power and discharging all filter capacitors with the grounding stick.
- Never perform troubleshooting work alone; another person should be present.
- Whenever a wire is disconnected, temporarily wrap it with electrical tape and secure the connector so it will not arc over to ground or another component when power is applied.

- Analyze each planned test before it is conducted. Because every test in the troubleshooting process requires time, steps should be arranged to provide the greatest amount of information about the problem.
- Check with the transmitter manufacturer to find out what testing procedures the company recommends. Ask what special precautions should be taken.

Troubleshooting the high-voltage plate supply usually is done under the following conditions:

- Exciter off
- Plate and screen IPA voltages off
- PA screen voltage off

Individual transmitters may require different procedures. Check with the manufacturer first.

Figure 12.1 shows a typical transmitter high-voltage power supply. Begin the troubleshooting process by breaking the circuit at point *A*. If the overload condition persists, the failure is caused by a problem in the power supply itself, not in the PA compartment. If, on the other hand, the overload condition disappears, a failure in

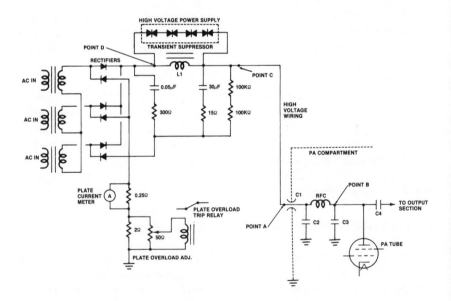

Figure 12.1 A typical transmitter high-voltage three-phase power-supply circuit. (Courtesy of *Broadcast Engineering* magazine, Intertec Publishing, Overland Park, Kan.)

the feedthrough capacitor (C1), decoupling capacitors (C2, C3), or blocking capacitor (C4) is indicated.

If a problem is indicated in the PA compartment, reconnect the high-voltage supply line at point *A* and break the circuit at point *B*. A return of the overload problem would indicate a failure in one of the decoupling capacitors or feedthrough capacitor.

To avoid unnecessary effort and time in troubleshooting, use the process of elimination to identify sections of the circuit to be examined. If, for example, the test at point *A* indicated the problem was not in the load, but in the power supply, a logical spot to perform the next test would be at point *C* (for long high-voltage cable runs). This test would identify or eliminate the interconnecting cable as a cause of the fault condition. If the cable run from the high-voltage supply to the PA compartment is short, point *D* might be the next logical point to check. Breaking the connection at the input to the power-supply filter allows the rectifiers and interconnecting cables to be checked. Transient-suppression devices (as shown above L1) should be considered a part of the component they are designed to protect. If a choke is removed from the circuit for testing, its protective device also must be removed. Failure to remove both connections usually results in damage to the protective device.

To prevent the creation of a new problem while trying to correct the original failure, break the circuit in only one point at a time. Also, study the possible adverse effects of each planned step in the process. Disconnecting certain components from the circuit may cause overvoltages or power-supply ripple that may damage other components in the transmitter. Consult the manufacturer on each planned step.

Use extreme care when performing any troubleshooting work on a transmitter. Transmitter high voltages can be lethal. Work inside the transmitter only after removing all ac power and using the grounding stick provided with the transmitter to discharge all capacitors. Remove primary power from the unit by tripping the appropriate power-distribution circuit breakers in the transmitter building. Do not rely on internal contactors or SCRs to remove all dangerous ac. Do not defeat protective interlock circuits. Although defeating an access-panel interlock switch may seem a good way to decrease work time, the consequences can be tragic.

12.3   RF SYSTEM FAULTS

Although RF troubleshooting may seem intimidating, there is no secret to it. Patient examination of the circuit and careful study of the schematic diagram will go a long way toward locating the problem. The first step in troubleshooting an RF problem is to determine whether the fault is RF-based or dc-based (as outlined in Sec. 12.2).

12.3.1   Troubleshooting Procedure

Check the load by examining the transmitter overload indicators. Most transmitters monitor reflected power from the antenna and will trip off if excessive VSWR is detected. If the VSWR fault indicator is not lit, the load probably is not the cause of the problem. A definitive check of the load can be made by switching the transmitter output to a dummy load and bringing up the high voltage. The PA tube may be checked by substituting one of known quality. When the tube is changed, carefully inspect the contact fingerstock for signs of overheating or arcing. Be careful to protect the socket from damage when removing and inserting the PA tube. Do not change the tube unless there is good reason to suspect that it may be defective. Do not touch the tube until it has cooled completely.

If problems with the PA stage persist, examine the grid circuit of the tube. Figure 12.2 shows the input stage of a grounded-screen FM transmitter. A short circuit in any of the capacitors in the grid circuit (C1 to C5) will effectively ground the PA grid. This will cause a dramatic increase in plate current, because the PA bias supply will be short-circuited to ground along with the RF signal from the IPA stage.

The process of finding a defective capacitor in the grid circuit begins with a visual inspection of the suspected components. Look for signs of discoloration caused by overheating or loose connections, and evidence of package rupture. The voltage and current levels found in a transmitter PA stage often are sufficient to rupture a capacitor if an internal short circuit occurs. Check for component overheating right after shutting down the transmitter. (As mentioned previously,

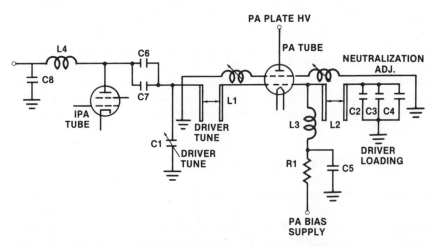

Figure 12.2   The PA grid input circuit of a grounded-screen transmitter. (Courtesy of *Broadcast Engineering* magazine, Intertec Publishing, Overland Park, Kan.)

remove all ac power, and discharge all capacitors first.) A defective capacitor often will overheat. However, such heating also may be the result of improper tuning of the PA or IPA stage, or a defective component elsewhere in the circuit.

Before replacing any components, study the transmitter schematic diagram to determine which parts in the circuit could cause the failure condition that exists. By knowing how the transmitter works, the engineer can keep from spending hours checking components that an examination of the fault condition and the transmitter design would show to be unlikely causes of the problem.

Check blocking capacitors C6 and C7. A breakdown in either component would have serious consequences. The PA tube would be driven into full conduction, and it could arc internally. The working voltages of capacitors C1 to C5 also could be exceeded, damaging one or more of the components. Because most of the wiring in the grid circuit of a PA stage consists of wide metal straps (required because of the skin effect), it is not possible to view stress points in the circuit to narrow the scope of the troubleshooting work. Areas of the system that are interconnected using components that have low power-dissipation capabilities, however, should be examined closely. For example, the grid bias decoupling components shown in Figure 12.2 (R1, L3, and C5) include a low-wattage (2 W) resistor and a small RF choke. Because of the limited power-dissipation ability of these two devices, a failure in decoupling capacitor C5 would be likely to cause R1, and possibly L3, to burn out. The failure of C5 in a short circuit would pull the PA grid to near ground potential, causing the plate current to increase and trip off the transmitter high voltage. Depending on the sensitivity and speed of the plate overload sensor, L3 could be damaged or destroyed by the increased current it would carry to C5, and therefore, to ground.

If L3 were able to survive the surge currents that resulted in PA plate overload, the choke would continue to keep the plate supply off until C5 was replaced. Bias supply resistor R1, however, probably would burn out because the bias power supply generally is switched on with the transmitter filament supply. Therefore, unless the PA bias power-supply line fuse opened, R1 would overheat and probably fail.

Because of the close spacing of components in the input circuit of a PA stage, carefully check for signs of arcing between components or sections of the tube socket. Keep all components and the socket itself clean at all times. Inspect all interconnecting wiring for signs of damage, arcing to ground, or loose connections.

12.3.2 Component Substitution

Substituting a new component for a suspected part can save valuable time in troubleshooting. With some components, it is cost-effective to replace a group of parts that may include one defective component because of the time that would be involved in gaining access to the damaged device. For example, the grid circuit of the PA stage shown in Figure 12.2 includes three *doorknob* capacitors (C2 to C4) formed into a single assembly. If one device were found to be defective, it might be advantageous simply to replace all three capacitors. These types of components often are integrated into a single unit that may be difficult to reach. Because doorknob capacitors are relatively inexpensive, it probably would be best to replace the three as a group. This way, the entire assembly could be eliminated as a potential cause of the fault condition.

A good supply of spare parts is invaluable for troubleshooting. In high-power transmitting equipment, substitution is sometimes the only practical means of finding the cause of a problem. The manufacturer's factory service department usually can recommend a minimum spare parts inventory. Obvious candidates include components that are not available locally, such as high-voltage fixed-value capacitors, vacuum variable capacitors, and specialized semiconductors.

12.3.3   Inside the PA Cavity

One of the factors that make it difficult to troubleshoot a cavity-type power amplifier is the nature of the major component elements. The capacitors do not necessarily look like capacitors, and the inductors do not necessarily look like inductors. It is often difficult to relate the electrical schematic diagram to the mechanical assembly that exists within the transmitter output stage. At VHF and UHF frequencies—the domain of cavity PA designs—inductors and capacitors can take on unusual mechanical forms.

Consider the PA cavity schematic diagram shown in Figure 12.3. The grounded-screen stage is of conventional design. Decoupling of the high-voltage power supply is accomplished by C1, C2, C3, and L1. Capacitor C3 is located inside the PA chimney (cavity inner conductor). The RF sample lines provide two low-power RF outputs for a modulation monitor or other test instruments. Neutralization inductors L3 and L4 consist of adjustable grounding bars on the screen grid ring assembly. The combination of L2 and C6 prevents spurious oscillations within the cavity. Figure 12.4 shows the electrical equivalent of the PA cavity schematic diagram. The 1/4-wavelength cavity acts as the resonant tank for the PA. Coarse tuning of the cavity is accomplished by adjustment of the shorting plane. Fine tuning is performed by the PA tuning control, which acts as a variable capacitor to bring the cavity into resonance. The PA loading control consists of a variable capacitor that matches the cavity to the load. There is one value of plate loading that will yield optimum output power, efficiency, and PA tube dissipation. This value is dictated by the cavity design and values of the various dc and RF voltages and currents supplied to the stage.

The logic of a PA stage often disappears when the maintenance engineer is confronted with the actual physical design of the system. As illustrated in Figure 12.5, many of the components take on an unfamiliar form. Blocking capacitor C4

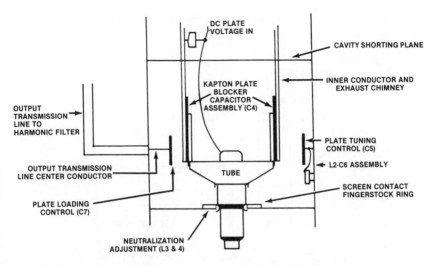

Figure 12.3 An FM transmitter PA output stage built around a 1/4-wavelength cavity with capacitive coupling to the load. (Courtesy of *Broadcast Engineering* magazine, Intertec Publishing, Overland Park, Kan.)

is constructed of a roll of *Kapton* insulating material sandwiched between two circular sections of aluminum. (Kapton is a registered trademark of DuPont.) PA plate tuning control C5 consists of an aluminum plate of large surface area that can be moved in or out of the cavity to reach resonance. PA loading control C7 is constructed in much the same way as the PA tuning assembly, with a large-area paddle feeding the harmonic filter, located external to the cavity. The loading

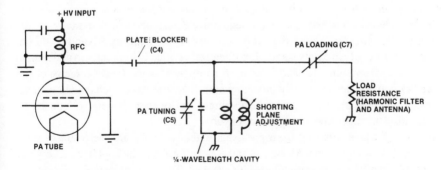

Figure 12.4 The equivalent electrical circuit of the PA stage shown in Figure 12.3. (Courtesy of *Broadcast Engineering* magazine, Intertec Publishing, Overland Park, Kan.)

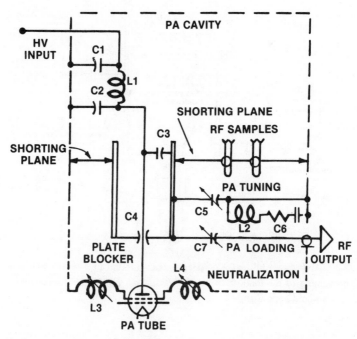

Figure 12.5 The mechanical equivalent of the PA stage shown in Figure 12.3. (Courtesy of *Broadcast Engineering* magazine, Intertec Publishing, Overland Park, Kan.)

paddle may be moved toward the PA tube or away from it to achieve the required loading. The L2-C6 damper assembly actually consists of a 50 Ω noninductive resistor mounted on the side of the cavity wall. Component L2 is formed by the inductance of the connecting strap between the plate tuning paddle and the resistor. Component C6 is the equivalent stray capacitance between the resistor and the surrounding cavity box. From this example, it can be seen that many of the troubleshooting techniques that work well with low-frequency RF and dc do not necessarily apply in cavity stages. It is, therefore, critically important to understand how the system operates and what each component does. Because many of the cavity components (primarily inductors and capacitors) are mechanical elements more than electrical ones, troubleshooting a cavity stage generally focuses on checking the mechanical integrity of the box.

Most failures resulting from problems within a cavity are the result of poor mechanical connections. All screws and connections must be kept tight. Every nut and bolt in a PA cavity was included for a reason. There are no insignificant screws that do not need to be tight. However, do not overtighten, either. Stripped threads and broken component connection lugs only cause additional grief.

When a problem occurs in a PA cavity, it is usually difficult to determine which individual element (neutralization inductor, plate tuning capacitor, loading capacitor, etc.) is defective, from the symptoms the failure will display. A fault within the cavity is usually a catastrophic event that will take the transmitter off the air. It is often impossible to bring the transmitter up for even a few seconds to assess the fault situation. The only way to get at the problem is to shut down the transmitter and take a look inside.

Closely inspect every connection, using a trouble light and magnifying glass. Look for signs of arcing or discoloration of components or metal connections. Check the mechanical integrity of each element in the circuit. Be certain the tuning and loading adjustments are solid, without excessive mechanical play. Look for signs of change in the cavity. Check areas of the cavity that may not seem to be vital parts of the output stage, such as the maintenance access door fingerstock and screws. Any failure in the integrity of the cavity, whether at the base of the PA tube or on part of the access door, will cause high circulating currents to develop and may prevent proper operation of the stage. If a problem is found that involves damaged fingerstock, replace the affected sections. Failure to do so probably will result in future problems because of the currents that can develop at any discontinuity in the cavity inner or outer conductor.

12.3.4   VSWR Overload

A VSWR overload in transmission equipment may result from a number of different problems. The following checklist presents some common problems and solutions:

1. VSWR overloads usually are caused by an improper impedance match external to the transmitter. The first step in the troubleshooting procedure is to substitute a dummy load for the entire antenna and transmission-line system. Connect the dummy load at the transmitter output port, thereby eliminating all coax, external filters, and other RF hardware that might be present in the system.
2. If the VSWR trip fault is eliminated in step 1, the problem lies somewhere in the transmission line or antenna. Next, the dummy load can be moved to the point at which the transmission line leaves the building and heads for the tower (if different than the point checked in step 1). This test will check any RF plumbing, switches, or filter assemblies. If the VSWR overload condition still is absent, the problem is centered in the transmission line or the antenna.
3. If a standby antenna is not available, operation still may be possible at reduced power on a temporary basis. For example, if arcing occurs in the antenna or line at full power, emergency operation may be possible at half

power. Inspect the antenna and line for signs of trouble. Repair work beyond this point normally requires specialized equipment and a tower crew. This discussion assumes that the problem is not caused by ice buildup on the antenna, a problem that can be alleviated by reducing the transmitter power output until VSWR trips do not occur.

4. If step 1 shows the VSWR overload to be internal to the transmitter, determine whether the problem is caused by an actual VSWR overload or by a failure in the VSWR control circuitry. Check for this condition by disabling the transmitter exciter and bringing up the high voltage. Under these conditions, RF energy will not be generated. (It is assumed that the transmitter has proper bias on all stages and is neutralized properly.) If a VSWR overload is indicated, the problem is centered in the VSWR control circuitry, not in the RF chain. Possible explanations for control circuitry failure include loose connections, dirty switch contacts, dirty calibration potentiometers, poor PC board edge connector contacts, defective IC amplifiers or logic gates, and intermittent electrolytic capacitors.

5. If step 4 shows that the VSWR overload is real, and not the result of faulty control circuitry, check all connections in the output and coupling sections of the final stage. Look for signs of arcing or loose hardware, particularly on any movable tuning components. Inspect high-voltage capacitors for signs of overheating, which might indicate failure; and check coils for signs of dust buildup, which might cause a flashover. In some transmitters, VSWR overloads can be caused by improper final stage tuning or loading. Consult the equipment instruction book for this possibility. Also, certain transmitters include glass-enclosed spark-gap lightning protection devices (*gas-gaps*) that can be disconnected for testing.

6. If VSWR overload conditions resulting from problems external to the transmitter are experienced at an AM radio station, check the following items:

Component dielectric breakdown. If a normal (near zero) reflected power reading is indicated at the transmitter under carrier-only conditions, but VSWR overloads occur during modulation, component dielectric breakdown may be the problem. A voltage breakdown could be occurring within one of the capacitors or inductors at the antenna tuning unit (ATU) or phaser. Check all components for signs of damage. Clean insulators as required. Carefully check any open-air coils or transformers for dust buildup or loose connections.

Narrowband antenna. If the overload occurs with any modulating frequency, the probable cause of the fault is dielectric breakdown. If, on the other hand, the overload seems particularly sensitive to high-frequency modulation, then narrow antenna bandwidth is indicated. Note the action of the transmitter forward/reflected power meter. An upward deflection of reflected power with modulation is a symptom of limited antenna bandwidth. The greater the upward deflection, the more limited the bandwidth. If these symptoms are observed, conduct an impedance sweep of the antenna system.

Static buildup. Tower static buildup is characterized by a gradual increase in reflected power as shown on the transmitter front panel. The static buildup, which usually occurs before or during thunderstorms and other bad weather conditions, continues until the tower base ball gaps arc over and neutralize the charge. The reflected power reading then falls to zero. A static drain choke at the tower base to ground usually will prevent this problem.

Guy wire arc-over. Static buildup on guy wires is similar to a nearby lightning strike in that no charge is registered during the buildup of potential on the reflected power meter. Instead, the static charge builds on the guys until it is of sufficient potential to arc across the insulators to the tower. The charge then is removed by the static drain choke and/or ball gaps at the base of the tower. To prevent static buildup on guy wires, place RF chokes across the insulators, or use nonmetallic guys. Arcing across the insulators also may be reduced or eliminated by regular cleaning.

12.4 POWER CONTROL FAULTS

A failure in the transmitter ac power control circuitry can result in problems ranging from zero RF output to a fire inside the unit. Careful, logical troubleshooting of the control system is mandatory. Two basic types of primary ac control are used in transmitters today:

1. SCR (thyristor) system
2. Relay logic

12.4.1 Thyristor Control System

A failure in the thyristor power control system of a transmitter is not easy to overlook. In the worst case, no high voltage at all will be produced by the transmitter. In the best case, power control may be erratic or uneven when the continuously variable power adjustment mode is used. Understanding how the servo circuit works and how it is interconnected is the first step in correcting such a problem.

Figure 12.6 shows a block diagram of a typical thyristor control circuit. Three gating cards are used to drive back-to-back SCR pairs, which feed the high-voltage power transformer primary windings. Although the applied voltage is three-phase, the thyristor power control configuration simulates a single-phase design for each phase-to-phase leg. This allows implementation of a control circuit that consists

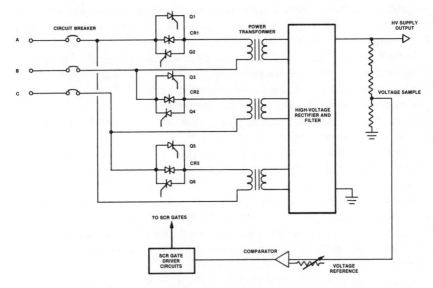

Figure 12.6 Block diagram of a three-phase thyristor power control system. (Courtesy of *Broadcast Engineering* magazine, Intertec Publishing, Overland Park, Kan.)

basically of a single-phase gating card duplicated three times (one for each load phase). This approach has advantages from the standpoint of design simplicity, and also from the standpoint of field troubleshooting. In essence, each power control circuit is identical, allowing test voltages and waveforms from one gating card to be compared directly with a gating card experiencing problems.

If the high-voltage supply will not come up at all, the problem involves more than a failure in just one of the three gating cards. The failure of any one gating board would result in reduced power output (and other side effects), but not in zero output. Begin the search with the interlock system.

12.4.2 Interlock Failures

Newer transmitters provide the engineer with built-in diagnostic readouts on the status of the transmitter interlock circuits. These may involve discrete LEDs or a microcomputer-driven visual display of some type. If the transmitter has an advanced status display, the process of locating an interlock fault is relatively simple. For an older transmitter that is not equipped this way, substantially more investigation will be needed.

Make a close observation of the status of all fuses, circuit breakers, transmitter cabinet doors, and access panels. Confirm that all doors are fully closed and secured. If it is operated remotely, switch the transmitter from *remote* to *local control* to eliminate the remote-control system as a possible cause of the problem. Observe the status of all control-panel indicator lamps. Some transmitters include an *interlocks open* lamp; other units provide an indication of an open interlock through the *filament on* or *plate off* pushbutton lamps. These indicators can save valuable minutes or even hours of troubleshooting, so pay attention to them. Replace any burned-out indicator lamps as soon as they are found to be defective.

If the front-panel indicators point to an interlock problem, pull out the schematic diagram of the transmitter, get out the DMM, and shut down the transmitter. If the transmitter interlock circuit operates from a low-voltage power supply, such as 24 V dc, use a voltmeter to check for the loss of continuity. Remove ac power from all sections of the transmitter except the low-voltage supply by tripping the appropriate front-panel circuit breakers. Be extremely careful when working on the transmitter to avoid any line voltage ac. If the layout of the transmitter does not permit safe troubleshooting with only the low-voltage power supply active, remove all ac from the unit by tripping the wall-mounted main breaker. Then, use an ohmmeter to check for the loss of continuity.

If the transmitter interlock circuit operates from 120 V ac or 220 V ac, remove all power from the transmitter by tripping the wall-mounted main breaker. Use an ohmmeter to locate the problem. Many older transmitters use line voltages in the interlock system. Do not try to troubleshoot such transmitters with ac power applied.

Finding a problem such as an open control circuit interlock is basically a simple procedure, in spite of the time involved. Do not rush through such work. When searching for a break in the interlock system, use a methodical approach to solving the problem. Consider the circuit configuration shown in Figure 12.7. The most logical approach to finding a break in the control ladder is to begin at the source of the 24 V dc input and, step by step, work toward the input of the power controller. Although this approach may be logical, it also can be time-consuming. Instead, eliminate stages of the interlock circuit. For example, make the first test at terminal *A*. A correct voltage reading at this point in the circuit will confirm that all of the interlock door switches are operating properly.

With the knowledge that the problem is after terminal *A*, move on to terminal *B*. If the 24 V supply voltage disappears, check the fault circuit overload relays to determine where the control signal is lost. Often, such interlock problems can be attributed to dirty contacts in one of the overload relays. If a problem is found with one set of relay contacts, clean all of the other contacts in the overload interlock string for good measure. Be sure to use the proper relay-contact cleaning tools. If sealed relays are used, do not attempt to clean them. Instead, replace the defective unit.

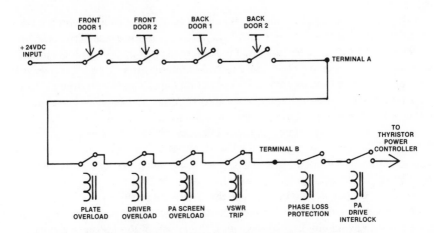

Figure 12.7 A typical transmitter interlock circuit. Terminals A and B are test points used for troubleshooting the system in the event of an interlock system failure. (Courtesy of *Broadcast Engineering* magazine, Intertec Publishing, Overland Park, Kan.)

12.4.3 Step-Start Faults

The high-voltage power supply of any medium- or high-power transmitter must include provisions for in-rush current-limiting upon the application of a *plate-on* command. The filter capacitor(s) in the power supply will appear as a virtual short circuit during a sudden increase in voltage from the rectifier stacks. To avoid excessive current surges through the rectifiers, capacitor(s), choke, and power transformer, nearly all transmitters use some form of *step-start* arrangement. Such circuits are designed to limit the in-rush current to a predictable level. This can be accomplished in a variety of ways.

For transmitters using a thyristor power control system, the step-start function can be designed easily into the gate firing control circuit. An R-C network at the input point of the gating cards is used to ramp the thyristor pairs from a zero conduction angle to full conduction (or a conduction angle preset by the user). This system provides an elegant solution to the step-start requirement, allowing plate voltage to be increased from zero to full value within a period of approximately 5 seconds.

Transmitters employing a conventional ac power control system usually incorporate a step-start circuit consisting of two sets of contactors: the *start contactor* and the *run contactor*, as illustrated in Figure 12.8. Surge-limiting resistors provide sufficient voltage drop upon application of a *plate-on* command to limit the surge current to a safe level. Auxiliary contacts on the start contactor cause the run contactor to close as soon as the start contacts have seated.

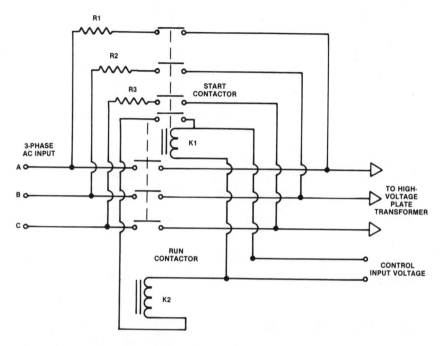

Figure 12.8 A typical three-phase step-start power control system. (Courtesy of *Broadcast Engineering* magazine, Intertec Publishing, Overland Park, Kan.)

A fault in the step-start circuit of a transmitter often is evidenced, at least initially, by random tripping of the plate supply circuit breaker at high-voltage turn-on. If left uncorrected, this condition can lead to problems such as failed power rectifiers or filter capacitors.

Troubleshooting a step-start fault in a system employing thyristor power control should begin at the R-C ramp network. Check the capacitor to see whether it has opened. Monitor the control voltage to the thyristor gating cards to confirm that the output voltage of the controller slowly increases to full value. If it does, and the turn-on problem persists, the failure involves one or more of the gating cards.

When troubleshooting a step-start fault in a transmitter employing the dual contactor arrangement, begin with a close inspection of all contact points on both contactors. Pay careful attention to the auxiliary relay contacts of the start contactor. If the contacts fail to close properly, the full load of the high-voltage power supply will be carried through the resistors and start contactor. These devices normally are sized only for intermittent duty. They are not intended to carry the full load current for any length of time. Look for signs of arcing or overheating of the contact pairs and current-carrying connector bars. Check the current-limiting resistors for excessive dissipation and continuity.

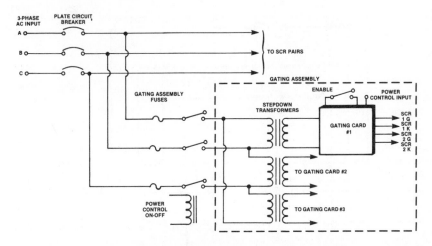

Figure 12.9 A common three-phase thyristor servo ac power control system. (Courtesy of *Broadcast Engineering* magazine, Intertec Publishing, Overland Park, Kan.)

12.4.4 Protection Circuits

Many engineers enjoy a false sense of security with transmission equipment because of the protection devices included in most designs. Although conventional overload circuits provide protection against most common failure modes, they are not foolproof. The first line of defense in the transmitter—the ac power system circuit breakers—may allow potentially disastrous currents to flow under certain fault conditions.

Consider the thyristor ac power servo system shown in Figure 12.9. This common type of voltage-regulator circuit adjusts the condition angle of the SCR pairs to achieve the desired dc output from the high-voltage power supply. An alternative configuration could have the output voltage sample derived from a transmission line RF pickup and amplifier/detector. In this way, the primary power control is adjusted to match the desired RF output from the transmitter. If one of the high-voltage rectifier stacks of this system were to fail in a short-circuit condition, the output voltage (and RF output) would fall, causing the thyristor circuit to increase the conduction period of the SCR pairs. Depending on the series resistance of the failed rectifier stack and the rating of the primary side circuit breaker, the breaker may or may not trip. Remember that the circuit breaker was chosen to allow operation at full transmitter power with the necessary headroom to prevent random tripping. The primary power system, therefore, can dissipate a significant amount of heat under reduced-power conditions, such as those that would be experienced with a drop in the high-voltage supply output. The difference between the maximum designed power output of the supply (and, therefore, the transmitter) and the failure-induced power demand of the system can be dissipated as heat without tripping the main breaker.

Operation under such fault conditions, even for 20 seconds or less, can cause considerable damage to power-supply components, such as the power transformer, rectifier stack, thyristors, or system wiring. Damage can range from additional component failures to a fire in the affected section of the transmitter.

Case in Point: The failure mode just outlined represents a real threat to high-power systems. The author is aware of a case in which just such a failure resulted in a fire inside a 20 kW FM transmitter. The following sequence of events led to the destruction of the unit:

- One or more transient overvoltages hit the transmitter site, causing an arc to occur within the driver stage plate transformer. The arcing continued until particles from the secondary winding broke free from the transformer.
- The failure of the driver transformer caused the driver output voltage to drop significantly, which decreased the RF output of the transmitter to about 25 percent of normal. Because the failure occurred between windings of the secondary of the driver plate transformer (and before the driver stage over-current sensor), plate voltage remained on. Also, because of the point where the secondary winding short circuit occurred, the transformer primary did not draw sufficient current to initially trip the driver circuit breaker. As a result, ac power continued to flow to the damaged transformer.
- Small pieces of molten metal continued to drop from the driver transformer, landing on the PA plate transformer. These particles dropped into the windings, causing the plate transformer to short-circuit, which started a localized fire.
- When the smoke finally cleared, the entire high-voltage power supply had been damaged. In addition to the two ruined transformers, logic relays were melted away, rectifier stacks were fried, and most of the wiring harness was destroyed. The transmitter was determined to be damaged beyond repair.

Disasters such as this are rare, but they do occur. Be prepared to respond to any emergency condition by understanding thoroughly how the transmitter works and by identifying potential weak points in the system. Troubleshooting is far too important to be left to chance, or to the inexperienced.

13

TEST EQUIPMENT FOR RF SYSTEMS

13.1 INTRODUCTION

The goal of every maintenance engineer is to ensure top-quality transmission standards and to provide the most reliable equipment possible for the end-user. These objectives do not just happen. They are the result of a carefully planned maintenance program. Such a program involves correct setup of the equipment whenever maintenance is required. Having the proper RF test equipment is an integral part of this maintenance effort. Setting up a maintenance shop for an RF facility can be an expensive proposition. However, good-quality test instruments are a requirement to protect the investment in the transmission system. Acquiring the proper test equipment is much the same as having an insurance policy on that investment.

The actual test equipment requirements to perform most maintenance and adjustment procedures on transmission equipment are relatively modest. A technician, armed with the following instruments, will be able to deal with most common RF problems:

- Digital multimeter (DMM), including a high-voltage probe and a clamp-on current probe.
- Logic probe.
- Wideband oscilloscope.
- SMPTE (Society of Motion Picture and Television Engineers) standard color-bar generator (for video transmission systems).
- Waveform monitor and vectorscope (for video transmission systems).
- Wattmeter and a selection of power-level slugs.
- FM and/or AM modulation monitor (or deviation meter).
- Spectrum analyzer.

13.2 VIDEO TRANSMISSION MEASUREMENTS[1]

Although there are now a number of computer-based television signal monitors capable of measuring video, sync, chroma, and burst levels (as well as pulse widths and other timing factors), the best way to see what a signal is doing is to monitor it visually. Probably the most useful and flexible single test instrument for television is a precision wideband, dual-channel oscilloscope with a dual time base and delayed sweep. The unit must have sufficiently high vertical frequency response for the fast rise times of pulse and digital signals. The sweep speed range must go low enough to view fields and frames, and high enough to resolve the color subcarrier (3.579545 MHz for NTSC) and beyond.

In some instances, the point of the waveform that must be observed is not visible because of the requirements of synchronizing the display to the signal. In such cases, a time base system with delayed sweep can be used to sync the scope to a stable point on the waveform. The signal is routed through a delay line before it is displayed, making the point of interest visible.

The waveform monitor and vectorscope are oscilloscopes especially adapted for the video environment. The waveform monitor, like a traditional oscilloscope, operates in a voltage-vs.-time mode. Although an oscilloscope time base can be set over a wide range of intervals, the waveform monitor time base triggers automatically on sync pulses in the television signal, producing line- and field-rate sweeps, as well as multiple lines, multiple fields, and shorter time intervals. Filters, clamps, and other circuits process the video signal for specific monitoring needs. The vectorscope operates in an x-y voltage-vs.-voltage mode to display chrominance information. It decodes the signal in much the same way as a television receiver or a video monitor to extract color information and to display phase relationships. These two instruments serve separate, distinct purposes. Some models combine the functions of both types of monitors in one chassis with a single CRT display. Others include a communications link between two separate instruments.

Beyond basic signal monitoring, the waveform monitor and vectorscope provide a means to identify and analyze signal aberrations. If the signal is distorted, these instruments allow a technician to learn the extent of the problem and to locate the offending equipment.

13.2.1 Visual Transmission Measurements

The primary parameters of a broadcast television signal that should be measured, monitored, and maintained within certain operational limits are detailed in vol. 3,

---

1 Portions of Sec. 13.2 were contributed by Carl Bentz, Special Projects Editor, Intertec Publishing, Overland Park, Kan.

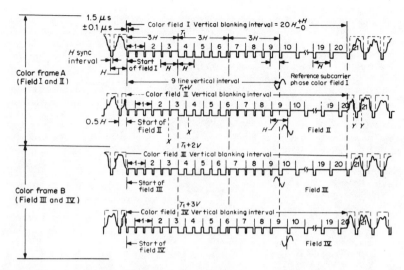

Figure 13.1   Components of the NTSC color television waveform.

Part 73 of the FCC rules (for the NTSC color transmission system). The modulation signals of a visual transmitter are more complicated than television aural or radio signals. For example, the sound broadcasting signal is quite narrow, compared with a visual signal. The visual channel is 6 MHz wide for NTSC, and the aural carrier is offset by 4.5 MHz from the visual. The visual information includes luminance, chroma, and pulse trains for horizontal and vertical synchronization. For the picture to appear correctly on the home receiver, all parts of the signal must pass through the transmitter-receiver chain with exact timing in relation to one another. They also must not be degraded individually by the transmission path or be influenced by other signals in the process.

Visual Transmission Format. The content of the visual television signal is embedded in a pulse-based framework that services both horizontal (line) and vertical (field) synchronization. (See Figure 13.1.) Also contained within the framework is a reference signal for the color subcarrier, or *burst.* The maximum amplitudes of the visual content of the signal are defined by the pulse framework. The framework is defined in terms of time and amplitude at the input and output of the visual transmitter chain. The system is designed to maintain the time and amplitude relationships from input to output. Most television transmission standards work on the basis of *negative modulation* in the visual transmitter; the largest value of RF amplitude corresponds to the zero level of video modulation (sync). The minimum value (residual carrier) of the RF amplitude corresponds to the maximum value of video modulation (white level).

There should be no fluctuations in the output power of the system as long as no low-frequency transients are allowed to pass through the power supply and control

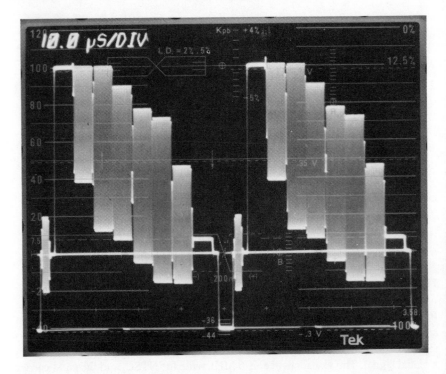

Figure 13.2   Waveform monitor display of a color-bar signal at the two-line rate. (Courtesy of Tektronix)

circuitry of the transmitter. When transients do exist, they are recognizable on a waveform monitor or oscilloscope.

13.2.2   Basic Waveform Measurements

Waveform monitors are used to evaluate the amplitude and timing of video signals and to show timing relationships between two or more signals. The familiar color-bar pattern is the only signal required for these basic tests. Figure 13.2 shows a typical waveform display of color bars.

It is important to realize that all color bars are not created equal. Some generators offer a choice of 75 percent or 100 percent amplitude bars. Sync, burst, and setup amplitudes remain the same in the two color-bar signals, but the peak-to-peak amplitudes of high-frequency chrominance information and low-frequency luminance levels change. The saturation of color, a function of chrominance and luminance amplitudes, remains constant at 100 percent in both modes. The 75 percent bar signal has 75 percent amplitude with 100 percent saturation. In 100 percent bars, amplitude and saturation are both 100 percent.

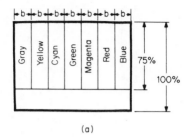

(a)

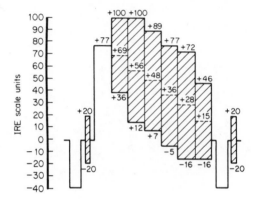

Figure 13.3   EIA RS189A color-bar displays: (a) Color monitor display of gray and color bars. (b) Waveform display of reference gray and primary/complementary colors, plus sync and burst.

Chrominance amplitudes in 100 percent bars exceed the maximum amplitude that should be transmitted. Therefore, 75 percent amplitude color bars, with no chrominance information exceeding 100 IRE, are the standard amplitude bars for NTSC. In the 75 percent mode, a choice of 100 IRE or 75 IRE white reference level may be offered. Figure 13.3 shows 75 percent amplitude bars with a 100 IRE white level. Either white level can be used to set levels, but operators must be aware of which signal has been selected. SMPTE bars have a white level of 75 IRE as well as a 100 IRE white flag.

To make precise evaluations of signals, the waveform monitor itself must be functioning properly. It should be periodically taken to a service center for calibration. The internal calibration signal for precise gain and sweep adjustments should be used regularly. The calibration signal is selected on the front panel and, if necessary, the vertical gain calibration control is adjusted until the resulting square-wave signal is exactly 140 IRE units in amplitude. (Some instruments may require settings of 100 IRE units. Consult the manual.)

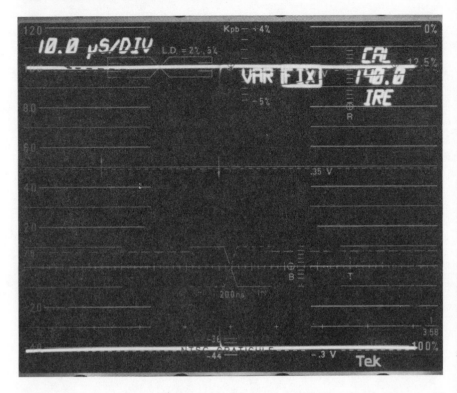

Figure 13.4   A waveform monitor display of a typical calibration signal. (Courtesy of Tektronix)

Many waveform monitors include a sweep calibrator. With the calibration signal enabled, adjust the horizontal calibration control until the square wave crosses the graticule base line at the major division marks. Figure 13.4 shows a typical calibrator signal. The adjustments generally are made with a screwdriver. (Do not confuse the calibration adjustment controls with the variable gain knobs.)

Before making measurements with the waveform monitor, check the setting of the *dc restorer*. The dc restorer normally should be on, to stabilize the display from variations in *average picture level* (APL). If the instrument offers slow or fast dc restorer speeds, remember that the slow speed stabilizes the display at an average dc level, but permits the observation of low-frequency abnormalities such as 60 Hz hum. The fast speed removes most of the hum.

The vertical response of a waveform monitor depends upon filters that process the signal in order to display certain components. The *flat response* mode displays all components of the signal. A *chroma response* filter removes luminance and displays only chrominance. The low-pass filter removes chrominance, leaving only low-frequency luminance levels in the display. Some monitors include an IRE

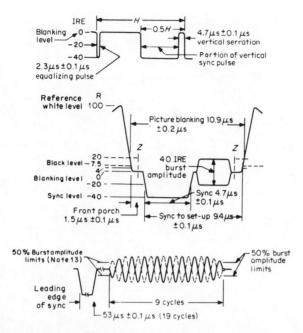

Figure 13.5   Sync pulse widths for NTSC as defined by the Federal Communications Commission. (Courtesy of Tektronix)

filter, designed to average out high-level, fine-detail peaks on a monochrome video signal. The IRE filter aids the operator in setting brightness levels. The IRE response removes most, but not all, of the chrominance.

If the waveform monitor has a dual filter mode, the operator can observe luminance levels and overall amplitudes at the same time. The instrument switches between the flat and low-pass filters. With a 2H sweep, the display on the left is low-pass-filtered, while information to the right is unfiltered. The line select mode is another useful feature for monitoring live signals.

Sync pulses. Duration and frequency of the sync pulses must be monitored closely. Most waveform monitors include 0.5 µs or 1 µs per division magnification (MAG) modes, which can be used to verify H-sync width between 4.4 µs and 5.1 µs. The width is measured at the - 4 IRE point. On waveform monitors with good MAG registration, sync appearing in the middle of the screen in the two-line mode remains centered when the sweep is magnified. Check the rise and fall times of sync, and the widths of the front porch and entire blanking interval. Examine burst and verify that there are 8 to 11 cycles of subcarrier.

Check the vertical intervals for correct format, and measure the timing of the equalizing pulses and vertical sync pulses. The acceptable limits for these parameters are shown in the FCC pulse width specification, reproduced in Figure 13.5.

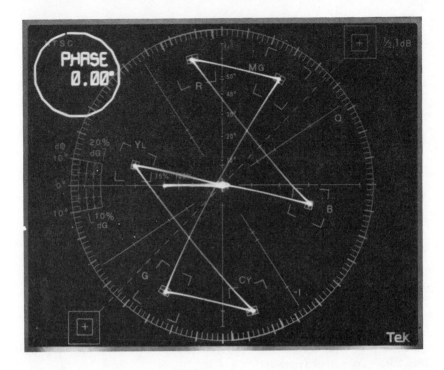

Figure 13.6 Vectorscope display of a color-bar signal. (Courtesy of Tektronix)

13.2.3 Basic Vectorscope Measurements

The vectorscope displays chrominance amplitudes, aids hue adjustments, and simplifies the matching of burst phases of multiple signals. These functions require only the color-bar test signal as an input stimulus. To evaluate and adjust chrominance in the television signal, observe color bars on the vectorscope. The instrument should be in its calibrated gain position. Adjust the vectorscope phase control to place the burst vector at the 9 o'clock position. Note the vector dot positions with respect to the six boxes marked on the graticule. If everything is adjusted correctly, each dot will fall on the crosshairs of its corresponding box, as shown in Figure 13.6.

The chrominance amplitude of a video signal determines the intensity or brightness of color. If the amplitudes are correct, the color dots fall on the crosshairs in the corresponding graticule boxes. If vectors overshoot the boxes, chrominance amplitude is too large; if they undershoot, it is too small. The boxes at each color location can be used to quantify the error. In the radial direction, the small boxes

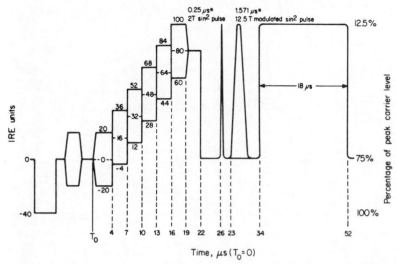

Figure 13.7   Composite vertical-interval test signal (VITS) inserted in field 1, line 18. The video level in IRE units is shown on the left. The radiated carrier signal is shown on the right.

indicate a 2.5 IRE error from the standard amplitude. The large boxes indicate a 20 percent error.

Other test signals, including a modulated staircase or multiburst, are used for more advanced tests. It is important to take a good look at how these signals appear immediately after they come out of the generator. Knowing what the undistorted signal looks like simplifies the identification of distortions.

13.2.4   Line Select Features

Some waveform monitors and vectorscopes have line select capability. They can display one or two lines out of the entire video frame of 525 lines. (In the normal display, all the lines are overlaid on top of one another.) The principal use of the single line feature is to monitor VITS (vertical interval test signals). The use of VITS allows in-service testing of the transmission system. A typical VITS waveform is shown in Figure 13.7. A full-field line selector drives a picture monitor output with an intensifying pulse. The pulse causes a single horizontal line on a picture monitor to be highlighted. This indicates where the line selector is within the frame to correlate the waveform monitor display with the picture.

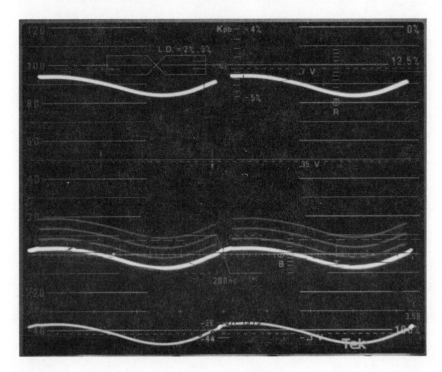

Figure 13.8 Waveform monitor display showing additive 60 Hz degradation. (Courtesy of Tektronix)

13.2.5 Distortion Mechanisms

The television system should respond uniformly to signal components of different frequencies. This parameter generally is evaluated with a waveform monitor. Checking the various parts of the frequency spectrum requires different signals. If the signals all are faithfully reproduced on the waveform monitor screen after passing through the video system, it is safe to assume that there are no serious frequency-response problems.

At very low frequencies, look for externally introduced distortions, such as power-line hum or power-supply ripple, and distortions resulting from inadequacies in the video processing equipment. Low-frequency distortions usually appear on the television screen as flickering or slowly varying brightness. Low-frequency interference can be seen on a waveform monitor when the dc restorer is set to the slow mode and a two-field sweep is selected. Sine-wave distortion from ac power-line hum may be observed in Figure 13.8. A *bouncing APL* signal can be used to detect distortion in the video chain. Vertical shifts in the blanking and sync levels indicate the possibility of low-frequency distortion.

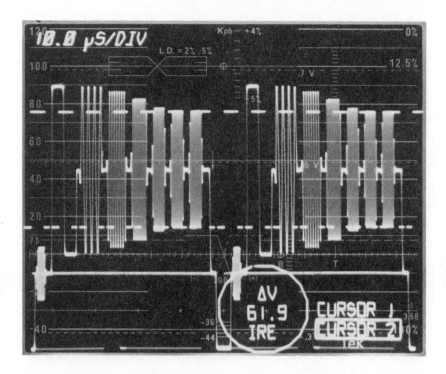

Figure 13.9 Waveform monitor display of a multiburst signal showing poor high-frequency response. (Courtesy of Tektronix)

Field-rate distortions appear as a difference in shading from the top to the bottom of the picture. A field-rate 60 Hz square wave is best suited for measuring field-rate distortion. Distortion of this type is observed as a tilt in the waveform in the two-field mode with the dc restorer off.

Line-rate distortions appear as streaking, shading, or poor picture stability. To detect such errors, look for tilt in the bar portion of a pulse-and-bar signal. The waveform monitor should be in the 1H or 2H mode with the fast dc restorer selected for the measurement.

The *multiburst* signal is used to test the high-frequency response of a system. The multiburst includes packets of discrete frequencies within the television passband, with the higher frequencies toward the right of each line. The highest frequency packet is at about 4.2 MHz, which is the upper frequency limit of the NTSC system. The next packet to the left is near the color subcarrier frequency (3.58 MHz) for checking the chrominance transfer characteristics. Other packets are included at intervals down to 500 kHz. The most common distortion is high-frequency rolloff, seen on the waveform monitor as reduced-amplitude packets at higher frequencies. This type of problem is shown in Figure 13.9. The

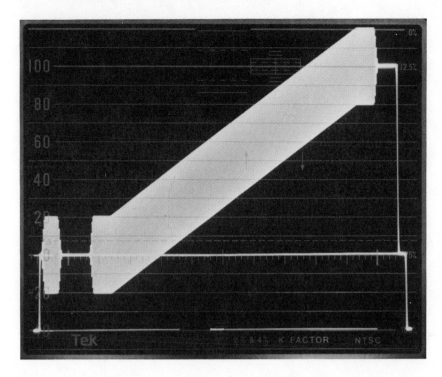

Figure 13.10   Waveform monitor display of a modulated ramp signal. (Courtesy of Tektronix)

television picture exhibits loss of fine detail and color intensity when such impairments are present. High-frequency peaking, appearing on the waveform as higher-amplitude packets at the higher frequencies, causes ghosting on the picture.

Differential Phase. Differential phase (dθ) distortion occurs if a change in luminance level produces a change in the chrominance phase. If the distortion is severe, the hue of an object will change as its brightness changes. A modulated staircase or ramp is used to quantify the problem. Either signal places chrominance of uniform amplitude and phase at different luminance levels. Figure 13.10 shows a 100 IRE modulated ramp. Because dθ may vary with changes in APL, measurements at the center and at the two extremes of the APL range are necessary for full evaluation of system response.

To measure dθ with a vectorscope, increase the gain control until the vector dot is on the edge of the graticule circle. Use the phase shifter to set the vector to the 9 o'clock position. Phase error appears as circumferential elongation of the dot. The vectorscope graticule has a scale marked with degrees of dθ error. Figure 13.11 shows a dθ error of 5°.

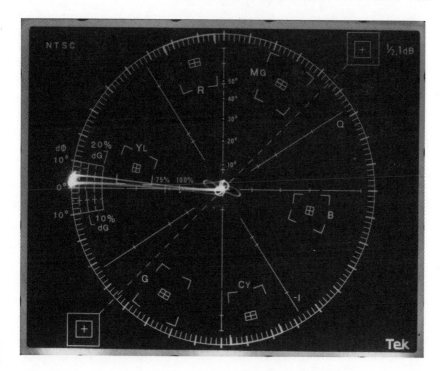

Figure 13.11 Vectorscope display showing 5° differential phase error. (Courtesy of Tektronix)

More information can be obtained from a swept R-Y display, which is a common feature of waveform monitor and vectorscope systems. If one or two lines of demodulated video from the vectorscope are displayed on a waveform monitor, differential phase appears as tilt across the line. In this mode, the phase control can be adjusted to place the demodulated video on the baseline, which is equivalent in phase to the 9 o'clock position of the vectorscope. Figure 13.12 shows a dθ error of 5.95° with the amount of tilt measured against a vertical scale. This mode is useful in troubleshooting. By noting the point along the line at which the tilt begins, it is possible to determine at what dc level the problem starts to occur. In addition, field-rate sweeps enable the operator to look at dθ over the field.

A variation of the swept R-Y display may be available in some instruments for precise measurement of differential phase. Highly accurate measurements can be made with a vectorscope that includes a precision phase shifter and a double-trace mode. This method involves nulling the lowest part of the waveform with the phase shifter, and then using a separate calibrated phase control to null the highest end of the waveform. A readout in tenths of a degree is possible.

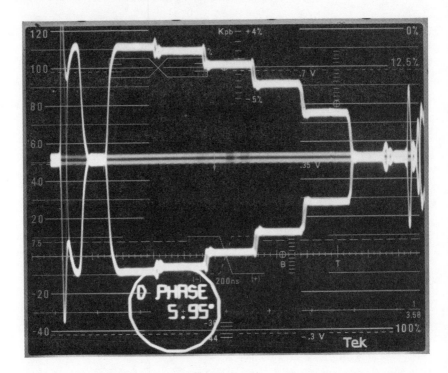

Figure 13.12 Waveform monitor display showing a differential phase error of 5.95° as a tilt on the vertical scale. (Courtesy of Tektronix)

Differential Gain. Differential gain (dG) distortion refers to a change in chrominance amplitude with changes in luminance level. The vividness of a colored object changes with variations in scene brightness. The modulated ramp or staircase is used to evaluate this impairment with the measurement taken on signals at different APL points.

To measure differential gain with a vectorscope, set the vector to the 9 o'clock position, and use the variable gain control to bring it to the edge of the graticule circle. Differential gain error appears as a lengthening of the vector dot in the radial direction. The dG scale at the left side of the graticule can be used to quantify the error. Figure 13.13 shows a dG error of 10 percent.

Differential gain can be evaluated on a waveform monitor by using the chroma filter and examining the amplitude of the chrominance from a modulated staircase or ramp. With the waveform monitor in 1H sweep, use the variable gain to set the amplitude of the chrominance to 100 IRE. If the chrominance amplitude is not uniform across the line, there is dG error. With the gain normalized to 100 IRE, the error can be expressed as a percentage. Also, dG can be evaluated precisely with

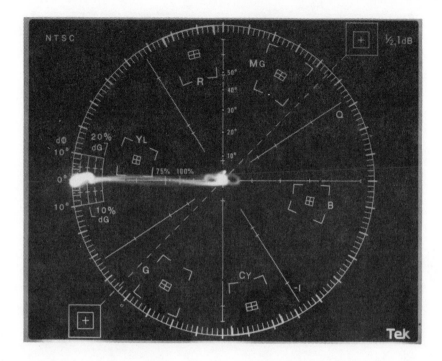

Figure 13.13 Vectorscope display of a 10 percent differential gain error. (Courtesy of Tektronix)

a swept display of demodulated video. This is similar to the single trace R-Y methods for differential phase. The B-Y signal is examined for tilt when the phase is set so that the B-Y signal is at its maximum amplitude. The tilt can be quantified against a vertical scale.

ICPM. Television receivers use a method known as *intercarrier sound* to reproduce audio information. Sound is recovered by beating the audio carrier against the video carrier, producing a 4.5 MHz IF signal, which is demodulated to produce the sound portion of the transmission. From the interaction between audio and video portions of the signal, certain distortions in the video at the transmitter can produce audio buzz at the receiver. Distortions of this type are referred to as *incidental carrier phase modulation,* or ICPM. Stereo audio for television has increased the importance of measuring this parameter at the transmitter, because the buzz is more objectionable in stereo broadcasts. Usually, it is suggested that less than 3° of ICPM be present.

ICPM is measured using a high-quality demodulator with a synchronous detector mode and an oscilloscope operated in a high-gain *x-y* mode. Some waveform

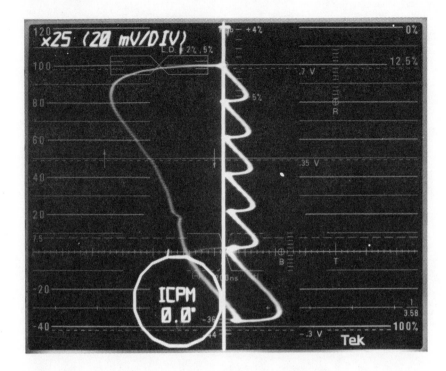

Figure 13.14 Waveform monitor display using the ICPM graticule of a five-level stairstep signal with no distortion. (Courtesy of Tektronix)

and vector monitors have such a mode as well. Video from the demodulator is fed to the y input of the scope, and the quadrature output is fed to the x input terminal. Low-pass filters make the display easier to resolve.

An unmodulated five-step staircase signal produces a polar display, shown in Figure 13.14, on a special graticule developed for this purpose. Notice that the bumps all rest in a straight vertical line, if there is no ICPM in the system. Tilt indicates an error, as shown in Figure 13.15. The graticule is calibrated in degrees per radial division for differential gain settings. Adjustment, but not measurement, can be performed without a graticule.

Tilt and Rounding. Good picture quality requires that the characteristics of the visual transmitter be as linear as possible. The transmitted channel bandwidth (6 MHz for NTSC and even greater for PAL and SECAM) introduces various obstacles to achieving a high degree of linearity. The wideband amplifiers of television must have a flat response over the entire frequency range of interest. The design process considers solutions for the problems of spurious harmonics resulting from stray component parameters, as well as distortions to signals from all sources.

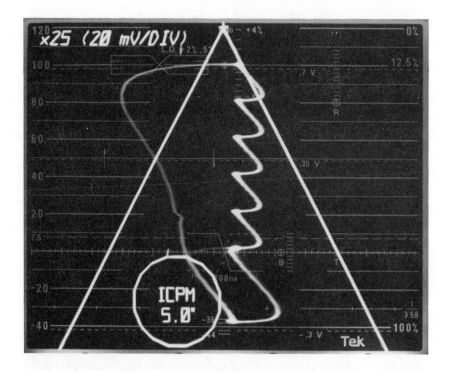

Figure 13.15   Waveform monitor display using the ICPM graticule of a five-level stairstep signal with 5° ICPM distortion. (Courtesy of Tektronix)

Frequency distortion in television systems, ranging from 15 kHz to several hundred kilohertz, is more visible on test equipment if a line-frequency square-wave signal (with a rise time of approximately 200 ns) modulates the visual transmitter. Distortion characteristics of this range generally fall into the category of *rounding* and *tilt*.

It is possible to run tests on a transmitter using a number of specific frequencies, monitoring the response to each at various points in the system. However, a more efficient method involves using square-wave test signals with rise times selected to avoid overshoots in transmission. During monitoring of the signals at various test points, tilt and rounding of the square-wave corners may be observed as impaired frequency response. Two square-wave frequencies prove to be useful for this work. The lower frequency of 60 Hz (50 Hz in PAL) aids in detecting response errors in the frequency range below approximately 15 kHz. A 15 Hz square wave serves to determine response to several hundred kilohertz. These two test signals identify low-frequency response problems when observed on an oscilloscope as a tilt of the flat part of the square wave (top or bottom) and a rounding of the corners.

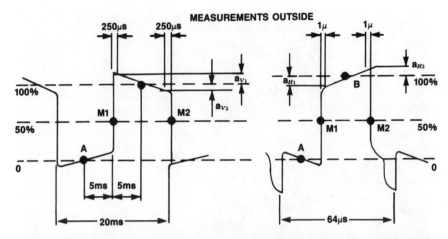

Figure 13.16   Measurement of tilt on a visual signal with 50 Hz and 15 kHz signals. (Source: Carl Bentz, "Inside the Visual PA, Part 2," *Broadcast Engineering* magazine, Intertec Publishing, Overland Park, Kan., November 1988)

Ideal response would produce flat tops and bottoms without any rounding at the corners.

Tilt can be determined for both test signal ranges. The measurement is made on a portion of the square wave derived from a demodulator. Particular start and end points of the measurement are suggested to avoid ringing as a result of pulse rise times. To ensure the best measurements, set the sweep range and vertical sensitivity of the oscilloscope as wide as possible to accommodate the entire length of the trace segment of interest. Tops and bottoms of the square waves should be measured separately. (See Figure 13.16.)

Rounding is determined from the 15 kHz signal. It is realistic to delay the start point and to define a duration along the scope trace for the measurement. Separate observations should be made on rising and falling sides of the square wave, as shown in Figure 13.17.

The effects of tilt and rounding on a video signal in the transmission system can be seen in the demodulated image. Unless it is excessive, tilt is least obvious. It appears partly as a flickering in the picture and partly as a variation in luminance level from top to bottom and/or side to side of the image. Rounding, in this case, manifests itself as a reduction of detail in medium-frequency parts of the image. In cases of excessive tilt or rounding error, it may be difficult for some television receivers to lock on to the degraded signal.

Group Delay. The transients of higher-frequency components (15 kHz to 250 kHz), produce overshoot on the leading edge of transitions and, possibly, ringing at the top and bottom of the waveform area. The shorter duration of the

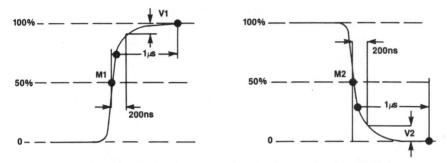

Figure 13.17   Measurement of rounding of a 15 kHz square-wave test signal to identify high-frequency-response problems. (Source: Carl Bentz, "Inside the Visual PA, Part 2," *Broadcast Engineering* magazine, Intertec Publishing, Overland Park, Kan., November 1988)

"flat" top and bottom lowers concern about tilt and rounding. A square wave contains a fundamental frequency and numerous odd harmonics. The number of the harmonics determines the rise or fall times of the pulses. Square waves with rise times of 100 ns (T) and 200 ns (2T) are particularly useful for television measurements. In the 2T pulse, significant harmonics approach 5 MHz; a T pulse includes components approaching 10 MHz. Because the television system should carry as much as information as possible, and its response should be as flat as possible, the T pulse is a common test signal for determination of group delay.

Group delay is the effect of time- and frequency-sensitive circuitry on a range of frequencies. Time, frequency, phase shift, and signal delay all are related and can be determined from circuit component values. Excessive group delay in a video signal appears as a loss of image definition. Group delay is a fact of life that cannot be avoided, but its effect can be reduced through *predistortion* of the video signal. Group delay adjustments can be made before the modulator stage of the transmitter, during monitoring of the signal from a feedline test port.

Group delay may be monitored using a special scope or waveform graticule. The goal in making adjustments is to fit all excursions of the signal between the smallest tolerance markings of the graticule, as illustrated in Figure 13.18. Because quadrature phase errors are caused by the vestigial sideband transmission system, synchronous detection is needed to develop the display.

13.2.6   Automated Video Signal Measurement

Video test instruments based on microcomputer systems provide the maintenance engineer with the capability for rapid measurement of a number of parameters with exceptional accuracy. Automated instruments offer a number of benefits, including

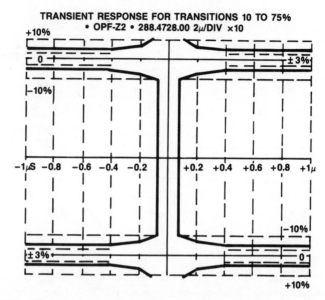

Figure 13.18 Tolerance graticule mask for measuring transient response of a visual transmitter. A complete oscillation is first displayed on the screen. The time base then is expanded, and the signal *x-y* position controls are adjusted to shift the trace into the tolerance mask. (Source: Carl Bentz, "Inside the Visual PA, Part 3," *Broadcast Engineering* magazine, Intertec Publishing, Overland Park, Kan., December 1988)

reduced setup time, test repeatability, waveform storage and transmission capability, and remote control of instrument functions. Typical features include:

- Waveform monitor functions
- Vectorscope monitor functions
- Picture display capability
- Automatic analysis of input signals
- RS-232 and/or GPIB I/O ports

Figure 13.19 shows a block diagram of a representative automated video test instrument. A sample output waveform is shown in Figure 13.20.

13.3 SPECTRUM ANALYZER

An oscilloscope-type instrument displays voltage levels referenced to time, and a spectrum analyzer indicates signal levels referenced to frequency. The frequency components of the signal applied to the input of the analyzer are detected and

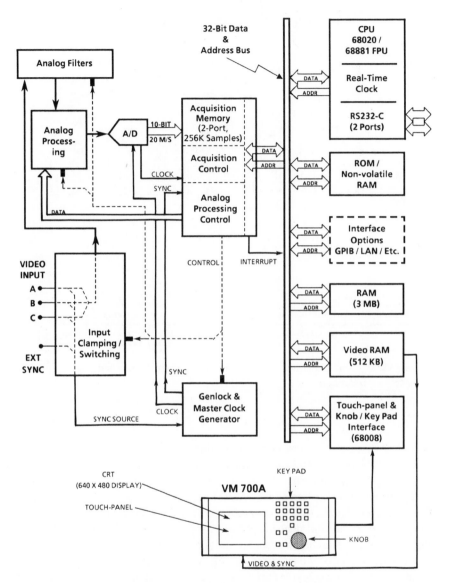

Figure 13.19 Block diagram of an automated video test instrument. (Courtesy of Tektronix)

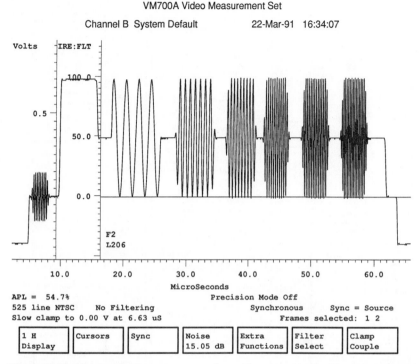

Figure 13.20 Representative output waveform of an automated video test instrument. (Courtesy of Tektronix)

separated for display against a frequency-related time base. Spectrum analyzers are available in a variety of ranges with some models designed for use with audio or video frequencies, and others intended for use with RF frequencies.

The primary application of a spectrum analyzer is the measurement and identification of RF signals. When connected to a small receiving antenna, the analyzer can measure carrier and sideband power levels. Expanding the sweep width of the display makes it possible to observe offset or multiple carriers. By increasing the vertical sensitivity of the analyzer and adjusting the center frequency and sweep width, it is possible to observe the occupied bandwidth of the RF signal. Convention dictates that the vertical axis displays amplitude, and the horizontal axis displays frequency. This frequency-domain presentation allows the user to glean more information about the characteristics of an input signal than would be possible with an oscilloscope. Figure 13.21 compares the oscilloscope and spectrum analyzer display formats.

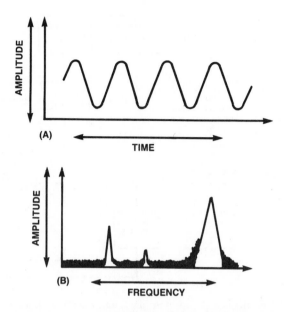

Figure 13.21   Comparison of waveform displays: (a) Oscilloscope. (b) Spectrum analyzer. (Data from: Harold Kinley, "Using Service Monitor/Spectrum Analyzer Combos," *Mobile Radio Technology* magazine, Intertec Publishing, Overland Park, Kan., July 1987)

13.3.1   Principles of Operation

A spectrum analyzer intended for use at RF frequencies is shown in block diagram form in Figure 13.22. The instrument includes a superheterodyne receiver with a swept-tuned local oscillator (LO) that feeds a CRT display. The tuning control determines the center frequency (F_c) of the spectrum analyzer, and the *scan-width* selector determines how much of the frequency spectrum around the center frequency will be covered. Full-feature spectrum analyzers also provide front-panel controls for scan-rate selection and bandpass filter selection. Key specifications for a spectrum analyzer include:

- *Resolution:* The frequency separation required between two signals so that they may be resolved into two distinct and separate displays on the CRT screen. Resolution usually is specified for equal-level signals. When two signals differ greatly in amplitude and are close together in frequency, greater resolution is required to separate them on the display.
- *Scan width:* The amount of frequency spectrum that may be scanned and shown on the CRT display. Scan width usually is stated in kilohertz or

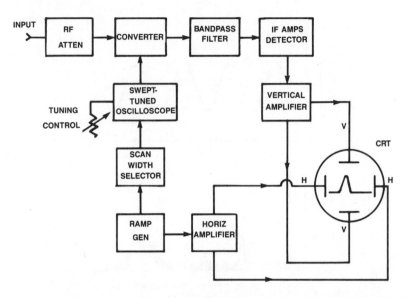

Figure 13.22 Block diagram of a spectrum analyzer. (Data from: Harold Kinley, "Using Service Monitor/Spectrum Analyzer Combos," *Mobile Radio Technology* magazine, Intertec Publishing, Overland Park, Kan., July 1987)

megahertz per division. The minimum scan width available usually is equal to the resolution of the instrument.

- *Dynamic range:* The maximum amplitude difference that two signals may have and still be viewed on the CRT display. Dynamic range usually is stated in decibels.
- *Sensitivity:* The minimum signal level required to produce a usable display on the CRT screen. If low-level signal tracing is planned, as in receiver or off-air monitoring, the sensitivity of the spectrum analyzer is important.

When using the spectrum analyzer, take care not to overload the front end with a strong input signal. Overloading can cause "false" signals to appear on the display. These false signals are the result of nonlinear mixing in the front end of the instrument. False signals may be identified by changing the RF attenuator setting to a higher level. The amplitude of false signals (caused by overloading) will drop much more than the amount of increased attenuation.

The spectrum analyzer is useful in troubleshooting receivers as well as transmitters. As a tuned signal tracer, it is well-adapted to stage-gain measurements and other tests. There is one serious drawback, however. The 50 Ω spectrum analyzer input can load many receiver circuits too heavily, especially high-impedance

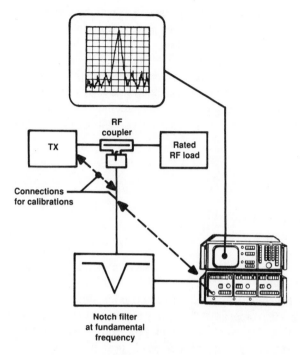

Figure 13.23   Typical test setup for measuring the harmonic and spurious output of a transmitter. The notch filter is used to remove the fundamental frequency to prevent overdriving the spectrum analyzer input. (Data from: Carl Pepple, "How to Use a Spectrum Analyzer at the Cell Site," *Cellular Business* magazine, Intertec Publishing, Overland Park, Kan., March 1989)

circuits such as FET amplifiers. Isolation probes are available to overcome loading problems. Such probes, however, also attenuate the input signal, and unless the spectrum analyzer has enough reserve gain to overcome the loss caused by the isolation probe, the instrument will fail to provide useful readings. Isolation probes with 20 dB to 40 dB of attenuation are typical. As a rule of thumb, probe impedance should be at least 10 times the impedance of the circuit to which it is connected.

13.3.2   Applications

The primary application for a spectrum analyzer centers on measuring the occupied bandwidth of an input signal. Harmonics and spurious signals may be checked and potential causes investigated. Figure 13.23 shows a typical test setup for making transmitter measurements.

The spectrum analyzer also is well-suited to making accurate transmitter FM deviation measurements. This is accomplished using the *Bessel null* method, a mathematical function that describes the relationship between spectral lines in frequency modulation. The Bessel null technique is highly accurate; it forms the basis for modulation monitor calibration. The concept behind the Bessel null method is to drive the carrier spectral line to zero by changing the modulating frequency. When the carrier amplitude is at zero, the modulation index is given by a Bessel function. Deviation may be calculated by using the following equation:

$$\Delta F_c = MI \; F_m$$

where: ΔF_c = deviation frequency

MI = modulation index
F_m = modulating frequency

The carrier frequency "disappears" at the Bessel null point, with all power remaining in the FM sidebands.

A tracking generator may be used in conjunction with the spectrum analyzer to check the dynamic response of frequency-sensitive devices, such as transmitter isolators, cavities, ring combiners, duplexers, and antenna systems. A tracking generator is a frequency source that is locked in step with the spectrum analyzer horizontal trace rate. The resulting display shows the relationship of the amplitude vs. frequency response of the device under test. The spectrum analyzer also may be used to perform gain-stage measurements. The combination of a spectrum analyzer and a tracking generator makes filter passband measurements possible. As measurements are made along the IF chain of a receiver, the filter passbands become increasingly narrow, as illustrated in Figure 13.24.

On-Air Measurements. The spectrum analyzer is used for three primary on-air tests:

- Measuring unknown signal frequencies. A spectrum analyzer may be coupled to the output of a transmitter to determine its exact operating frequency.
- Intermod and interference signal tracking. A directional yagi antenna may be used to identify the source of an interfering signal. If the interference is on-frequency, but carries little intelligence, chances are good that it is an intermod being produced by another transmitter. Use the wide-dispersion display mode of the analyzer, and note signal spikes that appear simultaneously with the interference. A troubleshooter, armed with a spectrum analyzer, a directional antenna, and the knowledge of how intermod signals are generated, usually can locate a suspected transmitter rapidly. Each suspected unit then can be tested individually by inserting an isolator between

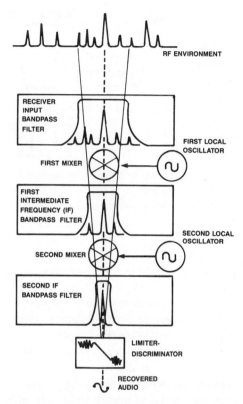

Figure 13.24 Using a spectrum analyzer to measure the tunnel effect of bandpass filters in a receiver. (Data from: Carl Pepple, "How to Use a Spectrum Analyzer at the Cell Site," *Cellular Business* magazine, Intertec Publishing, Overland Park, Kan., March 1989)

the transmitter PA and duplexer (or between the PA and antenna). When the intermod signal amplitude drops the equivalent of the reverse insert loss of the isolator, the offending transmitter has been located.

- Field-strength measurements. With an external antenna, a spectrum analyzer can be used for remote field-strength measurements. Omnidirectional, broadband antennas are the most versatile because they allow several frequency bands to be checked simultaneously. Obviously, the greater the external antenna height, the greater the testing range.

Spurious Harmonic Distortion. An incorrectly tuned or malfunctioning transmitter may produce spurious harmonics. The spectrum analyzer provides the best way to check for spectral purity. Couple a sample of the transmitter output signal to the analyzer input, either by loop coupling or RF sampling, as illustrated in

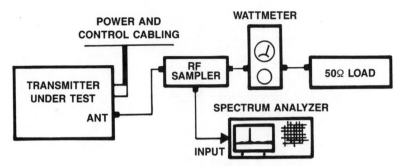

Figure 13.25 Common test setup to measure transmitter harmonic distortion with a spectrum analyzer. (Source: Richard J. Wolf, "Spectrum Analyzer Uses for Two-Way Technicians," *Mobile Radio Technology* magazine, Intertec Publishing, Overland Park, Kan., July 1987)

Figure 13.25. For low power levels from portable and mobile units, transmit into a dummy load, and use a flexible rubber antenna on the analyzer input. For maximum accuracy when measuring larger amounts of power, use an RF sampler to control the input level to the analyzer. Use maximum RF attenuation initially on the analyzer front end to prevent overload damage and internal intermod. False signals on the display also may be observed if covers or shields are not in place on the unit under test. The oscillator, doubler, or tripper levels may radiate sufficient signals to register on the analyzer display.

Key the transmitter, and observe any spurious harmonics on the analyzer CRT. After centering the main signal of interest and adjusting the input attenuation for maximum display amplitude, calculate the spurious radiation. Spurious signal attenuation is measured in decibels, referenced to the amplitude of the transmitter fundamental. When a radio transmitter has a harmonic distortion problem, the defective stage usually can be isolated by tuning each stage and observing the spurious harmonics on the analyzer display. When the defective stage is tuned, the harmonics either shift frequency or change in amplitude. Signal tracing, with a probe and heavy input attenuation, also helps to determine where the distortion first occurs. Overdriven stages are prime culprits.

Selective-Tuned Filter Alignment. The high input sensitivity and visual, frequency-selective display of the spectrum analyzer provide the technician with an efficient analog tuning instrument. Injection circuits can be peaked quickly by loop-coupling the analyzer input to the mixer section of a receiver. This is most helpful in tuning older radios when a service manual is not available. Radios that do not have test points or that require elaborate, specialized test sets can be tuned in a similar manner. Portable radio transmitters can be tuned directly into their flexible rubber antennas for maximum signal strength. This reveals problems such as improper signal transfer to the antenna or a defective antenna.

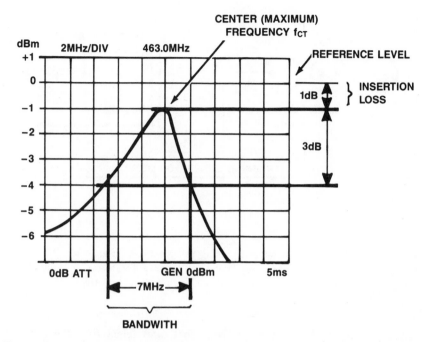

Figure 13.26 Typical spectrum analyzer display of a single-cavity filter. Bandwidth (BW) = 458 MHz to 465 MHz = 7 MHz. Filter quality factor (Q) = f_{CT}/BW = 463 MHz/7 MHz = 66. The trace shows 1 dB insertion loss. (Data from: Richard J. Wolf, "Spectrum Analyzer Uses for Two-Way Technicians," *Mobile Radio Technology* magazine, Intertec Publishing, Overland Park, Kan., July 1987)

Cavities, combiners, and duplexers all are selective-tuned filters. They are composed of three types of passive filters:

- Bandpass
- Band-reject
- Combination pass/reject

The spectrum analyzer is well-suited to tuning these types of filters, particularly duplexers. Couple the cavity between the tracking generator and the spectrum analyzer. Calibrate the tuning frequency in the center of the CRT display, and adjust the generator output and analyzer input level for an optimal display trace. Then tune the cavity onto frequency. Measure the various filter characteristics, and compare them with the manufacturer's specifications. Figure 13.26 shows a typical spectrum analyzer display of a single-cavity device. Important filter characteristics

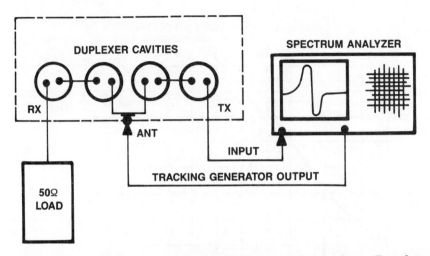

Figure 13.27 Test setup for duplexer tuning using a spectrum analyzer. (Data from: Richard J. Wolf, "Spectrum Analyzer Uses for Two-Way Technicians," *Mobile Radio Technology* magazine, Intertec Publishing, Overland Park, Kan., July 1987)

include frequency, bandwidth, insertion loss, and selectivity. Measure the bandwidth at the points 3 dB down from maximum amplitude. Insertion loss is equal to the cavity's attenuation of the center-tune frequency. Calculate filter Q (quality factor), which is directly proportional to selectivity, by dividing the filter center frequency by the bandwidth of the filter.

Duplexer tuning is similar to cavity tuning, but is complicated by interaction between the multiple cavities. Using the test setup shown in Figure 13.27, couple the tracking generator output to the duplexer antenna input. Couple the receive and transmit ports alternately to the spectrum analyzer while tuning the pass or reject performance of each cavity. When adjusting a combination cavity, tune the reject last because it will tend to follow the pass tuning. Alternate tuning between the receive and transmit sides of the duplexer until no further improvement can be gained. To assure proper alignment, terminate all open ports into a 50 Ω load while tuning.

The duplexer insertion loss of each port can be measured as the difference between a reference amplitude and the pass frequency amplitude. The reference signal is measured by short-circuiting the generator output cable to the analyzer input cable. This reference level nulls any cable losses to prevent them from being included in duplexer insertion loss measurements.

Small-Signal Troubleshooting. The spectrum analyzer is well-suited for use in small-signal RF troubleshooting. Stage gain, injection level, and signal loss can be checked easily.

Receiver sensitivity loss of less than 6 dB usually is a difficult problem to trace with most test instruments. Such losses usually are associated with the front end of the radio, either as a stage gain or injection amplifier deficiency. Before the spectrum analyzer was commonly available, there was little to aid a technician in tracing such a problem. An RF millivolt meter, with typical usable sensitivity of 1 mV, was of little help in troubleshooting such microvolt-level signals. Oscilloscopes are restricted in the same way, as well as having inadequate bandwidth for the high frequencies used in the front-end circuits of most receivers. A spectrum analyzer, with a typical 1 V sensitivity and broad RF bandwidth, is well-adapted for RF troubleshooting. Microvolt-level signals can be traced accurately, allowing signal losses to be examined at each individual stage.

To conduct measurements, inject a test signal generated by a communications monitor into the antenna input of the radio under test. Use an appropriate probe to trace the signal from stage to stage. Stage gain, mixer output, injection level, and filter loss can be checked from the antenna to the discriminator. Isolate the problem by comparing measurements taken with those of a correctly functioning unit. A standard test signal level of 100 V usually works well. It is large enough to allow testing of the RF amplifier without causing overload. If the radio includes an AGC circuit, disable it temporarily during testing to permit measurement of the true stage gain.

When tracing a sensitivity loss, start with the RF mixer. The mixer provides a junction point for narrowing the direction of troubleshooting. If the output of the mixer is correct, the defective stage lies beyond the mixer, probably in the IF amplifier. If the mixer output is low, check the input signal to the mixer. Check the RF amplifier, input cavities, and receive/transmit (RX/TX) switching circuit (if used). If the mixer IF carrier injection is low, check the multipliers and oscillators along that path.

The bandwidth of IF filters can be measured in-circuit by varying the receiver input frequency and measuring the bandwidth of the 3 dB rolloff points. A crystal bandpass filter can be tested for flaws by injecting an overdeviated signal into the receiver. A cracked filter crystal will produce a sharp, deep notch on the analyzer display. The key to comparative troubleshooting is accurate recording of test signal levels of a unit known to be working correctly before troubleshooting.

Remember to protect the front end of the analyzer from dc voltages and overload when tracing signals with a probe. When testing unknown signal levels, use maximum input attenuation and external attenuators as needed to avoid analyzer damage. Isolate the analyzer input from dc voltages by using a capacitively coupled probe.

13.4 NETWORK ANALYZER[2]

Antenna and transmission-line performance measurements are among the most neglected and least understood parameters at most transmission facilities. Many facilities do not have the equipment to perform useful measurements. Experience is essential, because much of the knowledge obtained from such tests is derived by interpreting the raw data. In general, transmission systems measurements should be made:

- Before and during installation of the antenna and transmission line. Barring unforeseen operational problems, this will be the only time that the antenna is at ground level. Ready access to the antenna allows the engineer to perform a variety of key measurements without climbing the tower.
- During system troubleshooting when attempting to locate a problem. Following installation, these measurements usually concern the transmission line itself.
- To ensure that the transmission line is operating normally. Many facilities check the transmission line and antenna system on a regular basis. A quick sweep of the line with a network analyzer and a time-domain reflectometer may disclose developing problems before they can cause a transmission-line failure.

Ideally, the measurements should be used to confirm a good impedance match, which can be interpreted as minimum VSWR or maximum *return loss*. Return loss is related to the level of signal that is returned to the input connector after the signal has been applied to the transmission line and reflected from the load. A line perfectly matched to the load would transfer all energy to the load. No energy would be returned, resulting in an infinite return loss, or an ideal VSWR of 1:1. The benefits of matching the transmission-line system for minimum VSWR include:

- Most efficient power transfer from the transmitter to the antenna system.
- Best performance with regard to overall bandwidth.
- Improved transmitter stability with tuning following accepted procedures more closely.
- Minimum transmitted signal distortions.

---

2 Portions of Sec. 13.4 were adapted from J. Whitaker and D. Markley, "Test Equipment for RF Troubleshooting," *Broadcast Engineering* magazine, Overland Park, Kan., November 1990.

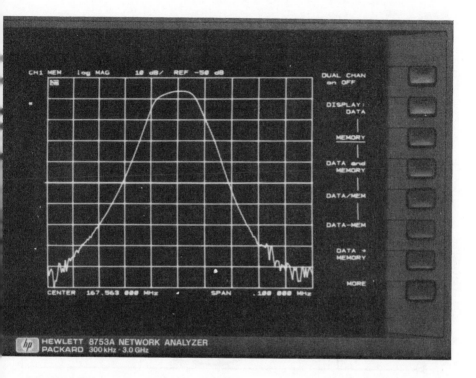

Figure 13.28   Conventional Cartesian display of swept frequency vs. signal level on a network analyzer. (Courtesy of Hewlett-Packard)

The network analyzer allows the maintenance engineer to perform a number of critical measurements in a short period of time. The result is an antenna system that is tuned as closely as practical for uniform impedance across the operating bandwidth. A well-matched system increases operating efficiency by properly coupling the signal from the transmitter to the antenna. Figure 13.28 shows a typical Cartesian display of antenna system performance. Figure 13.29 shows an actual network analyzer plot of an FM broadcast antenna.

13.4.1   Measuring VSWR

Historically, a *slotted line* device was used to measure VSWR on a transmission line. A slotted line includes a probe that penetrates the outer conductor of the line through a slot. The probe, in close proximity with the inner conductor, measures the voltage or samples the field along the center conductor. The sample is detected, which results in a voltage proportional to the actual signal on the center conductor.

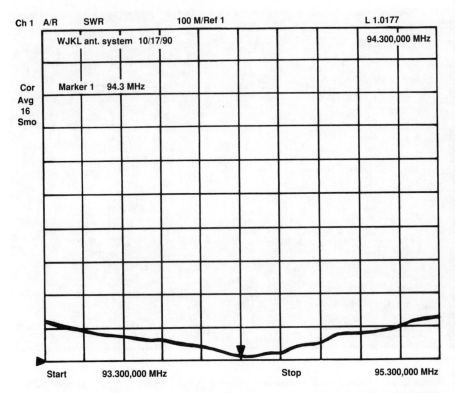

Figure 13.29   Network analyzer plot of an FM broadcast antenna operating at 94.3 MHz. (Source: Jerry Whitaker and Don Markley, "Test Equipment for RF Systems," *Broadcast Engineering* magazine, Intertec Publishing, Overland Park, Kan., November 1990)

It is an accurate, reliable instrument. However, the slotted line procedure takes a considerable amount of time to sweep a transmission line accurately over a wide bandwidth, and then to plot the resulting data.

A network analyzer incorporating a return loss bridge performs antenna measurements more quickly. The analog network analyzer consists of a sweep generator coupled to a tracking receiver. A sample of the signal applied to the transmission line is compared with the return signal through a return loss bridge or directional coupler. By adding a storage or *normalizer* device to store the signal digitally, the instrument can provide a stable display while sweeping the line at low speed to find all irregularities that may exist at discrete frequencies.

Rather than a sweep generator, digital designs use an integral synthesizer. In this way, the return loss is measured at discrete frequencies. Software-calibration procedures correct each measurement for system frequency- and phase-response errors, delay irregularities, and directivity errors in the return loss bridge or directional coupler. If a software-controlled unit is calibrated at the top of the transmission line, measurements will accurately show antenna characteristics without effects of the transmission line. Results are plotted on an x-y plotter or defined and stored for later printout.

One particularly desirable feature of a network analyzer is its capability to display either a *Smith chart* (discussed in Sec. 13.5) or a more simple Cartesian x-y presentation of return loss vs. frequency. (Some units may provide both displays simultaneously.) The Smith chart is useful, but interpretation can be confusing. The Cartesian presentation usually is easier to interpret, but is not technically better.

13.4.2 Calibration

Calibration methods vary for different instruments. For one method, a short circuit is placed across the network analyzer terminals, producing a return loss of zero (the short circuit reflects all signals applied to it). The instrument then is checked with a known termination. This step often causes the inexperienced technician to go astray. The termination should have known characteristics and full documentation. It is acceptable procedure to check the equipment by examining more than one termination, where the operator knows the characteristics of the devices used. Significant changes from the known characteristics suggest that additional tests should be performed. After the test unit is operating correctly, check to ensure that the adapters and connectors to be used in the measurement do not introduce errors of their own. An accepted practice for this involves the use of a piece of transmission line of known quality. A 20 ft section of line should sufficiently separate the input and output connectors. The results of any adjustment at either end will be noticeable on the analyzer. Also, the length allows adjustments to be made fairly easily. The section of line used should include tuning stubs or tuners to permit the connectors to be matched to the line across the operating channel.

Next, the station's dummy load must be matched to the transmission line. Do not assume that the dummy load is an appropriate termination by itself, or a station reference. The primary function of a dummy load is to dissipate power in a manner that allows easy measurement. It is neither a calibration standard nor a reference. Experience proves it is necessary to match dummy and transmission-line sections to maintain a good reference. The load is matched by looking into the transmission line at the patch panel (or other appropriate point). Measurements then are taken at locations progressively closer to the transmitter, until the last measurement is made at the output connection of the transmitter. After the dummy load is checked, it serves as a termination.

13.4.3 Antenna Measurements

An antenna should be tuned properly before it is placed into service. Any minor tuning adjustments to the antenna should be made at its base, not by compensation at the input to the transmission line. Impedance adjustments typically are made with tuning rings on the center conductor or with an impedance-matching section. Adjustments are performed while observing the return loss on the network analyzer. Transmission-line rings are less convenient, but less expensive than an impedance-matching section. The rings can be used for matching short runs or the overall line between the transmitting equipment and the antenna. Either tuning method may be used at the antenna.

Both tuning systems operate by introducing a discontinuity into the transmission line. The ring effectively changes the diameter of the center conductor, causing an impedance change at that point on the line. This introduces a reflection into the line, the magnitude of which is a function of the size of the ring. The phase of the reflection is a function of the location of the ring along the length of the center conductor.

Installing the ring usually is a cut-and-try process. It may be necessary to open, adjust, close, and test the line several times. However, after a few cuts, the effect of the ring will become apparent. It is not uncommon to need more than one ring on a given piece of transmission line for a good match over the required bandwidth. When a match is obtained, the ring normally is soldered into place.

Impedance-matching hardware also is available for use with waveguide. A piece of material is placed into the waveguide, and its location is adjusted to create the desired mismatch. For any type of line, the goal is to create a mismatch equal in magnitude, but opposite in phase, to the existing undesirable mismatch. The overall result is a minimum mismatch and minimum VSWR.

A tuner alters the line characteristic impedance at a given point by changing the distance between the center and outer conductors by effectively moving the outer conductor. In reality, it increases the capacity between the center and outer conductors to produce a change in the impedance, and introduce a reflection at that point.

13.5 THE SMITH CHART

The Smith chart is acknowledged to be the most universal tool for RF design work. RF problems can be represented by complicated mathematics, such as hyperbolic or quadratic functions, or by polynomial functions and differential equations. These equations involve complex numbers, and they must be solved for each specific frequency at each characteristic impedance, a time-consuming and error-prone process. The Smith chart simplifies these equations into a chart format. Several complications must be overcome to achieve a general solution for RF design problems. First is the variation of basic parameters caused by frequency. Each frequency reacts differently to a given impedance. Second is the wide range of possible fundamental impedances. For example, no two antennas of different design have exactly the same characteristic impedance, although most fall into the range of 30 Ω to 120 Ω. The Smith chart provides a method to "normalize" the frequency and characteristic impedance of an RF problem to permit solution on a standardized chart. Two quantities must be determined before a Smith chart is used: *normalized impedance* and *propagation constant*. In addition, the expected or measured characteristic impedance of at least one part of the RF circuit must be defined.

13.5.1 Normalized Impedance

Whenever an RF signal enters a transmission line or RF amplifier, there is a brief time during which energy flows only forward into the line (or circuit). When the RF wave has traveled to the end or termination of the transmission line, any mismatch, no matter how small, will cause some energy to be reflected. During the brief time when RF energy is only flowing forward into the circuit, the signal "sees" the characteristic impedance (ideal impedance) of the transmission-line circuit. For transmission lines, this number is specified by the manufacturer over a range of frequencies. Four properties are used to define the characteristic impedance:

1. The pure wire resistance of the transmission line per linear unit (R).
2. The inductive reactance of the wire at the frequency of interest (L).
3. The capacitive susceptance of the wire/shield combination at the frequency of interest (C).
4. The conductance per unit length of the transmission line (G).

For applications in which the length of the transmission line is short, the resistance of the wire is negligible, and so is the conductance. Even for long lengths of transmission line, such as antenna feedlines, the resistance can be factored into the solution after the rest of the graphical work has been done. Thus, the general equation for characteristic impedance (Z) of a typical transmission line is:

$$Z = \sqrt{(L/C)}$$

Because the inductive reactance and capacitive susceptance are calculated at the frequency of circuit operation, the frequency-related terms drop out, simplifying the process. For RF applications, the frequency-dependent values of inductance and capacitance of a transmission line result from:

- Physical dimensions of the cable, such as diameter and thickness of the inner conductor.
- Distance between the inner conductor and the outer shield.
- Dielectric constant between the inner and outer conductors.

For waveguide, the physical dimensions are the width and breadth of a cross section of the guide expressed in fractions of a wavelength of the operating frequency. Note that these parameters relate to operation at a single frequency. Modulated signals present further complications.

13.5.2   Propagation Constant

The propagation constant of a transmission line is defined as the natural logarithm of the ratio between the input current and output current for the forward-traveling portion of the wave (assuming that the length of the transmission line is exactly 1 wavelength at the frequency of interest). The propagation constant is a complex number, with the *real* portion representing the resistive attenuation of the transmission line, and the *complex* portion representing the phase constant. The phase constant includes both the wavelength within the transmission line and the velocity of propagation. The characteristic impedance and propagation constant of a transmission line usually are provided by the manufacturer. They represent information that defines the operating efficiency of the transmission line in and of itself, disregarding the ultimate circuit application.

Although the characteristic impedance and propagation constant are given by the manufacturer, and define the ideal performance of forward-traveling waves in a unit length of transmission line, the *normalized characteristic impedance* provides the actual specifics that result from using the transmission line in an RF application. Connecting a generator and a receiver to the transmission line introduces several circuit mismatches. The generator does not have much impact on the problem, as long as the output impedance is similar to the characteristic impedance of the selected transmission line. The receiver, or load that is placed on the end of the transmission line, and the length of the transmission line in wavelengths exert a much greater influence on the circuit. For example, if the receiver is a totally open circuit, Kirchoff's Law can be applied to determine that almost 100 percent of the energy injected into the transmission line will be reflected. The Smith chart can be used to identify the best way to provide maximum energy transfer from the transmission line into the load. In addition to transmission-line problems, Smith chart techniques apply to RF amplifiers, attenuators, directional couplers, antennas, antenna-matching networks, and signal-distribution networks.

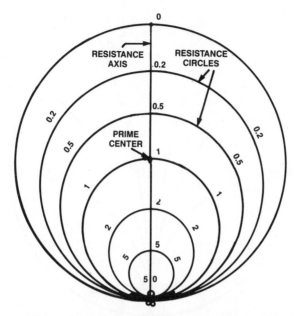

Figure 13.30   Resistance circles of the Smith chart coordinate system. (Source: Gerry Kaufhold, "The Smith Chart, Part 3," *Broadcast Engineering* magazine, Intertec Publishing, Overland Park, Kan., February 1990)

13.5.3   Applying the Smith Chart

The Smith chart is built around two families of curves. The first is the *resistance circle*, as illustrated in Figure 13.30. Resistance circles are centered on the only straight line on the chart, the vertical line (called the *resistance axis*). Notice that the concentric circles are given values starting with zero at the top, and ending with infinity at the bottom. The circles represent ratios of whatever value is assigned to the center point, called the *prime center*, which has a value of 1.0. In operation, the prime center usually is set equal to the characteristic impedance of the transmission line, and the ratios of each circle serve as multipliers. This normalizes the chart for any value assigned the prime center. All points along a resistance circle have the same resistance.

The second family of curves, the *reactance curves*, is shown in Figure 13.31. Only segments of these curves are plotted. As with the resistance family, these curves are labeled with normalizing multipliers, with respect to prime center. Also,

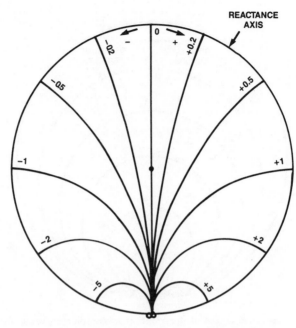

Figure 13.31   Reactance curves of the Smith chart coordinate system. (Source: Gerry Kaufhold, "The Smith Chart, Part 3," *Broadcast Engineering* magazine, Intertec Publishing, Overland Park, Kan., February 1990)

all points along any one curve have the same reactance value. Combining the two curves yields a precursor to the Smith chart, illustrated in Figure 13.32.

One other family of useful circles, usually not printed on the chart, is plotted with a compass, as needed. These are the standing wave ratio (SWR) circles. They occur as concentric circles centered on prime center, as shown in Figure 13.33. In typical use, Smith charts are drawn with the resistance component line horizontal. The SWR and attenuation scales are *radial* scales. They are used by aligning a straightedge at right angles to the scale of interest, finding the desired value, and tracing a line up the Smith chart. The bottom transmission coefficient scale is a *magnitude* scale, which must be applied differently. A compass or divider is used to transfer linear distances from portions of the Smith chart.

The Smith chart can be used to efficiently design a filter and/or impedance-matching network. The technician can solve each part of each problem by using the chart, taking advantage of the graphical nature of the tool to obtain possible solutions quickly. Because power components for RF work come in limited size and value ranges, impedance-matching problems often must be reworked to obtain

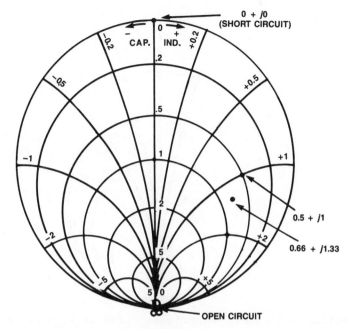

Figure 13.32 Complete Smith chart coordinate system, with sample values plotted. (Source: Gerry Kaufhold, "The Smith Chart, Part 3," *Broadcast Engineering* magazine, Intertec Publishing, Overland Park, Kan., February 1990)

final component values that can be realized with off-the-shelf parts. The flexibility of the Smith chart in solving a complex problem is a great help in this respect. When using a Smith chart, perform the following steps (in sequence):

- Write down the pertinent actual resistance and reactance values.
- Choose a convenient denominator that brings the normalized resistance of the characteristic impedance close to 1.0. Normalize all other values by dividing by this number. Note which values are impedances for series circuits and which are admittances for shunt circuits.
- Plot the normalized values onto the Smith chart.
- Resolve the problem by applying the rules of the Smith chart.
- Denormalize circuit values by multiplying.
- Solve the reactance equations at the frequency of interest to obtain correct final component values for the circuit.

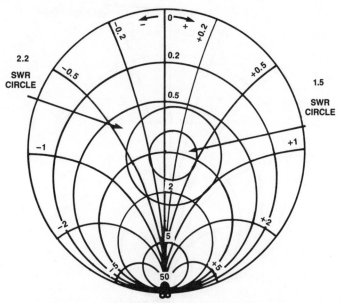

Figure 13.33 SWR circles drawn on a Smith chart. (Source: Gerry Kaufhold, "The Smith Chart, Part 3," *Broadcast Engineering* magazine, Intertec Publishing, Overland Park, Kan., February 1990)

13.6 BIBLIOGRAPHY

Bentz, Carl: "Inside the Visual PA, Part 1," *Broadcast Engineering* magazine, Intertec Publishing, Overland Park, Kan., October 1988.

_____: "Inside the Visual PA, Part 2," *Broadcast Engineering* magazine, Intertec Publishing, Overland Park, Kan., November 1988.

_____: "Inside the Visual PA, Part 3," *Broadcast Engineering* magazine, Intertec Publishing, Overland Park, Kan., December 1988.

Kaufhold, Gerry: "The Smith Chart, Part 1," *Broadcast Engineering* magazine, Intertec Publishing, Overland Park, Kan., November 1989.

_____: "The Smith Chart, Part 2," *Broadcast Engineering* magazine, Intertec Publishing, Overland Park, Kan., January 1990.

_____: "The Smith Chart, Part 3," *Broadcast Engineering* magazine, Intertec Publishing, Overland Park, Kan., February 1990.

_____: "The Smith Chart, Part 4," *Broadcast Engineering* magazine, Intertec Publishing, Overland Park, Kan., March 1990.

Kinley, Harold: "Using Service Monitor/Spectrum Analyzer Combos," *Mobile Radio Technology* magazine, Intertec Publishing, Overland Park, Kan., July 1987.

Lewis, John: "A New Era in Television Test and Measurement," *SMPTE Journal*, SMPTE, White Plains, NY, November 1988.

Pepple, Carl: "How to Use a Spectrum Analyzer at the Cell Site," *Cellular Business* magazine, Intertec Publishing, Overland Park, Kan., March 1989.

Whitaker, Jerry, and Don Markley: "Test Equipment for RF Systems," *Broadcast Engineering* magazine, Intertec Publishing, Overland Park, Kan., November 1990.

Wolf, Richard J.: "Spectrum Analyzer Uses for Two-Way Technicians," *Mobile Radio Technology* magazine, Intertec Publishing, Overland Park, Kan., July 1987.

Technical staff: "Rigs and Recipes: How to Measure and Monitor," Rohde & Schwarz application note, Rohde & Schwarz.

14

TRANSMITTER PERFORMANCE CHARACTERISTICS

14.1 INTRODUCTION

Many shortcomings in the received audio, video, and/or datastream of a transmission resulting from noise, reflections, and terrain limitations cannot be controlled by the maintenance engineer. However, other problems can be prevented through proper equipment adjustment. The critical performance parameters for a transmission system include:

- Output power.
- Transmission line/antenna VSWR.
- Frequency stability/accuracy.
- Occupied bandwidth.
- Distortion mechanisms, including total harmonic distortion (THD), intercarrier phase distortion (ICPM), and quadrature distortion.
- Differential phase and gain.
- Frequency-response errors.
- Envelope delay.

14.1.1 Maintaining RF System Performance

The preventive maintenance schedule required to maintain a satisfactory level of performance depends upon a number of factors, including:

- Operating environment
- Equipment quality

- Age of the equipment
- Type of service being performed

A base station transmitter in an air-conditioned building usually will have fewer problems than one subject to wide ranges of temperature and humidity. Abrupt temperature changes can stress electronic components and cause premature failure. Older equipment requires more frequent testing than newer equipment.

DC input power to the final RF output stage is a function of the applied plate or collector voltage and the plate or collector current of the final tubes or transistors. Never allow power to exceed final stage voltage and current ratings. Test points or built-in meters usually are provided for checking the dc voltage and current levels of the final amplifier.

Carrier frequency should be measured during initial installation, and periodically thereafter. An off-frequency transmitter degrades system performance and may cause interference to other services on neighboring channels. A service monitor or a frequency counter can be used to measure transmitter frequency without a direct physical connection. Any frequency-measuring instrument must be calibrated properly. Accuracy should be greater than half the frequency tolerance required in the particular radio service. For example, if the rules of a given radio service require that the transmitter frequency be held within 0.0005 percent, then the frequency-measuring instrument should have an accuracy of at least 0.00025 percent. Check the instrument against the short-wave frequency standard station WWV at regular intervals to maintain proper calibration.

For transmitters employing a phase-locked loop (PLL) frequency synthesizer configured to operate on many different frequencies, it usually is sufficient to measure the frequency at three points: the lowest, middle, and highest operating frequencies. If an adjustment is necessary, make it at the middle frequency. This three-point measurement procedure has one potential flaw: It does not catch errors caused by incorrect programming. Such faults can result from defective switch contacts, trouble on programming lines, defective PROMs or EPROMs, and defective programmable frequency dividers. To detect such defects, check each channel individually.

Deviation. Transmitter deviation or modulation must be set properly. A deviation level set too high or too low adversely affects radio range. If the deviation is set too low, the signal-to-noise ratio of the overall system is diminished. If the deviation is set too high, the transmitter signal may exceed the receiver's modulation acceptance bandwidth, causing severe distortion on modulation peaks and out-of-band spurious radiation. Although dedicated deviation meters are available, most technicians use a service monitor to set deviation. The service monitor may have an analog meter or a digital display for the deviation measurement. Almost all service monitors employ an oscilloscope with a display screen calibrated to indicate the deviation level. Compared with an analog meter or digital display, a

CRT graphic display provides much more information about system performance. For example, a CRT can display the degree of clipping, the peak deviation level, and deviation symmetry. Most dedicated FM deviation meters also provide a *scope* output connection for use with a conventional oscilloscope to view the modulation waveform.

The difference between the positive and negative excursions of the deviated waveform is defined as *modulation dissymmetry*. Theoretically, positive and negative excursions should be perfectly symmetrical. In practice, however, there is always some dissymmetry. In a multichannel phase-modulated transmitter, modulation dissymmetry may be worse on some channels than others as a result of modulator tuning. If the modulator tuning is peaked on one channel, response on another channel may not be as good. Modulator tuning may have to be compromised to achieve a balanced response on all channels.

14.2 VACUUM TUBE OPERATING PARAMETERS[1]

As discussed briefly in Sec. 10.4.4, each type of transmitter will tune up in its own way. Still, generalizations can be made that apply to most systems, depending on the basic stage design.

14.2.1 Stage Tuning

In a typical transformer-coupled audio amplifier, the plate-to-plate load impedance required is given in the technical data sheet for the tube type under consideration, or it can be calculated relatively easily. The secondary load impedance normally is defined by the application. It only remains for the design engineer to specify the turns ratio of the transformer. After the output transformer and secondary load are optimized, the proper excitation is determined by the plate current. If a means is available to measure the grid voltage swing, this can also be used to indicate proper excitation.

When optimizing a tetrode or pentode RF amplifier for proper excitation and loading, the procedure is different, depending upon whether the screen voltage is taken from a fixed supply or a dropping resistor supply with poor regulation. If both the screen supply and grid bias are from fixed sources with good regulation, the plate current is almost entirely controlled by RF excitation. The loading then is varied until the maximum power output is obtained. Following these adjustments, the excitation is trimmed, along with the loading, until the desired control and

---

1 Section 14.2 was adapted with permission from: "The Care and Feeding of Power Grid Tubes," Varian Associates, San Carlos, Calif., 1982.

screen grid currents are obtained. In the case of an RF amplifier where both the screen and grid bias are taken from sources with poor regulation, the stage will tune very much like a triode RF power amplifier. The plate current will be controlled principally by varying the loading, and the excitation will be trimmed to give the desired control grid current. In this case, the screen current will be set almost entirely by the choice of the dropping resistor. It will be found that excitation and loading will vary the screen voltage considerably, and these should be trimmed to yield a desired screen voltage.

The grounded-grid amplifier has been used for many years, and with the advent of new high-power *zero bias* triodes, this configuration has become more common. Adjusting the excitation and loading of a grounded-grid RF amplifier requires a slightly different procedure. The plate voltage (plate and screen voltage in the case of a tetrode or pentode) must be applied before the excitation. If this precaution is not followed, there is a good chance that the control grid will be damaged. The loading is increased as the excitation is increased. When the desired plate current is reached, note the power output. The loading can be reduced slightly and the excitation increased until the plate current is the same as before. If the power output is less than before, a check can be made with increased loading and less excitation. By proper trimming, the proper grid current, plate current, and optimum power output can be attained. In a grounded-grid circuit, the cathode or input circuit is in series with the plate circuit. Because of this arrangement, any change made in the plate circuit will have an effect on the input circuit. Therefore, the driver amplifier does not see its designed load until the driven stage is up to full plate current.

14.2.2   Amplifier Balance

In a push-pull RF amplifier, lack of balance in the plate circuit or unequal plate dissipation usually is caused by a lack of symmetry in the RF circuit. Normally, the tubes are similar enough that such imbalance is not associated with the tubes and their characteristics. This point can be checked readily by interchanging the tubes in the sockets (provided both tubes have common dc voltages to plate, screen, and grid), and observing whether the unbalanced condition remains with the socket location or moves with the tube. If it remains with the socket location, the circuit requires adjustment. If appreciable imbalance is associated with the tube, it is likely that one tube is abnormal and should be replaced.

The basic indicators of balance are the plate current and plate dissipation of each tube. It is assumed that the circuit applies the same dc plate voltage, dc screen voltage (if a tetrode or pentode), and dc grid bias to each tube from common supplies. Also, it is initially assumed that the plate circuit is mechanically and electrically symmetrical (or approximately so).

Imbalance in a push-pull RF amplifier usually is caused by unequal RF voltages applied to the grids of the tubes, or by the RF plate circuit applying unequal RF voltages to the plates of the tubes. First balance the grid excitation until equal dc plate currents flow in each tube. Then balance the RF plate circuit until equal plate dissipation appears on each tube, or equal RF plate voltage is observed. The balance of plate current is a more important criterion than equality of screen current (in a tetrode or pentode) or grid current. This is because tubes tend to be more uniform in plate current characteristics. However, the screen current also is sensitive to a lack of voltage balance in the RF plate circuit and may be used as an indicator. If the tubes differ somewhat in screen current characteristics, and the circuit has common dc supply voltages, the final trimming of the plate circuit balance may be made by interchanging tubes and adjusting the circuit to give the same screen current for each tube, regardless of its location.

Note that the dc grid current has not been used as an indicator of balance of the RF power amplifier. It is probable that following the foregoing procedure, the grid currents will be fairly well-balanced, but this condition in itself is not a safe indicator of grid excitation balance.

14.2.3   Parallel Tube Amplifiers

The previous discussion has been oriented toward the RF push-pull amplifier. The same comments can be directed to parallel tube RF amplifiers. The problem of balance—to be certain each tube carries its fair share of the load—still must be considered. In audio frequency power amplifiers operating in class AB_1 or class AB_2, the idle dc plate current per tube should be balanced by separate bias adjustments for each tube. In many cases, some lack of balance of the plate currents will have a negligible effect on the overall performance of the amplifier.

When tubes are operating in the idle position close to cutoff, operation is in a region where the plate current cannot be held to a close percentage tolerance. At this point, the action of the plate and screen voltages is in delicate balance with the opposing negative grid voltage. The state of balance is indicated by the plate current. Minor variations of individual grid wires or variations in the diameter of grid wires can upset the balance. It is practically impossible to control such minor variations in manufacture.

14.2.4   Harmonic Energy

A pulse of plate current delivered by the tube to the output circuit contains components of the fundamental and most harmonic frequencies. To generate output power that is a harmonic of the exciting voltage applied to the control grid, it is necessary merely to resonate the plate circuit to the desired harmonic frequency. To optimize the performance of the amplifier, it is necessary to adjust the angle of plate current flow to maximize the desired harmonic. The shorter the length of the current pulse, in the case of a particular harmonic, the higher the plate efficiency.

However, the bias, exciting voltage, and driving power must be increased. Also, if the pulse is too long or too short, the output power will decrease appreciably. The plate circuit efficiency of tetrode and pentode harmonic amplifiers is quite high. In triode amplifiers, if feedback of the output harmonic occurs, the phase of the voltage feedback usually reduces the harmonic content of the plate pulse, thereby lowering the plate circuit efficiency. Because tetrodes and pentodes have negligible feedback, the efficiency of a harmonic amplifier usually is comparable to that of other amplifiers. Also, the high-amplification factor of a tetrode or pentode causes the plate voltage to have little effect on the flow of plate current. A well-designed tetrode or pentode also permits large RF voltages to be developed in the plate circuit while still passing high peaks of plate current in the RF pulse. These two factors further help to increase plate efficiency.

Controlling Harmonics. The previous discussion of harmonics has been applicable to the case in which harmonic power in the load is the objective. The generation and radiation of harmonic energy must be kept to a minimum in a *fundamental frequency* RF amplifier. It generally is not appreciated that the pulse of a grid circuit also contains energy on the harmonic frequencies, and control of these harmonic energies may be quite important. The ability of the tetrode and pentode to isolate the output circuit from the input circuit over a very wide range of frequencies is important in avoiding feedthrough of harmonic voltages from the grid. An important part of this shielding is the result of the basic physical design of the devices. The following steps facilitate the reduction of unwanted harmonic energy in the output circuit:

- Keep the circuit impedance between plate and cathode as low as practical for high harmonic frequencies. This may be accomplished by having some or all of the tuning capacitance of the resonant circuit close to the tube.
- Provide complete shielding of the input and output compartments.
- Use inductive output coupling from the resonant plate circuit, and possibly a capacitive or Faraday shield between the coupling coil and the tank coil. A high-frequency attenuating circuit, such as a Pi or Pi-L network, also may be used.
- Use low-pass filters on all supply leads and wires leading to the output and input compartments.
- Use resonant traps for particular frequencies as required.
- Use a low-pass filter in series with the output transmission line.

14.2.5   Intermodulation Distortion

In general, the criteria used in the selection of operating parameters for tubes in high-fidelity audio frequency amplifier service are applicable for selecting the operating conditions for linear RF amplifiers. In the case of the sideband linear amplifier, the degree of linearity of the stage is of considerable importance. Intermodulation distortion products in linear power-amplifier circuits can be caused by either amplitude gain nonlinearity or phase shift with changes in input signal level. Intermodulation distortion products appear only when the RF signal has a varying envelope amplitude. A single continuous-frequency wave will be amplified by a fixed amount and shifted in phase by a fixed amount. The nonlinearity of the amplifier will produce only harmonics of the input wave. If the input RF wave changes at an audio rate, however, the nonlinearity of the amplifier will cause undesirable intermodulation distortion.

When an RF signal with varying amplitude is passed through a nonlinear device, many new products are produced. The frequency and amplitude of each component can be determined mathematically. The nonlinear device can be represented by a power series expanded about the zero-signal operating point. Intermodulation distortion power is wasted and serves no purpose other than to cause interference to adjacent channels. Intermodulation distortion in a power amplifier tube is principally the result of its transfer characteristics. An ideal transfer characteristic curve is shown in Figure 14.1.

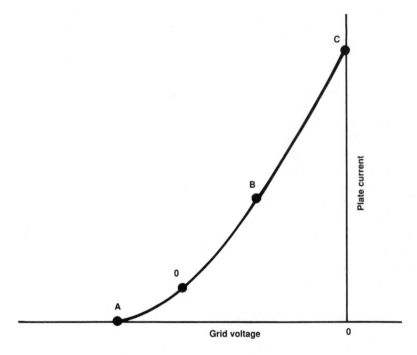

Figure 14.1   Ideal grid-plate transfer curve for class AB operation. (Courtesy of Varian Associates)

Even-order products do not contribute to the intermodulation distortion problem because they fall outside the amplifier passband. Therefore, if the transfer characteristic produces an even-order curvature at the small-signal end of the curve (from point A to point B) and the remaining portion of the curve (point B to point C) is linear, the tube is considered to have an ideal transfer characteristic. If the operating point of the amplifier is set at point 0 (located midway horizontally between point A and point B), there will be no distortion in a class AB amplifier. However, no tube has this idealized transfer characteristic. It is possible, by clever manipulation of the electron ballistics within a given tube structure, to alter the transfer characteristic and minimize the distortion products.

14.3   TRANSMISSION-LINE/ANTENNA VSWR

Among all the possible antenna and feedline measurements that can be made, VSWR is the most common. Antenna defects that affect gain and directivity usually (but not always) cause an increase in VSWR. A practical understanding of VSWR can aid in troubleshooting antenna and feedline problems. An in-line directional wattmeter commonly is used to measure both the forward and reflected power in a feedline. It is the comparison of reflected power (P_r) to forward power (P_f) that provides a qualitative measure of the match between the transmitter and the feedline, or between the antenna and the feedline.

14.3.1   Operating Parameters

On a theoretical, lossless feedline, VSWR is the same at all points along the line. Actually, all feedlines have some attenuation distributed uniformly along their entire lengths. Any attenuation between the reflection point and the source reduces VSWR at the source. This concept is illustrated in Figure 14.2. Standing waves on the feedline increase feedline attenuation to a level greater than normal attenuation under matched-line conditions. The amount of feedline attenuation increase depends on the VSWR and how great the losses are under matched-line conditions. The higher the matched-line attenuation, the greater the additional attenuation caused by a given VSWR. Figure 14.3 correlates additional attenuation with matched-line attenuation for various VSWR levels.

When a feedline is terminated by an impedance equal to the feedline *characteristic impedance*, the feedline impedance at any point is equal to the feedline characteristic impedance. When the feedline is terminated by an impedance not equal to the feedline characteristic impedance, the feedline input impedance will vary with changes in feedline length. The greater the impedance mismatch at the load, the wider the feedline input impedance variation with changes in line length. The greatest feedline impedance change occurs with the worst mismatches: open- or short-circuits.

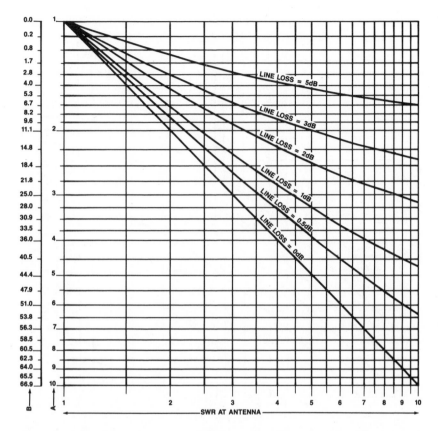

Figure 14.2 Illustration of how the VSWR on a transmission line affects the attenuation performance of the line. Scale A represents the VSWR at the transmitter; scale B represents the corresponding reflected power. (Source: Harold Kinley, "Base Station Feedline Performance Testing," *Mobile Radio Technology* magazine, Intertec Publishing, Overland Park, Kan., April 1989)

A short circuit at the end of a feedline appears as an open circuit at a point 1/4-wavelength down the feedline. It is the greatest impedance transformation that can occur on a feedline. Less severe mismatches cause smaller impedance transformations. At a point 1/2-wavelength down the line, the impedance equals the load impedance. Feedlines with high VSWR sometimes are called *resonant* or *tuned* feedlines. Matched lines—those with little VSWR—are called *nonresonant* or *untuned* lines.

This phenomenon means that high feedline VSWR can cause the feedline input to present a load impedance to the transmitter that differs greatly from the feedline's characteristic impedance. Some transmitters are equipped with an output matching

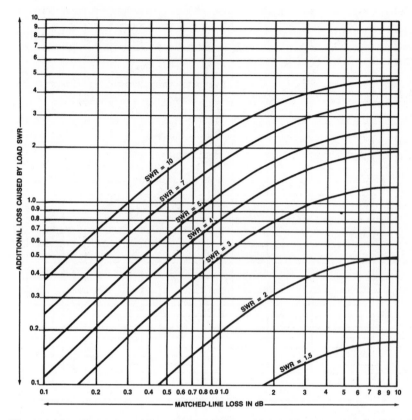

Figure 14.3 Illustration of how VSWR causes additional loss in a length of feedline. (Source: Harold Kinley, "Base Station Feedline Performance Testing," *Mobile Radio Technology* magazine, Intertec Publishing, Overland Park, Kan., April 1989)

network that provides a *conjugate match* to a broad range of impedances. Other transmitters offer more limited load matching provisions.

14.4 RADIO TRANSMITTER PERFORMANCE

Assessing the performance of an amplitude- or frequency-modulated transmitter has been simplified greatly by recent advancements in RF system design. The overall performance of most transmitters can be determined by measuring the audio performance of the system as a whole. If the audio tests indicate good performance, it is likely that all of the RF parameters also meet specifications. Both AM and FM transmission have certain limitations that prevent them from ever being completely transparent media. At the top of the list for FM is multipath distortion. In many locations, some degree of multipath is unavoidable. Running a close second is the practical audio bandwidth limitation of both AM and FM systems.

14.4.1   Key System Measurements

The procedures for measuring AM and FM transmitter performance vary widely, depending on the type of equipment being used. Certain basic measurements, however, can be applied to characterize the overall performance of most systems:

- *Frequency response:* The actual deviation from a constant amplitude across a given span of frequencies.
- *Total harmonic distortion:* The creation by a nonlinear device of spurious signals harmonically related to the applied waveform. THD is sensitive to the noise floor of the system under test. If the system has a signal-to-noise ratio of 60 dB, the distortion analyzer's best possible reading will be greater than 0.1 percent (60 dB = 0.001 = 0.1 percent).
- *Intermodulation distortion:* The creation by a nonlinear device of spurious signals not harmonically related to the audio waveform. These distortion components are *sum-and-difference (beat note)* mixing products. The IMD measurement is relatively impervious to the noise floor of the system under test.
- *Signal-to-noise ratio (S/N):* The amplitude difference, expressed in decibels, between a reference level audio signal and the residual noise and hum of the overall system.
- *Crosstalk:* The amplitude difference, expressed in decibels, between two or more channels of a multichannel audio transmission system.

14.4.2   Synchronous AM in FM Systems

The output spectrum of an FM transmitter contains many sideband frequency components, theoretically an infinite number. The practical considerations of transmitter design and frequency allocation make it necessary to restrict the bandwidth of all FM broadcast signals. Bandwidth restriction brings with it the undesirable side effects of phase shifts through the transmission chain, the generation of *synchronous AM* components, and distortion in the demodulated output of certain receivers. In most medium- and high-power FM transmitters, the primary out-of-band filtering is performed in the output cavity of the final stage. Previous stages in the transmitter (the exciter and IPA) are designed to be broadband, or at least more broadband than the PA. As the bandwidth of an FM transmission system is reduced, synchronous amplitude modulation increases for a given carrier devia-

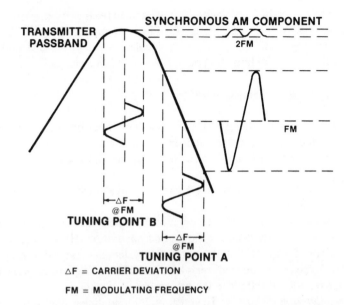

Figure 14.4 The generation of synchronous AM in a bandwidth-limited FM system. Note that minimum synchronous AM occurs when the system operates in the center of its passband. (Courtesy of *Broadcast Engineering* magazine, Intertec Publishing, Overland Park, Kan.)

tion (modulation). Synchronous AM is produced as tuned circuits with finite bandwidth are swept by the frequency of modulation. The amount of synchronous AM generated is dependent on tuning, which determines (to a large extent) the bandwidth of the system. Figure 14.4 illustrates how synchronous AM is generated through modulation of the FM carrier.

Bandwidth Limiting. The design goal of most medium- and high-power FM transmission equipment is to limit the bandwidth of the transmitted RF signal at one stage only. In this way, control over system bandwidth can be maintained tightly, and the tradeoffs required in any practical FM system can be optimized. If, on the other hand, more than one narrowband stage exists within the transmitter, adjustment for peak efficiency and performance can be a difficult proposition. The following factors can affect the bandwidth of an FM system:

- Total number of tuned circuits in the system.
- Amplitude and phase response of the total combination of all tuned circuits in the RF path.
- Amount of drive to each RF stage (saturation effects).
- Nonlinear transfer function within each RF stage.

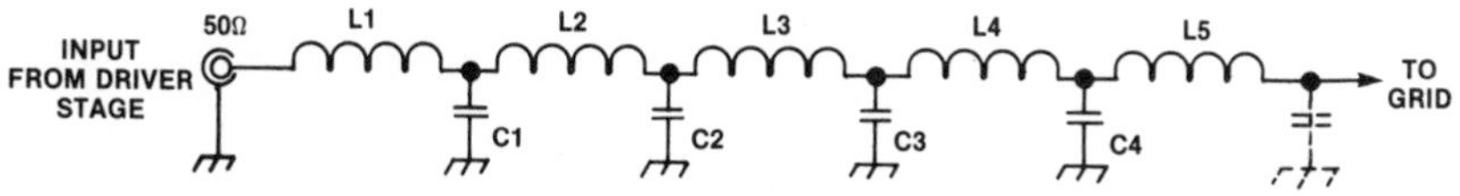

Figure 14.5   IPA-to-PA grid matching network designed for wide-bandwidth operation. (Courtesy of *Broadcast Engineering* magazine, Intertec Publishing, Overland Park, Kan.)

Although bandwidth generally is the domain of the transmitter design engineer, the following techniques can be used to improve the bandwidth of a given system:

- Maintain wide bandwidth until the final RF stage. This can be accomplished through the use of a broadband exciter and broadband (usually solid-state) IPA.
- Minimize the number of interactive-tuned networks within the system. Through proper matching network design, greater bandwidth and simplified tuning can be accomplished.
- Use a single tube design (for medium- and high-power transmitters) or a broadband solid-state design (where practical). The objective is to maintain wide bandwidth until the last possible point in the RF chain.

Bandwidth affects the gain and efficiency of any FM RF amplifier. The bandwidth is determined by the load resistance across the tuned circuit and the output or input capacitance of the RF stage. Because grid-driven PA tubes typically exhibit high input capacitance, the PA input circuit generally has the greatest effect on bandwidth limiting in an FM transmitter. The grid circuit can be broadened through resistive swamping in the input circuit, or through the use of a broadband input impedance matching network. One approach involves a combination of series inductor and shunt capacitor elements implemented on a printed wiring board, as illustrated in Figure 14.5. In this design, the inductors and capacitors are etched into the copper-clad laminate of the PWB. This approach provides the necessary impedance transformation from a 50 Ω solid-state driver to a high-impedance PA grid in a series of small steps, rather than more conventional L, Pi, or T matching networks.

Effects of Synchronous AM. The effect of restricted bandwidth on synchronous AM performance for an example FM RF amplifier is plotted in Figure 14.6. Notice that as the -3 dB points of the RF system passband are narrowed, a dramatic increase in synchronous AM occurs. The effect shown in the figure is applicable to any bandwidth-limited FM system, including FM broadcast transmitters, television aural transmitters, microwave relay systems, diplexers, and transmitting antennas.

Bandwidth also affects the distortion floor of the demodulated baseband signal of an FM system, as plotted in Figure 14.7. The data shown applies to a test setup

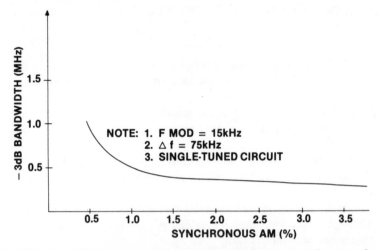

Figure 14.6 Typical synchronous AM content of an FM system as a function of system bandwidth. (Courtesy of *Broadcast Engineering* magazine, Intertec Publishing, Overland Park, Kan.)

involving a single-tuned circuit fed by an FM generator with a deviation of 75 kHz at a modulating frequency of 15 kHz (no deemphasis is applied to the output signal). Remember that FM is a nonlinear process, and that interpretation of distortion numbers for multiple-carrier signals is not accurate. The example shown, however, illustrates how the bandwidth of the RF channel can set a minimum performance limit on system total harmonic distortion. By understanding the mechanics of synchronous AM, the technician can adjust an FM system for optimum performance.

14.4.3 Incidental Phase Modulation

Incidental phase modulation (IPM) is produced by an AM transmitter as a result of amplitude modulation. In other words, as an AM transmitter develops the amplitude-modulated signal, it also produces a phase-modulated, or PM, version of the audio as well. In theory, IPM is of little consequence for monophonic broadcast AM because FM and PM do not affect the carrier amplitude. An envelope detector will, in theory at least, ignore IPM. This is true to a point, but stations attempting to broadcast AM stereo must pay close attention to IPM. Stereo AM broadcasting in the United States utilizes a combination of AM and PM modulation. Receivers designed to decode stereo broadcasts will not perform properly if excessive IPM is produced by the transmitter. IPM is, in fact, an important parameter for any type of AM radio service that codes intelligence into a PM

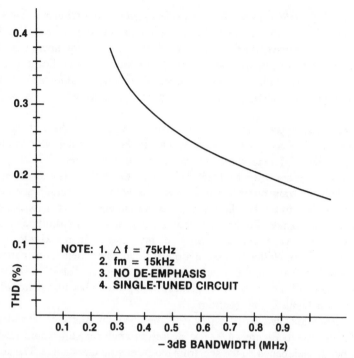

NOTE: 1. Δf = 75kHz
      2. fm = 15kHz
      3. NO DE-EMPHASIS
      4. SINGLE-TUNED CIRCUIT

Figure 14.7 Typical total harmonic distortion (THD) content of a demodulated FM signal as a function of transmission-channel bandwidth. (Courtesy of *Broadcast Engineering* magazine, Intertec Publishing, Overland Park, Kan.)

component. Although the effects of IPM vary from one system to another, optimum performance is realized when IPM is minimized.

As a general rule, because IPM is a direct result of the modulation process, it can be generated in any stage that is influenced by modulation. The most common cause of IPM in plate-modulated and pulse-modulated transmitters is improper neutralization of the final RF amplifier. Adjusting the transmitter for minimum IPM is an accurate way of achieving proper neutralization. The reverse is not always true, however, because some neutralization methods will not necessarily result in the lowest amount of IPM.

Improper tuning of the IPA stage is the second most common cause of IPM. As modulation changes the driver loading to the PA grid, the driver output also may vary. The circuits that feed the driver stage usually are isolated enough from the PA that they do not produce IPM. An exception is when the power supply for the IPA is influenced by modulation. Such a problem could be caused by a loss of capacitance in the filters of the high-voltage power supply.

Bandwidth Considerations. When a perfect AM transmitter is modulated with a tone, two sidebands are created, one above the carrier and one below it. Phase modulation of the same carrier, however, produces an infinite number of sidebands at intervals above and below the carrier. The number of significant sidebands depends on the magnitude of the modulation. With moderate amounts of IPM, the transmitted spectrum can be increased to the point at which the legal channel limits are exceeded.

Corrective Procedures. The spectrum analyzer is generally the best instrument for identifying IPM components. First, modulate the transmitter with a pure sine wave, and make a THD measurement with a distortion analyzer. This will provide an idea of what the spectrum analyzer should look like when it is hooked up to the transmitter. If the transmitter has an excessive amount of harmonic distortion, it will be impossible to tell the difference between the normal distortion sidebands and the IPM sidebands. Figure 14.8 shows what an ideal transmitted spectrum would look like on a spectrum analyzer when modulated with a 1 kHz tone at 100 percent modulation. Notice that there are just three components: one carrier and two sidebands. This transmitter would measure zero percent distortion, because no harmonic sidebands are present. Also, no IPM is present, because it would show up as extra sidebands in the spectrum.

If the transmitter had an IPM problem, the spectrum analyzer display might resemble the one in Figure 14.9. The distortion analyzer still would read zero percent, but the sidebands on the spectrum analyzer would not agree. The key is to start with a low harmonic distortion reading on the transmitter so that any sidebands will be recognizable immediately as the result of IPM. The most practical way to do this is to modulate the transmitter with a 1 kHz tone to a modulation level that

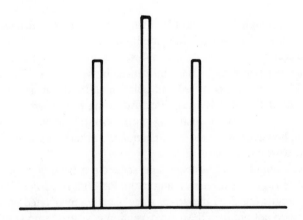

Figure 14.8 The spectrum of an ideal AM transmitter. THD measures zero percent; there are no harmonic sidebands. (Source: D. Bordonaro, "Reducing IPM in AM Transmitters," *Broadcast Engineering* magazine, Intertec Publishing, Overland Park, Kan., May 1988)

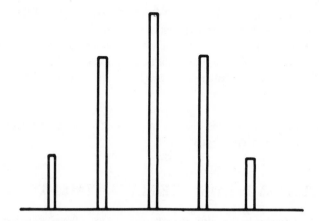

Figure 14.9   The spectrum of an ideal transmitter with the addition of IPM. (Source: D. Bordonaro, "Reducing IPM in AM Transmitters," *Broadcast Engineering* magazine, Intertec Publishing, Overland Park, Kan., May 1988)

is as close to 100 percent as possible, without exceeding 1 percent THD. If the distortion measures less than 1 percent, no distortion sideband will be greater than 40 dB below the 1 kHz sidebands. Any sidebands that are greater than this value are the result of IPM.

14.5   VIDEO TRANSMITTER PERFORMANCE

There are two broad classifications of signal distortions in television systems: linear and nonlinear. Linear distortions occur independent of picture level, and nonlinear distortions vary with the amplitude of the video signal. Linear distortions usually occur in systems with nonuniform frequency response, and are divided into four general categories:

- Short-time distortions affecting horizontal sharpness.
- Line-time distortions causing horizontal streaking.
- Field-time distortions causing errors in low-frequency response and problems with vertical shading of the picture.
- Long-time distortions in the dc-to-extremely-low frequency range, causing flicker and slow changes in picture brightness.

Nonlinear distortion occurs with changes in the average picture level. The primary types of nonlinear distortion include:

- *Luminance nonlinear distortion:* Occurs when the gain of the system varies as the picture changes from black to white.
- *Differential gain distortion:* Change in the amplitude of the chrominance signal as luminance amplitude shifts from black to white.
- *Differential phase distortion:* Change in color subcarrier phase as the luminance signal varies in amplitude.

See Sec. 13.2 for a detailed discussion of waveform and vectorscope measurement techniques for video signals.

14.6   TRANSMISSION SITE MANAGEMENT

Uncontrolled transmission site growth can reduce communications efficiency significantly. Without a coordinated effort on the part of the site manager and the users, the site may eventually "grow deaf" as more transmitters are added. This problem is the result of two conditions:

1. Intermodulation products
2. Elevated noise floor

Both conditions affect the ability of equipment at the site to receive signals from fixed and mobile transmitters.

14.6.1   Intermodulation Products

An intermodulation product, or intermod, is a signal created when two or more transmitter frequencies are mixed with one another unintentionally. Mixing can occur in transmitter circuitry, in receiver circuitry, or elsewhere. The apparent effect is the same: the creation of one or more low-level signals strong enough to interfere with desired signals and prevent reception. Sometimes the intermod falls on unused frequencies. Sometimes it falls on a channel in use or near enough to the channel to cause interference.

Mathematics can be used to calculate the intermod that may result from frequencies in use at a site. Normally produced by a computerized analysis, this *intermod study* is helpful in several ways. If the study is conducted as one or more new frequencies are under consideration for future use at an existing site, the study may reveal which frequencies are more likely to produce harmful intermod and which are not. If the study is conducted as part of an investigation into an interference problem, it may help in identifying transmitters that play a role in generating the intermod.

Intermod can be generated by any nonlinear device or circuit. The most common place for intermod to be generated is in the power amplifier of an FM transmitter. Such an amplifier does not need to preserve waveform integrity; the waveform is restored by the tuned output circuit. Thus, the amplifier is operated in an efficient, but nonlinear, mode: class C. In contrast, amplifiers for amplitude companded single-sideband (ACSSB) and other AM services must be operated in a linear mode. Such amplifiers normally do not act as mixers.

Generally, solid-state amplifiers are better mixers than their electron tube counterparts, a fact that may explain why intermod may appear suddenly when older equipment is replaced. A mix occurs when RF energy from one transmitter (Tx 1) finds its way down the transmission line and into the final amplifier of another transmitter (Tx 2), as illustrated in Figure 14.10. The energy mixes in the final amplifier with the prefiltered second-harmonic of Tx 2 to produce another in-band signal, a third-order intermod product. The reverse mix also may occur in the final amplifier of Tx 1, producing another intermod signal.

Once discovered, a mix generated in a transmitter final amplifier is relatively easy to eliminate: Increase the isolation between the offending transmitters. This may be accomplished by increasing the space between the antennas or by placing an isolator at the output of each transmitter.

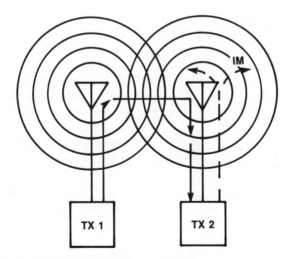

Figure 14.10   The mechanics of intermod generation in a two-transmitter installation. (Source: Al Dolgosh and Ron Jakubowski, "Controlling Interference at Communications Sites," *Mobile Radio Technology* magazine, Intertec Publishing, Overland Park, Kan., February 1990)

Receiver Intermod. Intermod also may be generated in the front end of a receiver. High-level RF energy from transmitters at a site can generate a bias voltage in the receiver sufficient to place the receiver RF amplifier into a nonlinear state. In such a mode, the stage acts as a mixer and creates unwanted frequencies. Filters may be used to solve this problem. The filters should notch out the offending transmitted signals. When another transmitter is added to the site, another notch filter usually must be added to the receiver transmission line.

Hardware Intermod. At a large transmission site, the entire area—including towers, guy wires, antennas, clamps, and obstruction lighting hardware—is saturated by high-level RF. Any loose or moving hardware can cause broadband noise. Corroded connections, tower joints, or clamps can form semiconductor junctions that may generate harmful intermod. Because the antenna system is exposed to a changing external environment, the degree of semiconductivity may vary with moisture or temperature. Changing semiconductivity causes intermittent problems, whose severity may change as well.

14.7   BIBLIOGRAPHY

Bordonaro, Dominic: "Reducing IPM in AM Transmitters," *Broadcast Engineering* magazine, Intertec Publishing, Overland Park, Kan., May 1988.

Dolgosh, Al, and Ron Jakubowski: "Controlling Interference at Communications Sites," *Mobile Radio Technology* magazine, Intertec Publishing, Overland Park, Kan., February 1990.

Hershberger, David, and Robert Weirather: "Amplitude Bandwidth, Phase Bandwidth, Incidental AM, and Saturation Characteristics of Power Tube Cavity Amplifiers for FM," Harris Corporation application note, Quincy, Ill., 1982.

Kinley, Harold: "Base Station Feedline Performance Testing," *Mobile Radio Technology* magazine, Intertec Publishing, Overland Park, Kan., April 1989.

Mendenhall, Geoffrey N.: "Testing Television Transmission Systems for Multichannel Sound Compatibility," Broadcast Electronics application note, Quincy, Ill., 1985.

_____: "A Systems Approach to Improving Subcarrier Performance," *Proceedings* of the 1986 NAB Engineering Conference, National Association of Broadcasters, Washington, D.C., 1986.

_____: "Fine-Tuning FM Final Stages," *Broadcast Engineering* magazine, Intertec Publishing, Overland Park, Kan., May 1987.

15

SAFETY: THE KEY TO STAYING ALIVE

15.1 INTRODUCTION

Electrical safety is something every engineer thinks about at one time or another. Perhaps the last time you thought about electrical safety was when you reached into the driver cabinet of a high-power transmitter or a klystron power supply. Or, it may have been when you forgot that the computer power supply was still plugged in, and your finger accidentally contacted the primary voltage. Safety is critically important to maintenance personnel, because in addition to working around powered hardware, they often must meet stringent deadlines. Safety is not something to be taken lightly.

15.2 ELECTRICAL SHOCK[1]

It takes surprisingly little current to injure a person. Studies at Underwriters' Laboratories (UL) show that the electrical resistance of the human body varies with the amount of moisture on the skin, the muscular structure of the body, and the applied voltage. The typical hand-to-hand resistance ranges from 500 Ω to 600 kΩ, depending on the conditions. Higher voltages have the capability to break down the outer layers of the skin, which can reduce the overall resistance value. UL uses the lower value, 500 Ω, as the standard resistance between major extremities, as from the hand to the foot. This value generally is considered the minimum that

---

1 Sections 15.2, 15.3 and 15.6 were contributed by Brad Dick, editor, *Broadcast Engineering* magazine, Intertec Publishing, Overland Park, Kan.

Table 15.1   The Effects of Current on the Human Body

1 mA or less	No sensation, not felt
More than 3 mA	Painful shock
More than 10 mA	Local muscle contractions, sufficient to cause "freezing" to the circuit for 2.5 percent of the population
More than 15 mA	Local muscle contractions, sufficient to cause "freezing" to the circuit for 50 percent of the population
More than 30 mA	Breathing is difficult, can cause unconsciousness
50 mA to 100 mA	Possible ventricular fibrillation of the heart
100 mA to 200 mA	Certain ventricular fibrillation of the heart
More than 200 mA	Severe burns and muscular contractions; heart more apt to stop than fibrillate
More than a few amperes	Irreparable damage to body tissues

would be encountered. In fact, it may not be unusual, because wet conditions or a cut or other break in the skin significantly reduce human body resistance.

15.2.1   Effects on the Human Body

Table 15.1 lists some effects that typically result when a person is connected across a current source with a hand-to-hand resistance of 2.4 kΩ. The table shows that a current of 50 mA will flow between the hands, if one hand is in contact with a 120 V ac source and the other hand is grounded. The table indicates that even the relatively small current of 50 mA can produce *ventricular fibrillation* of the heart, and might even cause death. Medical literature describes ventricular fibrillation as very rapid, uncoordinated contractions of the ventricles of the heart, resulting in loss of synchronization between heartbeat and pulse beat. The electrocardiograms shown in Figure 15.1 compare a healthy heart rhythm with one in ventricular fibrillation. Unfortunately, once ventricular fibrillation begins, it will continue. Barring resuscitation techniques, death will ensue within a few minutes.

The route taken by the current through the body greatly affects the degree of injury. Even a small current, passing from one extremity through the heart to another extremity, is dangerous and capable of causing severe injury or electrocution. There are cases in which a person who has contacted extremely high current levels has lived to tell about it. However, when this happens, it is usually because the current passes only through a single limb, not through the entire body. In these instances, the limb is often lost, but the person survives.

Current is not the only factor in electrocution. Figure 15.2 summarizes the relationship between current and time on the human body. The graph shows that 100 mA flowing through a human adult body for 2 seconds will cause death by electrocution. An important factor in electrocution, the *let-go range*, also is shown

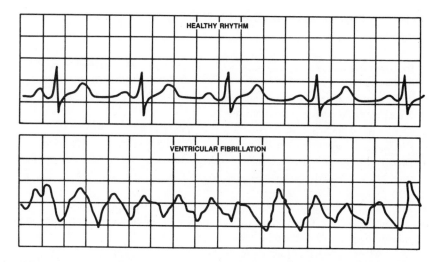

Figure 15.1   Electrocardiograms showing the healthy rhythm of a heart (top), and ventricular fibrillation of the heart (bottom). (Courtesy of *Broadcast Engineering* magazine, Intertec Publishing, Overland Park, Kan.)

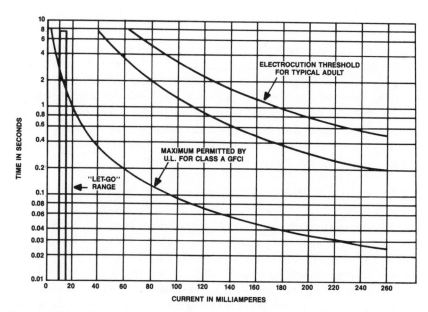

Figure 15.2   Effects of electrical current and time on the human body. Note the "let go" range. (Courtesy of *Broadcast Engineering* magazine, Intertec Publishing, Overland Park, Kan.)

Table 15.2   Required Safety Practices for Engineers Working Around High-Voltage Equipment

- Remove all ac power from the equipment. Do not rely on internal contactors or SCRs to remove dangerous ac.
- Trip the appropriate power-distribution circuit breakers at the main breaker panel.
- Place signs as needed to indicate that the circuit is being serviced.
- Switch the equipment being serviced to the *local control* mode as provided.
- Discharge all capacitors using the discharge stick provided by the manufacturer.
- Do not remove, short-circuit, or tamper with interlock switches on access covers, doors, enclosures, gates, panels, or shields.
- Keep away from live circuits.
- Allow any component to completely cool down before attempting to replace it.
- If a leak or bulge is found on the case of an oil-filled or electrolytic capacitor, do not attempt to service the part until it has completely cooled.
- Know which parts in the system contain PCBs. Handle them appropriately.
- Minimize exposure to RF radiation.
- Avoid contact with beryllium oxide (BeO) ceramic dust or fumes.
- Avoid contact with hot surfaces within the system.
- Know your equipment, and do not take chances.

on the graph. This is described as the amount of current that causes "freezing," or the inability to let go of the conductor. At 10 mA, 2.5 percent of the population would be unable to let go of a "live" conductor. At 15 mA, 50 percent of the population would be unable to let go of an energized conductor. It is apparent from the graph that even a small amount of current can "freeze" someone to an electrical conductor. The objective for those who must work around electrical equipment is to protect themselves from electrical shock. Table 15.2 lists required precautions for maintenance personnel working near high voltages.

15.2.2   Circuit-Protection Hardware

The typical primary panel or equipment circuit breaker or fuse will not protect you from electrocution. In the time that a fuse or circuit breaker takes to blow, you could be dead. However, two common devices, properly used, may help prevent electrocution. The *ground-fault interrupter* (GFI), shown in Figure 15.3, works by monitoring the current being applied to the load. The GFI uses a differential transformer and looks for an imbalance in load current. If a current (5 mA, ± 1 mA) begins flowing between the neutral and ground or between the hot and ground leads, the differential transformer detects the leakage and opens the primary circuit within 2.5 ms.

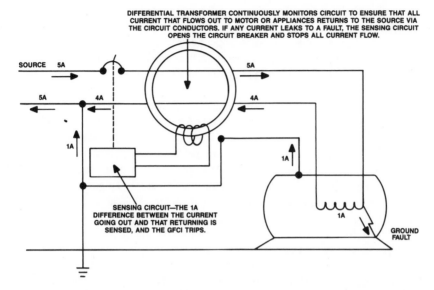

DIFFERENTIAL TRANSFORMER CONTINUOUSLY MONITORS CIRCUIT TO ENSURE THAT ALL
CURRENT THAT FLOWS OUT TO MOTOR OR APPLIANCES RETURNS TO THE SOURCE VIA
THE CIRCUIT CONDUCTORS. IF ANY CURRENT LEAKS TO A FAULT, THE SENSING CIRCUIT
OPENS THE CIRCUIT BREAKER AND STOPS ALL CURRENT FLOW.

SENSING CIRCUIT—THE 1A
DIFFERENCE BETWEEN THE CURRENT
GOING OUT AND THAT RETURNING IS
SENSED, AND THE GFCI TRIPS.

Figure 15.3 Basic design of a ground-fault interrupter (GFI). (Courtesy of *Broadcast Engineering* magazine, Intertec Publishing, Overland Park, Kan.)

OSHA (Occupational Safety and Health Administration) rules specify that temporary receptacles (those not permanently wired) and receptacles used on construction sites be equipped with GFI protection. Receptacles on two-wire, single-phase portable and vehicle-mounted generators of not more than 5 kW, where the generator circuit conductors are insulated from the generator frame and all other grounded surfaces, need not be equipped with GFI outlets. GFIs will not protect you from every type of electrocution. If you become connected to both the neutral and the hot wire, the GFI will treat you as part of the load, and will not open the primary circuit.

15.2.3   Working with High Voltage

Rubber gloves are a common safety measure used by engineers working on high-voltage equipment. High voltage gloves are designed to provide protection from hazardous voltages or RF when the wearer is working on "hot" ac or RF circuits. Although the gloves may provide some protection from these hazards, placing too much reliance on them poses the potential for disastrous consequences. Do not assume that they always provide complete protection. The gloves found in many facilities may be old and untested. Some may even have been "repaired" by users, perhaps with electrical tape. Few tools could be more hazardous than such a pair of gloves.

Know the voltage rating of the gloves. Gloves are rated differently for ac and dc voltages. For instance, a *class 0* glove has a minimum dc breakdown voltage of 35 kV; the minimum ac breakdown voltage, however, is only 6 kV. Furthermore, high-voltage rubber gloves are not tested at RF frequencies, and RF can burn a hole in the best of them. If you use rubber gloves for RF protection, you do so at your own risk.

How many engineers regularly have their high-voltage gloves tested? Not many. How do you know that the gloves are still safe for use at 20 kV? Have the gloves, because of age, wear, or microscopic holes, lost their ability to protect you? Always perform a visual inspection before using any pair of gloves. Look for cracks, and "roll" test them for leaks. The roll test is accomplished by flipping the glove while holding the cuff to capture air inside. Roll the cuff toward the fingers and listen for air leaks, or pass the glove near your face to detect air leaks. If you are not completely sure that the gloves are safe to use, replace them or have them tested. Your local electric utility may be able to provide the name of the company it uses to test its gloves. Untested gloves should not be used.

A good pair of gloves, however, could save your life. A consultant tells the story of working on a high-voltage fuse panel in a field. While he was bent over, looking into the cabinet, a stray goat wandered over and butted him into the equipment. Fortunately, the consultant was wearing high-voltage gloves. When, on reflex, he extended his hands to catch himself (inside the fuse panel), he was not electrocuted. Although it is probably safe to say that most facilities are not visited by live animals, you could encounter a similar situation if you were to slip on a tool or bolt carelessly left on the floor.

Working on "live" circuits calls for much more precaution than simply donning a pair of gloves. It demands a certain frame of mind—an acute awareness of everything in the area, especially ground points. Be conscious of where you are placing your feet, arms, and legs. Do not lean against a grounded surface or kneel while working. It is critical that you do not allow the use of gloves to create a false sense of security which can lead to the development of dangerous work habits. Recall the axiom of keeping one hand in your pocket while working on a device with current flowing? That advice actually is based on simple electricity. It is not the "hot" connection that poses danger, but the ground connection that lets the current begin to flow. A study in California showed that more than 90 percent of electrical equipment fatalities occurred when the grounded person contacted a live conductor. Line-to-line electrocution accounted for less than 10 percent of the deaths.

When working around high voltages, keep your hands, feet, and the rest of your body away from grounded surfaces. Even concrete can act as a ground if the voltage is high enough. If you must work in "live" cabinets, then consider using, in addition to rubber gloves, a rubber floor mat, rubber vest, and rubber sleeves. Although this may seem to be a lot of trouble, consider the consequences of making a mistake.

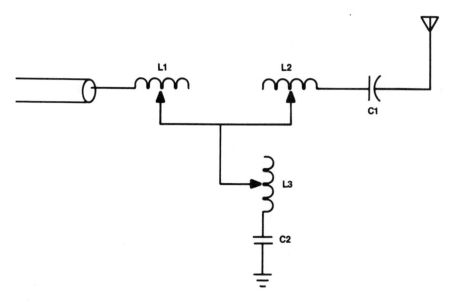

Figure 15.4   The voltages present on a typical antenna tuning unit can become extremely hazardous, especially if a coil tap such as L2 or L3 is disconnected. (Courtesy of *Broadcast Engineering* magazine, Intertec Publishing, Overland Park, Kan.)

Of course, the best troubleshooting methodology is never to work on any circuit unless you are positive that no hazardous voltages are present. In addition, any circuits or contactors that normally contain hazardous voltages should be grounded firmly before you begin work.

Another important safety rule is never to work alone. Who is going to shut down the equipment and call for help if you are knocked unconscious? Even if you cannot afford to hire a trained assistant, have someone accompany you while you are performing maintenance work.

RF Considerations. Broadcast engineers often rely on electrical gloves to make adjustments on live RF circuits. This practice, however, can be extremely dangerous. Consider the typical antenna tuning unit shown in Figure 15.4. In this configuration, disconnecting the coil from either L2 or L3 places the full transmitter output literally at your fingertips. Depending on the impedances involved, the voltages can become quite high, even in a circuit that normally is relatively tame.

If the station has an antenna of 120 electrical degrees, the base impedance might be approximately 106 Ω + j202. With 1 kW feeding into the antenna, the rms voltage at the tower base will be approximately 700 V. The peak voltage (which determines insulating requirements) will be close to 1 kV, and perhaps more than twice that if the carrier is being modulated. At the instant the output coil clip is disconnected, the current in the shunt leg will increase rapidly, and the voltage easily could more than double. What might happen if the engineer is wearing a pair of lineman's gloves that happen to be several years old, dirty, or worn?

Although many newer transmitters incorporate automatic shutdown circuits for high-VSWR conditions (as would occur when a coil tap is disconnected), many older transmitters or high-power units operating at lower power levels might continue to deliver power into this unusual load—even if the load happened to include the station engineer. In addition, some transmitters use dc in the final output network as an arc-detection device. If the blocking capacitor is short-circuited, leaky or missing, then that voltage is available at the network to help hold the engineer while the RF cooks him.

Of course, if luck is on his side, the engineer will only draw a large arc, escaping without burns or shock. This fact may provide some comfort to while he recovers from the broken arm he got falling off the ladder trying to get away from the arc. There is simply no substitute for safe work habits. It is important to treat every station as if it were a 50 kW facility and develop the work habits appropriate to such a plant.

15.2.4   A Case in Point

The following true story illustrates dramatically how even the most experienced engineer could be injured or killed while performing routine work. One morning, the engineer of a UHF-TV station responded to a trouble call from the studio. The 50 kW transmitter had developed a problem that required immediate attention. To remain on the air, the studio had bypassed the failed visual klystron amplifier and multiplexed the aural and visual signals on the single aural klystron.

When the engineer arrived at the mountaintop location, he found the transmitter to be operating properly. Normally, he would have returned the system to standard operating conditions. Instead, he elected to perform some preventive maintenance on the visual amplifier. From experience, the engineer knew the transmitter high-voltage contactor occasionally stuck in the *on* position. With this transmitter design, the stuck contactor could be found only through a visual inspection.

Before beginning his work, the engineer followed good safety practices. He shut down all power to the visual transmitter and associated high-power supply. Because he planned to work only on the visual amplifier, he thought no danger existed. However, the visual and aural power supplies were identical in appearance, and he mistakenly opened the aural supply cabinet instead of the visual supply. Thinking he was in the visual supply, he noticed the high-voltage connector stuck in the *on* position. As he reached down to free it manually, he touched an exposed conductor on the live aural supply.

He was thrown 5 ft across the concrete pad, but remained conscious. The transmitter overload circuits shut down the high voltage, probably saving his life. The engineer managed to return to the building to answer the ringing phone. It was the studio wondering why the transmitter was off the air. The paramedics arrived by helicopter almost 25 minutes later and found the engineer lying on the floor. Fortunately, he survived and was able to return to work a few days later. All he had to show for the accident was a half-dollar-size burn on one foot and a pinhole-size burn on a finger. This was one lucky engineer!

The moral of this story is to check, double check, and triple check power supplies, especially high-voltage supplies. Then use a shorting stick before touching anything. Also, leave that shorting stick across the potential source while working. You never know when someone will attempt to push the *on* button. Also be sure that all cabinets and components are labeled, shorting bars present, and interlocks working.

15.3   OPERATING HAZARDS

Particular hazards exist in the operation and maintenance of high-power RF equipment. Maintenance personnel must exercise extreme care around such hardware. Consider the following guidelines:

- Use caution around the high-voltage stages of the equipment. Many power tubes operate at voltages high enough to kill through electrocution. Always break the primary ac circuit of the power supply, and discharge all high-voltage capacitors.
- Minimize exposure to RF radiation. Do not permit personnel to be in the vicinity of open energized RF-generating circuits, RF transmission systems (waveguides, cables, or connectors), or energized antennas. High levels of radiation can result in severe bodily injury, including blindness. Cardiac pacemakers also may be affected. (See Sec. 15.5 for additional information on nonionizing radiation.)
- Avoid contact with beryllium oxide (BeO) ceramic dust or fumes. BeO ceramic material often is used as a thermal link to carry heat from a tube or transistor to the heat sink. Do not perform any operation on a BeO ceramic that might produce dust or fumes, such as grinding, grit blasting, or acid cleaning. Beryllium oxide dust and fumes are highly toxic, and breathing them can result in serious injury or death. BeO ceramics must be disposed of only in a manner prescribed by the device manufacturer.
- Avoid contact with hot surfaces within the equipment. The anode portion of most power tubes is air-cooled. The external surface normally operates at a high temperature (up to 250°C). Other portions of the tube also may reach high temperatures, especially the cathode insulator and the cathode/heater surfaces. All hot surfaces may remain hot for an extended time after the tube is shut off. To prevent serious burns, avoid bodily contact with these surfaces, both during tube operation and for a reasonable cooldown period afterward.

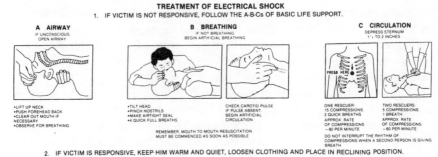

TREATMENT OF ELECTRICAL SHOCK
1. IF VICTIM IS NOT RESPONSIVE, FOLLOW THE A-B-Cs OF BASIC LIFE SUPPORT.

A AIRWAY	B BREATHING	C CIRCULATION
IF UNCONSCIOUS OPEN AIRWAY	IF NOT BREATHING BEGIN ARTIFICIAL BREATHING	DEPRESS STERNUM 1½ TO 2 INCHES

•LIFT UP NECK
•PUSH FOREHEAD BACK
•CLEAR OUT MOUTH IF NECESSARY
•OBSERVE FOR BREATHING

•TILT HEAD
•PINCH NOSTRILS
•MAKE AIRTIGHT SEAL
•4 QUICK FULL BREATHS

CHECK CAROTID PULSE
IF PULSE ABSENT
BEGIN ARTIFICIAL
CIRCULATION

REMEMBER, MOUTH TO MOUTH RESUSCITATION
MUST BE COMMENCED AS SOON AS POSSIBLE

ONE RESCUER
15 COMPRESSIONS
2 QUICK BREATHS
APPROX. RATE
OF COMPRESSIONS
— 80 PER MINUTE

TWO RESCUERS
5 COMPRESSIONS
1 BREATH
APPROX. RATE
OF COMPRESSIONS
— 60 PER MINUTE

DO NOT INTERRUPT THE RHYTHM OF
COMPRESSIONS WHEN A SECOND PERSON IS GIVING
BREATH

2. IF VICTIM IS RESPONSIVE, KEEP HIM WARM AND QUIET, LOOSEN CLOTHING AND PLACE IN RECLINING POSITION.
PLACE VICTIM FLAT ON HIS BACK ON A HARD SURFACE
CALL FOR MEDICAL ASSISTANCE AS SOON AS POSSIBLE

Figure 15.5   Basic first aid treatment for electrical shock. (Courtesy of *Broadcast Engineering* magazine, Intertec Publishing, Overland Park, Kan.)

15.3.1   First Aid Procedures

Be familiar with first aid treatment for electrical shock and burns. Always keep a first aid kit on hand at the facility. Figure 15.5 illustrates the basic treatment for electrical shock victims. Photocopy the information, and post it in a prominent location. Or, better yet, obtain more detailed information from your local heart association or Red Cross chapter. Personalized instruction on first aid usually is available locally. Table 15.3 lists basic first aid procedures for burns.

15.3.2   Aluminum Electrolytic Capacitors

Misapplication of aluminum electrolytic capacitors may be hazardous. Personal injury or property damage may result from explosion of a capacitor or from expulsion of electrolyte because of mechanical disruption of the component. A person whose skin, mouth, or eyes have been exposed to the electrolytes contained in an aluminum electrolytic capacitors must be treated immediately.

The most common forms of misapplication of an electrolytic capacitor include:

- Exposure of the component to a reverse voltage in excess of the specified limits.
- Exposure to temperatures above a specified limit.
- Application of a voltage beyond the specified surge voltage.
- Application of excessive ripple current or voltage.

If contact is made with the electrolyte of an aluminum electrolytic capacitor, following these guidelines:

Table 15.3   Basic First Aid Procedures

For extensively burned and broken skin:

- Cover affected area with a clean sheet or cloth.
- Do not break blisters, remove tissue, remove adhered particles of clothing, or apply any salve or ointment.
- Treat victim for shock as required.
- Arrange for transportation to a hospital as quickly as possible.
- If arms or legs are affected, keep them elevated.
- If medical help will not be available within an hour, and the victim is conscious and not vomiting, prepare a weak solution of salt and soda. Mix 1 teaspoon of salt and 1/2-teaspoon of baking soda to each quart of tepid water. Allow the victim to sip slowly about 4 oz. (half a glass) over a period of 15 minutes. Discontinue fluid intake if vomiting occurs. (Do not allow alcohol consumption.)

For less severe burns (first- and second-degree):

- Apply cool (not ice-cold) compresses, using the cleanest available cloth article.
- Do not break blisters, remove tissue, remove adhered particles of clothing, or apply salve or ointment.
- Apply clean, dry dressing if necessary.
- Treat victim for shock as required.
- Arrange for transportation to a hospital as quickly as possible.
- If arms or legs are affected, keep them elevated.

- Eye contact: If contact lenses are worn, remove them at once. Immediately flush the open eye(s) for 15 minutes with large amounts of water. If pain is experienced, apply two drops of 0.5 percent tetracaine (Pontocaine). Seek immediate medical attention.
- Skin or clothing contact: Flush the affected area thoroughly with running water as soon as possible after contact, then wash with soap and water or a mild detergent.
- Mouth contact or accidental swallowing: Drink large quantities of water or milk. Follow with Milk of Magnesia, beaten egg, or vegetable oil. Call a physician immediately.

15.3.3   OSHA Safety Considerations

Just following the rules is the best way to help prevent accidents. The federal government has taken a number of steps to help improve safety within the

Table 15.4 Sixteen Common OSHA Violations. Data from: National Electrical Code, NFPA No. 70.

FACT SHEET NUMBER	SUBJECT	NEC REFERENCE
1	Guarding of live parts	110-17
2	Identification	110-22
3	Uses allowed for flexible cord	400-7
4	Prohibited uses of flexible cord	400-8
5	Pull at joints and terminals must be prevented	400-10
6-1	Effective grounding, Part 1	250-51
6-2	Effective grounding, Part 2	250-51
7	Grounding of fixed equipment, general	250-42
8	Grounding of fixed equipment, specific	250-43
9	Grounding of equipment connected by cord and plug	250-45
10	Methods of grounding, cord and plug connected equipment	250-59
11	Alternating-current circuits and systems to be grounded	250-5
12	Location of overcurrent devices	240-24
13	Splices in flexible cords	400-9
14	Electrical connections	110-14
15	Marking equipment	110-21
16	Working clearances about electrical equipment	110-16

workplace. OSHA helps industries monitor and correct safety practices. Although OSHA may have received some bad press when it was conceived, the organization has developed a number of useful guidelines that could help prevent accidents within your facility. OSHA records show that electrical standards are among the most frequently violated of all safety standards. Table 15.4 lists 16 of the most common electrical violations.

Protective Covers. Exposure of live conductors is a common safety violation. All potentially dangerous electrical conductors should be covered with protective panels. How many equipment rooms have you seen with a breaker panel or disconnect switch cover removed? Have you ever operated a piece of high-voltage equipment with the cover panels removed? If so, this was in violation of OSHA regulations. The danger is that someone may come into contact with the exposed, current-carrying conductors. It also is possible for metallic objects such as ladders, cable, or tools to contact a hazardous voltage, creating a life-threatening condition. Open panels also represent a fire hazard.

Identification and Marking. Circuit breakers and switch panels should be identified and labeled properly. The labels on breakers and equipment switches may be years old, and may no longer accurately describe the equipment that is actually in use. This is a safety hazard. Although the facility engineer may know which breaker controls what equipment, other staff members or the fire department may not be able to discern that information from the labels. Improper labeling of the circuit panel may lead to casualties or unnecessary damage, if the only person who understands the system is unavailable in an emergency. If there are a number of devices connected to a single disconnect switch or breaker, provide a diagram or drawing for clarification. Label with brief phrases and use clear, permanent, and legible markings.

Equipment marking is a closely related area of concern. This is not the same thing as equipment identification. Marking equipment means labeling the equipment breaker panels and ac disconnect switches according to device rating. Breaker boxes should have nameplates showing the manufacturer, rating, and other pertinent electrical factors. The intent of this rule is to prevent devices from being subjected to excessive loads or voltages. If you are using old, unmarked switches or breakers, contact the manufacturer for rating information. Then, properly label them with the approved ratings.

Extension Cords. Extension (flexible) cords often are misused. Although it may be easy to connect a new piece of equipment with a flexible cord, be careful. The National Electrical Code (NEC) lists only eight approved uses for flexible cords.

Have you ever "installed" an additional ac outlet by using a flexible cord? Do not; this practice is not permitted. Use of a flexible cord where the cable passes through a hole in the wall, ceiling, or floor is an often-violated rule. Running the cord through doorways, windows, or similar openings also is prohibited. A flexible cord should not be attached to building surfaces or concealed behind building walls or ceilings. These common violations are summarized in Figure 15.6.

Along with improper use of flexible cords, failing to provide adequate strain relief on connectors is a common problem. Whenever possible, use manufactured

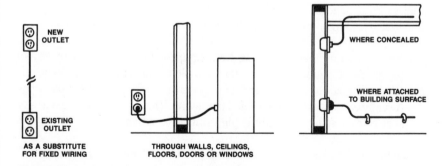

Figure 15.6   Using flexible cord for permanent installations and in concealed locations, as shown, is not permitted under NEC regulations. (Courtesy of *Broadcast Engineering* magazine, Intertec Publishing, Overland Park, Kan.)

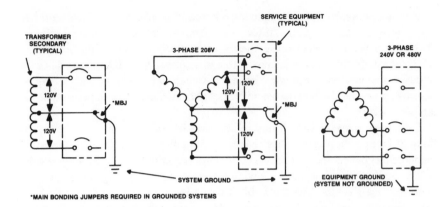

Figure 15.7 Even though regulations have been in place for many years, OSHA inspections still uncover violations in the grounding of primary electrical service systems. Note that the main bonding jumpers are required in only two of the designs. (Courtesy of *Broadcast Engineering* magazine, Intertec Publishing, Overland Park, Kan.)

cable connections. If you must make your own extension cords or attach power plugs, be sure to install a strain relief.

Grounding. OSHA regulations describe two types of grounding; *system grounding* and *equipment grounding.* System grounding actually connects one of the current-carrying conductors (such as the terminals of a supply transformer) to ground. (See Figure 15.7.) Equipment grounding connects all of the noncurrent-carrying metal surfaces together and to ground. From a grounding standpoint, the only difference between a grounded electrical system and an ungrounded electrical system is that the *main-bonding jumper* from the service equipment ground to a current-carrying conductor is omitted in the ungrounded system.

The system ground performs two functions:

1. It provides the final connection from equipment-grounding conductors to the grounded circuit conductor, thus completing the ground-fault loop.
2. It ties the electrical system and its enclosures solidly to their surroundings (usually earth, structural steel, and plumbing). This prevents voltages at any source from rising to harmfully high voltage-to-ground levels.

It should be noted that equipment grounding—bonding all electrical equipment to ground—is required whether or not the system is grounded. System grounding should be handled by the electrical contractor installing your power feeds.

The purpose of equipment grounding is twofold:

Table 15.5   Major Points to Consider When Developing a Facility Safety Program

- Management assumes the leadership role regarding safety policies.
- Responsibility for safety and health-related activities is clearly assigned.
- Hazards are identified, and steps are taken to eliminate them.
- Employees at all levels are trained in proper safety procedures.
- Thorough accident/injury records are maintained.
- Medical attention and first aid are readily available.
- Employee awareness and participation are fostered through incentives and an ongoing, high-profile approach to workplace safety.

1. It bonds all surfaces together so that there can be no voltage difference among them.
2. It provides a ground-fault current path from a fault location back to the electrical source, so that if a fault current develops, it will rise to a level high enough to operate the breaker or fuse.

The National Electrical Code is complex, and it contains numerous requirements concerning electrical safety. If your facility's electrical wiring system has gone through many changes over the years, it might be wise to have it inspected by a qualified consultant. You can perform many of the checks yourself by using the fact sheets shown in Table 15.4. The fact sheets are available from your local OSHA office.

15.3.4   Management Responsibility

The key to operating a safe facility is diligent management. If your facility does not yet have a safety program, now is the time to start one. A carefully thought-out plan ensures a coordinated approach to protecting staff members from injury and the station from potential litigation. Although the details and overall organization may vary, seven basic elements are present in workplaces that have good accident-prevention programs. These practices are summarized in Table 15.5. A more detailed examination of the seven points follows.

Management Leadership. The position of management is crucial to the success of any safety program. The attitude of managers toward job safety will be reflected by the employees. If the manager is not interested in preventing accidents, nobody else is likely to be. Following are some useful steps that might be taken to develop an effective safety program:

- If you have not yet done so, post the OSHA workplace poster, "Job Safety and Health Protection," where all employees can see it. This is an OSHA requirement.

- Hold a meeting to discuss job safety with employees and talk about mutual responsibilities. If necessary, write a policy statement on safety, and display it near the OSHA poster.
- Establish a management review process for all accident reports to ensure follow-up when needed. Managers should comment favorably on good practices and work to correct any unsafe work practices that are discovered.
- Set a good example for others in the safe way you go about your work.

Assign Responsibility. After a basic policy has been developed, it must be implemented. The key is to delegate the details for the policy to those who normally perform supervisory duties. These people are in the best position to oversee any safety program. If you perform primary supervision, then you should assume the task.

Identify Hazards. Identify any hazards that may exist within the facility. Install the necessary procedures to control or eliminate them as soon as possible. This step may require a careful tour of your operation. Sample checklists are available from your local OSHA office.

Training. Train those who will implement the program, and make them aware of the overall goals. Be sure that all employees, especially new or reassigned workers, receive adequate training. If there are situations or areas in which special protective equipment is required, be sure that employees know how to use it.

Maintain Good Records. From a legal standpoint, it is far better to have a complete record of all accidents and injuries on the job than to claim no knowledge of them. Obtain a report for every injury requiring medical treatment (other than basic first aid). Record each injury on OSHA form No. 200. If there are other types of occupational injuries and illness for recordable cases, use form No. 101. Workers' compensation forms also may be used for these reports. Prepare an annual summary, OSHA form No. 200, and post it no later than Feb. 1. This form must be remain posted until March 1, and all these records must be maintained for 5 years.

First Aid. Maintain sufficient medical supplies, and be sure someone on your staff knows basic first aid procedures. This step is not required if the facility is near an infirmary, clinic, or hospital.

Employee Awareness. If you demonstrate concern about safety, it is likely your employees also will show interest. Display safety pamphlets, and recruit employee help in identifying hazards. Reward your employees for proper safety performance. Often, an incentive program helps to maintain safe work practices. Eliminate any hazards identified, and obtain OSHA forms or first aid supplies that may be needed. The OSHA *Handbook for Small Business* outlines the legal requirements imposed by the Occupational Safety and Health Act of 1970. The handbook, which is available from your local OSHA office, also suggests ways in which a company can develop an effective safety program.

Free on-site consultations also are available. A consultant will tour your facility and offer practical advice about safety. These consultants do not issue citations, propose penalties or routinely provide information about you or your workplace conditions to the federal inspection staff. Contact the nearest OSHA office for additional information. Table 15.6 provides a basic checklist of safety points for consideration.

The key to a successful program is making safety a part of your regular business activity. Build the necessary standards into your daily operating practices. As an engineer, you should be concerned about safety. Your livelihood — and even your life — may depend on how effective you are in maintaining a safe work environment.

15.4   TOWER MAINTENANCE AND SAFETY

The transmitting tower is the most visible and expensive component of most transmission systems. A comprehensive maintenance program is essential to the safety of personnel who work in the vicinity of the tower, and to the long-term survival of the structure. The FCC addresses tower safety in Vol. 1, Part 17 of its rules. The commission specifies the painting and lighting requirements for towers ranging from less than 250 ft. to more than 2000 ft.

By specifying paints, colors, lamps, and lighting filters, the FCC addresses the safety concerns that relate to aircraft. OSHA focuses on personnel safety matters. The organization's main concern is for the people who must work on and around the tower structure. Many cities have statutes that apply to antennas and towers, and to the safety of the general public. In some cases, the local regulations may be far more stringent than any demands made by the FCC and OSHA.

To some, the various tiers of regulations may seem to be overkill. Yet, the tower represents a liability for which the facility owner is solely responsible. The regulatory efforts at each level are intended to aid industry in reducing such liability.

15.4.1   On the Tower

The majority of work to be done on existing towers will be limited to two projects: replacing lamps and/or colored lighting filters, and occasional painting. Make sure a reliable tower service company is contracted to perform these tasks. Replacement lamps should be kept in stock to assure that any failures can be corrected quickly. Painting is a long-term project. Even though a reputable service company is contracted to do the work, the facility engineer should oversee the project. Not only must the paint and its method of application answer FCC and FAA requirements, but painting may have a direct effect on the longevity of the structure. Unlike

Table 15.6   Sample Checklist of Important Safety Items

Refer regularly to this checklist to maintain a safe facility. For each category shown, be sure that:

ELECTRICAL SAFETY

- Fuses of the proper size have been installed.
- All ac switches are mounted in clean, tightly closed metal boxes.
- Each electrical switch is marked to show its purpose.
- Motors are clean and free of excessive grease and oil.
- Motors are properly maintained and provided with adequate overcurrent protection.
- Bearings are in good condition.
- Portable lights are equipped with proper guards.
- All portable equipment is double-insulated or properly grounded.
- The facility electrical system is checked periodically by a contractor competent in the NEC.
- The equipment grounding conductor or separate ground wire has been carried all the way back to the supply conductor.
- All extension cords are in good condition, and the grounding pin is not missing or bent.

EXITS AND ACCESS

- All exits are visible and unobstructed.
- All exits are marked with a readily visible, properly illuminated sign.
- There are sufficient exits to ensure prompt escape in case of an emergency.

FIRE PROTECTION

- Portable fire extinguishers of the appropriate type are provided in adequate numbers.
- All remote vehicles have proper fire extinguishers.
- Fire extinguishers are inspected monthly for general condition and operability, and noted on the inspection tag.
- Fire extinguishers are mounted in readily accessible locations.
- The fire alarm system is tested annually.

noncoated metal, galvanized metal usually does not require painting. Even so, the zinc coating of galvanized metal eventually will wear away. Seven points are suggested by the National Association of Broadcasters with regard to contracting a tower-painting project:

- Purchase the paint yourself, in accordance with FCC rules.
- Require brushing, scraping, and priming of the structure before the final coat of paint is applied.
- Stipulate that painting is to be done only when the tower is dry and the relative humidity is low.
- Allow painting to be done only if the tower temperature is at least 50°F.
- Apply paint as specified by the manufacturer.
- Inform the painters that the job will be inspected as the work progresses.
- Request evidence that the painting company has insurance coverage for its employees.

Insurance coverage should include workers' liability, contractor's public liability, contractor's protective liability, automobile liability, and direct-damage insurance. Ask your insurance representative to investigate the evidence of coverage.

15.4.2   Inspection

As with any other piece of equipment at a facility, preventive maintenance may avoid costly problems in regard to the tower. Insurance requirements often specify that complete inspection of the tower be made annually as a requisite for renewal of coverage. In some areas, local statutes also require inspections to be made and the results approved by a professional engineer. During such inspections, search for any signs of wear, stress or corrosion of the tower, all attachments, and hardware.

Check guy wires, guy attachments on the tower, guy anchors, and insulators (if used). Include all ladders, catwalks, ice bridges, and other ancillary parts of the tower in the inspection. Pay special attention to concrete foundations, the ball gap on AM towers, and weep holes on the base insulator (if used). In addition, check guy tensions, linearity, and plumb of the tower. These factors can be checked occasionally by facility personnel if the proper equipment is available. A tower structural consultant will be able to assist in developing a formal inspection plan. Checks throughout the year, especially after storms or high winds, can help in the detection and prevention of a potential tower failure.

15.5   NONIONIZING RADIATION

Nonionizing radio frequency radiation (RFR) resulting from high-intensity RF fields represents a growing concern to technicians who must work around high-power transmission equipment. The principal medical concern regarding nonionizing radiation involves heating of various body tissues, which can have serious effects, particularly if there is no mechanism for heat removal. Recent research also has noted subtle psychological and physiological changes, in some cases, in workers at radiation levels below the threshold for heat-induced biological changes. However, the consensus is that most effects are thermal in nature.

High levels of RFR can, according to some studies, affect one or more body systems or organs. Those areas identified to date as potentially sensitive involve:

- Metabolic effects on the central nervous system and cardiac system.
- Reproductive, developmental, and genetic effects (particularly at high exposure levels).
- The ocular (eye) system.
- The immunological system.

In spite of these studies, many of which are continuing, there is still no clear evidence in Western literature that exposure to medium-level nonionizing radiation results in detrimental effects. Soviet findings, on the other hand, suggest that occupational exposure to RFR at power densities above 1.0 mW/cm^2 does cause symptoms, particularly in the central nervous system.

Clearly, the jury is still out as to the ultimate biological effects of RFR. Until the situation is more clearly defined, however, the assumption must be made that potentially serious effects may result from excessive exposure. Compliance with existing standards should be the minimum goal, for members of the general public as well as for transmission facility employees.

15.5.1   NEPA Mandate

The National Environmental Policy Act of 1969 required the FCC to place controls on nonionizing radiation. The purpose was to prevent harm to the public at large and to those who must work near sources of the radiation. Action was delayed because no hard evidence existed that low- and medium-level RF energy was harmful to human life. Also, there was no evidence showing that radio waves from radio and television stations constituted a health hazard.

During the delay, many studies were carried out in an attempt to identify those levels of radiation that might be harmful. From the research, suggested limits were developed by the American National Standards Institute (ANSI) and stated in the document known as ANSI C95.1-1982. The protection criteria outlined in the standard are shown in Figure 15.8.

The energy level criteria were developed by representatives from a number of industries and educational institutions after performing research on the possible effects of nonionizing radiation. The projects focused on absorption of RF energy by the human body, based upon simulated human body models. In preparing the document, ANSI attempted to determine those levels of incidental radiation that

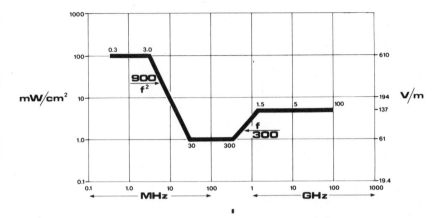

Figure 15.8 The power-density limits for nonionizing radiation exposure for humans. (Source: Donald Markley, "Complying with RF Emission Standards," *Broadcast Engineering* magazine, Intertec Publishing, Overland Park, Kan., May 1986)

would cause the body to absorb less than 0.4 W/kg of mass (averaged over the whole body) or peak absorption values of 8 W/kg over any 1 g of body tissue.

From the data, the researchers found that energy was absorbed more readily at some frequencies than at others. The absorption rates were found to be functions of the size of a specific individual and the frequency of the signal being evaluated. It was the results of these absorption rates that culminated in the shape of the *safe curve* shown in the figure. ANSI concluded that no harm would occur to individuals exposed to radio energy fields, as long as specific values were not exceeded when averaged over a period of 0.1 hours. Also, higher values for a brief period would not pose difficulties, if the levels shown in the standard document were not exceeded when averaged over the 0.1 hour time period.

The FCC adopted ANSI C95.1-1982 as a standard that would provide adequate protection to the public and to industry personnel who are involved in working around antenna structures. The regulation provides the communications industry with established criteria to protect the population at large, as well as facility operators. As long as a transmission facility complies with the ANSI document, a reasonable response can be offered to those who express fears about the possible effects of station operation. Unfortunately, not all communities have agreed with the ANSI standard. In several areas, specific communities have attempted to enforce their own standards, which involve less realistic and far more stringent values.

In adopting the standard, the commission has placed a requirement on all transmission stations to comply with ANSI C95.1-1982. Existing stations were grandfathered without having to show compliance until a *major action* occurred. A major action is defined as a renewal or modification that requires a new application to be filed with the commission. According to this definition, all stations eventually must show compliance.

The commission has ruled that compliance with the standard will be handled through self-certification. That is, if station personnel think that they have complied with the regulations, it is necessary only to mark a box on the application to indicate that it is not a major environmental action. A brief statement of explanation must be filed with the application. Then, unless an objection is filed, or the commission has reason to suspect that the operation does not comply, no further action will be necessary.

To assist individual stations, the commission has published a series of curves and graphs for engineers to use as rough guidelines to determine compliance status. Any assumptions that have been made in the preparation of the tables and graphs have been weighted toward greater amounts of protection. Compliance also can be determined by means other than the curves or tables in the guidelines. If station operation does not comply according to the tables, compliance may be shown through a more detailed analysis. Such an analysis requires calculations or measurements with an explanation of the assumptions or data upon which the calculations are based.

If should be noted that the standard concerns only areas that are accessible and where excessive levels may exist. For stations where problems are found, it may only be necessary to construct a fence around the tower at the required distance. In addition, no unauthorized personnel may be permitted inside that fence. For engineers who need to enter the area for a brief period, the exposure is averaged over 0.1-hour intervals. On the other hand, there may be locations where compliance with C95.1-1982 will prove to be expensive or difficult for licensees. In particular, multistation sites located on large towers and large buildings may require auxiliary antennas mounted at a second level to allow adjustment and maintenance of the primary systems. Also, problems may be encountered on mountaintop facilities where the tower is only tall enough to keep the antenna clear of the transmitter building.

The complete radiation report, prepared by the FCC Office of Science and Technology, is detailed and lengthy. However, it is recommended that maintenance engineers obtain and read a copy, as well as the ANSI C95.1-1982 document. By reviewing both documents, station engineers can gain an understanding of the thoroughness with which the standard was determined and the reasonable approach that the commission has taken in allowing the self-certification process.

Multiple-User Sites. At a multiple-user site, the responsibility for assessing the RFR situation (although officially triggered by either by the addition of a new user or the license renewal of all site tenants) is, in reality, a joint effort by all site tenants. In a multiple-user environment involving various frequencies, and thus various protection criteria, compliance is indicated when the fraction of the RFR limit within each pertinent frequency band is established and added to the sum of all the other fractional contributions. The sum must not be greater than one (1.0). Evaluating the multiple-user environment is not a simple matter. Corrective actions, if indicated, may be quite complex. In general, a group of stations sharing such an environment would be well-advised to secure the services of a consulting firm well-versed in such work (and, ideally, not the primary consultant to any member of the group) to perform the initial assessment as well as to provide recommendations for any necessary modifications to the facilities of the various users. An impartial "referee" often can prevent unnecessary conflict and arbitrary assignment of responsibility among the members of a group.

15.5.2 Operator Safety Considerations

RF energy must be contained properly by shielding and transmission lines. All input and output RF connections, cables, flanges, and gaskets must be RF-leakproof. Never operate a power tube without a properly matched RF energy absorbing load attached. Never look into or expose any part of the body to an antenna or open RF-generating tube, circuit or RF transmission system while it is energized. Monitor the RF system for radiation leakage at regular intervals and after servicing.

15.6 POLYCHLORINATED BIPHENYLS

The problem of *polychlorinated biphenyls* (PCBs) in the environment has received a significant amount of publicity. Yet, many maintenance engineers still do not recognize the problems or costs associated with mishandling of PCBs. PCBs belong to a family of organic compounds known as *chlorinated hydrocarbons*. Virtually all PCBs in existence today have been synthetically manufactured. PCBs have a heavy oil-like consistency, a high boiling point, a high degree of chemical stability, low flammability, and low electrical conductivity. These characteristics resulted in the past widespread use of PCBs in high-voltage capacitors and transformers. Commercial products containing PCBs were distributed widely from 1957 to 1977 under several trade names including:

- Aroclor
- Pyroclor
- Sanotherm
- Pyranol
- Askarel

Table 15.7   Commonly Used Names for PCB Insulating Material

Apirolio	Abestol	Askarel	Aroclor B	Chlorextol	Chlophen
Chlorinol	Clorphon	Diaclor	DK	Dykanol	EEC-18
Elemex	Eucarel	Fenclor	Hyvol	Inclor	Inerteen
Kanechlor	No-Flamol	Phenodlor	Pydraul	Pyralene	Pyranol
Pyroclor	Sal-T-Kuhl	Santothern FR	Santovac	Solvol	Therminal

Askarel also is a generic name used for nonflammable dielectric fluids containing PCBs. Table 15.7 lists some common trade names used for Askarel. These trade names typically will be listed on the nameplate of a PCB transformer or capacitor.

15.6.1   Health Risk

PCBs are harmful because, once they are released into the environment, they tend not to break apart into other substances. Instead, PCBs persist, taking several decades to slowly decompose. By remaining in the environment, they can be taken up and stored in the fatty tissues of all organisms, from which they are released slowly into the bloodstream. Therefore, because of the storage in fat, the concentration of PCBs in body tissues can increase with time, even though PCB exposure levels may be quite low. This process is called *bioaccumulation*. Furthermore, as PCBs accumulate in the tissues of simple organisms, and as they are consumed by progressively higher organisms, the concentration increases. This process is called *biomagnification*. These two factors are especially significant because PCBs are harmful even at low levels. Specifically, PCBs have been shown to cause chronic (long-term) toxic effects in some species of animals and aquatic life. Well-documented tests on laboratory animals show that various levels of PCBs can cause reproductive effects, gastric disorders, skin lesions, and cancerous tumors. PCBs may enter the body through the lungs, the gastrointestinal tract, and the skin. After absorption, PCBs circulate in the bloodstream and are stored in fatty tissues and a variety of organs, including the liver, kidneys, lungs, adrenal glands, brain, heart, and skin.

The health risk lies not only in the PCB itself, but also in the chemicals developed when PCBs are heated. Laboratory studies have confirmed that PCB by-products, including *polychlorinated dibenzofurans* (PCDFs) and *polychlorinated dibenzo-p-dioxins* (PCDDs), are formed when PCBs or chlorobenzenes are heated to temperatures ranging from approximately 900°F to 1300°F. Unfortunately, these products are more toxic than PCBs themselves.

The liability from a PCB spill or fire can be tremendous. For example, a Binghamton, NY, fire involving a large PCB transformer resulted in $20 million in cleanup expenses. The consequences of being responsible for a fire-related incident with a PCB transformer may be monumental.

15.6.2   Governmental Action

The U.S. Congress took action to control PCBs in October 1975, by passing the Toxic Substances Control Act (TSCA). A section of this law specifically directed the EPA to regulate PCBs. Three years later, the EPA issued regulations to implement a congressional ban on the manufacture, processing, distribution, and disposal of PCBs. Since, several revisions and updates have been issued by the EPA. One of these revisions, issued in 1982, specifically addressed the type of equipment used in industrial plants and broadcast stations. Failure to properly follow the rules regarding the use and disposal of PCBs has resulted in high fines and some jail sentences.

Although PCBs no longer are being produced for most electrical products in the United States, the EPA estimates that more than 107,000 PCB transformers and 350 million small PCB capacitors were in use or in storage in 1984. Approximately 77,600 of these transformers were used in or near commercial buildings. Approximately 3.3 million large PCB capacitors were in use as late as 1981. The threat of widespread contamination from PCB fire-related incidents is one reason behind the EPA's efforts to reduce the number of PCB products in the environment. The users of high-power equipment are affected by the regulations primarily because of the widespread use of PCB transformers and capacitors. These components usually are located in older (pre-1979) systems, so this is the first place to look for them. However, some facilities also maintain their own primary power transformers. Unless these transformers are of recent vintage, it is quite likely that they too contain PCB dielectrics. Table 15.8 lists the primary classifications of PCB devices.

15.6.3   PCB Components

The two most common PCB components are transformers and capacitors. A PCB transformer is one containing at least 500 ppm (parts per million) PCBs in the dielectric fluid. An Askarel transformer generally contains 600,000 ppm or more. A PCB transformer may be converted to a *PCB-contaminated device* (50 to 500 ppm) or a *non-PCB device* (less than 50 ppm) by having it drained, refilled and tested. The testing must not take place until the transformer has been in service for a minimum of 90 days. Note, this is *not* something that a maintenance technician can do. It is the exclusive domain of specialized remanufacturing companies.

PCB transformers must be inspected quarterly for leaks. If an impervious dike is build around the transformer sufficient to contain all the liquid material, the inspections can be conducted yearly. Similarly, if the transformer is tested and found to contain less than 60,000 ppm, a yearly inspection is sufficient. Failed PCB transformers cannot be repaired; they must be disposed of properly. If a leak develops, it must be contained, and daily inspections must begin. A cleanup must be initiated as soon as possible, but no later than 48 hours after the leak is discovered. Adequate records must be kept of all inspections, leaks, and actions

Table 15.8   Definition of PCB terms as Identified by the EPA

TERM	DEFINITION	EXAMPLES
PCB	Any chemical substance that is limited to the biphenyl molecule that has been chlorinated to varying degrees, or any combination of substances that contain such substances.	PCB dielectric fluids, PCB heat transfer fluids, PCB hydraulic fluids, 2,2',4-trichlorobiphenyl
PCB article	Any manufactured article, other than a PCB container, that contains PCBs and whose surface(s) has been in direct contact with PCBs.	Capacitors, transformers, electric motors, pumps, pipes
PCB container	A device used to contain PCBs or PCB articles and whose surface(s) has been in direct contact with PCBs.	Packages, cans, bottles, bags, barrels, drums, tanks
PCB article container	A device used to contain PCB articles or equipment, and whose surface(s) has not been in direct contact with PCBs.	Packages, cans, bottles, bags, barrels, drums, tanks
PCB equipment	Any manufactured item, other than a PCB container or PCB article container, which contains a PCB article or other PCB equipment.	Microwave ovens, fluorescent light ballasts, electronic equipment
PCB item	Any PCB article, PCB article container, PCB container, or PCB equipment that deliberately or unintentionally contains, or has as a part of it, any PCBs.	
PCB transformer	Any transformer that contains PCBs in concentrations of 500 ppm or greater.	
PCB contaminated	Any electrical equipment that contains more than 50, but less than 500 ppm, of PCBs. (Oil-filled electrical equipment other than circuit breakers, reclosers, and cable whose PCB concentration is unknown must be assumed to be PCB contaminated electrical equipment.)	Transformers, capacitors, circuit breakers, re-closers, voltage regulators, switches, cable, electromagnets

taken for 3 years after disposal of the component. Combustible materials must be kept a minimum of 5 m from a PCB transformer and its enclosure.

As of Oct. 1, 1990, the use of PCB transformers (500 ppm or greater) was prohibited in or near commercial buildings when the secondary voltages are 480 V ac or higher. Radial PCB transformers can continue to be used, provided certain electrical protection is provided. The EPA regulations also require that the operator notify others of the possible dangers. All PCB transformers (including PCB transformers in storage for reuse) must be registered with the local fire department. Supply the following information:

- The location of the PCB transformer(s).
- Address(es) of the building(s). For outdoor PCB transformers, the location.
- Principal constituent of the dielectric fluid in the transformer(s).
- Name and telephone number of the contact person in the event of a fire involving the equipment.

Any PCB transformers used in a commercial building must be registered with the building owner. All building owners within 30 m of such PCB transformers also must be notified. In the event of a fire-related incident involving the release of PCBs, the Coast Guard National Spill Response Center must be notified immediately (telephone 800-424-8802). Appropriate measures also must be taken to contain and control any possible PCB release into water.

Capacitors are divided into two size classes: *large* and *small*. A PCB small capacitor contains less than 1.36 kg (3 lbs) of dielectric fluid. A capacitor having less than 100 cubic inches is also considered to contain less than 3 lbs of dielectric fluid. A PCB large capacitor has a volume of more than 200 cubic inches and is considered to contain more than 3 lbs of dielectric fluid. Any capacitor having a volume between 100 and 200 cubic inches is considered to contain 3 lbs of dielectric, provided the total weight is less than 9 lbs. A PCB *large high-voltage capacitor* contains 3 lbs or more of dielectric fluid and operates at voltages of 2 kV or greater. A *large low-voltage capacitor* also contains 3 lbs or more of dielectric fluid, but operates at below 2 kV.

The use, servicing, and disposal of PCB small capacitors is not restricted by the EPA unless there is a leak. In that event, the leak must be repaired or the capacitor disposed of. Disposal may be performed by an approved incineration facility, or the component may be placed in a specified container and buried in an approved chemical waste landfill. Currently, chemical waste landfills are for disposal only of liquids containing 50 to 500 ppm PCBs, or for solid PCB debris. Items such as capacitors that are leaking oil containing greater than 500 ppm PCBs should be taken to an EPA-approved PCB disposal facility.

15.6.4   Identifying PCB Components

The most important task for the maintenance manager is to identify any PCB items on the premises. Equipment built after 1979 probably does not contain any PCB-filled devices. To be sure, however, inspect all capacitors, transformers, and power switches. A call to the manufacturer may help. Older equipment (pre-1979) is more likely to contain PCB transformers and capacitors. A liquid-filled transformer usually has cooling fins, and the nameplate may provide useful information about its contents. If the transformer is unlabeled or the fluid is not identified, it must be treated as a PCB transformer. Untested (not analyzed) mineral-oil-filled transformers are assumed to contain at least 50 ppm PCBs, but less than 500 ppm PCBs. This places them in the category of PCB-contaminated electrical equipment, which has different requirements than PCB transformers. Older high-voltage systems are likely to include both large and small PCB capacitors. Equipment rectifier panels, exciter/modulators, and power amplifier cabinets may contain a significant number of small capacitors. In older equipment, these capacitors often are Askarel-filled. Unless leaking, such devices pose no particular hazard. If a leak does develop, proper disposal techniques must be followed. Also, liquid-cooled rectifiers may contain Askarel. Even though their use is not regulated, treat them as PCB articles, with 50 ppm or more PCBs. Never make assumptions about PCB contamination; check with the manufacturer to be sure.

Any PCB article or PCB container being stored for disposal must be date-tagged when removed, and inspected for leaks every 30 days. It must be removed from storage and disposed of within 1 year from the date it was placed in storage. Items stored for disposal must be kept in a storage facility meeting the requirements of 40 CFR (Code of Federal Regulations), sec. 761.65(b)(1), unless they fall under alternate regulation provisions. There is a difference between PCB items stored for disposal and those stored for reuse. After an item has been removed from service and tagged for disposal, it cannot be returned to service. Disposal must conform to EPA guidelines.

15.6.5   Labeling PCB Components

After identifying PCB devices, proper labeling is the second step that must be taken. PCB article containers, PCB transformers, and large high-voltage capacitors must be marked with the standard 6-in x 6-in large marking label (ML) as shown in Figure 15.9. Equipment containing these transformers or capacitors also should be marked. PCB large low-voltage capacitors (less than 2 kV) need not be labeled until removed from service. If the capacitor or transformer is too small to hold the large label, a smaller 1-in x 2-in size is approved for use. Labeling each PCB small capacitor is not required. However, any equipment containing PCB small capacitors should be labeled on the outside of the cabinet or on access panels. Properly

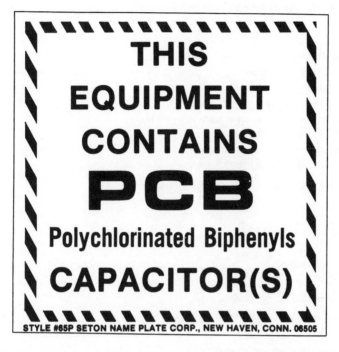

Figure 15.9   The large marking label (ML) is used to identify PCB transformers and PCB large capacitors. (Courtesy of *Broadcast Engineering* magazine, Intertec Publishing, Overland Park, Kan.)

label any spare capacitors and transformers that fall under the regulations. Use the large label to identify any doors, cabinet panels, or other means of access to PCB transformers. The label must be placed where it can be read easily by firefighters. All areas used for storage of PCBs and PCB items for disposal must be marked with the large 6-in x 6-in PCB label.

15.6.6   Record Keeping

Inspections are a critical component in the management of PCBs. EPA regulations specify a number of steps that must be taken and the information that must recorded. Table 15.9 summarizes the schedule requirements. Table 15.10 can be used as a checklist for each transformer inspection. This record must be retained for 3 years. In addition to the inspection records, some facilities may need to maintain an annual report. This report details the number of PCB capacitors, transformers, and other PCB items on the premises. The report must contain the dates the items were removed from service, their disposition, and detailed information regarding their

Table 15.9   Inspection Schedule Required for PCB Transformers and Other Contaminated Devices

PCB Transformers
- Standard PCB transformer . . . . Quarterly
- If full-capacity impervious dike added . . . . Annually
- If retrofitted to <60,000 ppm PCBs . . . . Annually
- If leak discovered, clean up as soon as possible . . . . Daily
  (retain these records for 3 years)
- PCB article or container stored for disposal . . . . Monthly
  (remove and dispose of within 1 year)

Retain all records for 3 years after disposing transformer.

Table 15.10   Inspection Checklist for PCB Components

Transformer location: _____

Date of visual inspection: _____

Leak discovered _____ (Yes/No)

If yes, date discovered (if different from inspection date): _____

Location of leak: _____

Name of person performing inspection: _____

Estimate of amount of dielectric fluid released from leak: _____

Date of cleanup, containment, repair, or replacement: _____

Description of cleanup, containment, or repair performed: _____

_____

Results of any containment and daily inspection required for uncorrected active leaks:

_____

characteristics. Such a report must be prepared if the facility uses or stores at least one PCB transformer containing greater than 500 ppm PCBs, 50 or more PCB large capacitors, or at least 45 kg of PCBs in PCB containers. Retain the report for 5 years after the facility ceases using or storing PCBs and PCB items in the prescribed quantities. Table 15.11 lists the required information in the annual PCB report.

15.6.7   Disposal

Disposal of PCBs is not a minor consideration. When contracting for PCB disposal, verify the company's license with the area EPA office. That office also can supply

Table 15.11   Required Information for PCB Annual Report

I.   PCB device background information:
   a.   Dates when PCBs and PCB items are removed from service.
   b.   Dates when PCBs and PCB items are placed into storage for disposal, and are placed into transport for disposal.
   c.   The quantities of the items removed from service, stored, and placed into transport are to be indicated using the following breakdown.
   - Total weight, in kilograms, of any PCB and PCB items in PCB containers, including identification of container contents (such as liquids and capacitors).
   - Total number of PCB transformers and total weight, in kilograms, of any PCBs contained in the transformers.
   - Total number of PCB large high- or low-voltage capacitors.
II.   The location of the initial disposal or storage facility for PCBs and PCB items removed from service, and the name of the facility owner or operator.
III.   Total quantities of PCBs and PCB items remaining in service at the end of calendar year according to the following breakdown:
   a.   Total weight, in kilograms, of any PCB and PCB items in PCB containers, including the identification of container contents (such as liquids and capacitors).
   b.   Total number of PCB transformers and total weight, in kilograms, of any PCBs contained in the transformers.
   c.   Total number of PCB large high- or low-voltage capacitors.

background information on the firm's compliance and enforcement history. In a worst-case scenario, your facility could be held liable for contamination damages and cleanup costs, even if the PCBs were handed off in an approved manner. Sometimes operators get into trouble simply because they don't recognize the potential hazards that might be present. In one case, a radio station that was demolished following a fire had the debris hauled to a county landfill recycling center. The waste contained some small PCB transformers, one of which broke upon being dumped. The result was that approximately 70 tons of garbage became PCB-contaminated. The recycling pit had to be scrubbed clean and three truckloads of garbage transported to and buried in a chemical waste landfill, at a cost of several hundred thousand dollars.

The fines levied in such cases are not mandated by federal regulations. Rather, the local EPA administrator, usually in consultation with local authorities, determines the cleanup procedures and costs. Civil penalties for administrative complaints issued for violations of the PCB regulations are determined according to a matrix provided in the PCB penalty policy. This policy, published in the Federal Register, considers the amount of PCBs involved and the potential for harm posed by the violation. Table 15.12 lists the costs to dispose of several PCB components

Table 15.12   Actual PCB Disposal Costs for One Radio Station

Phase 1:
- Deliver four approved drums for capacitor disposition.
- Furnish labor and materials to remove 18 oil samples, and send to laboratory for testing.

Total cost $3,823

Phase 2:
- Furnish necessary transportation of materials, and decommission and dispose of one transformer (25-in X 25-in X 44-in), and 30 gallons of Inerteen . $3,342
- Pick up and dispose of four drums of capacitors $2,538
- Extra drum of capacitors, if needed (disposal each) . . . . $ 472
- Replace transformer $2,021

Total project cost $12,196

at one facility. Although the disposal costs may seem high, they are small compared with the potential liability of cleaning up a PCB spill.

15.6.8   Proper Management

Properly managing the PCB risk is not difficult. The keys are to understand the regulations and to follow them carefully. A PCB management program should include the following steps:

- Locate and identify all PCB devices. Check all stored or spare devices.
- Properly label PCB transformers and capacitors according to EPA requirements.
- Perform the required inspections, and maintain an accurate log of PCB items, their location, inspection results, and actions taken. These records must be maintained for 3 years after disposal of the PCB component.
- Complete the annual report of PCBs and PCB items by July 1 of each year. This report must be retained for 5 years.
- Arrange for any necessary disposal through a company licensed to handle PCBs. If there are any doubts about the company's license, contact the EPA.
- Report the location of all PCB transformers to the local fire department and owners of any nearby buildings.

The importance of following the EPA regulations cannot be overstated.

15.7   BIBLIOGRAPHY

Hammar, Willie: "Occupational Safety Management and Engineering," Prentice-Hall, New York.

Markley, Donald: "Complying with RF Emission Standards," *Broadcast Engineering* magazine, Intertec Publishing, Overland Park, Kan., May 1986.

Smith, Milford K. Jr.: "RF Radiation Compliance," *Proceedings* of the 1989 Broadcast Engineering Conference, Society of Broadcast Engineers, Indianapolis, Ind.

Pfrimmer, Jack: "Identifying and Managing PCBs in Broadcast Facilities," *Proceedings* of the 1987 NAB Engineering Conference, National Association of Broadcasters, Washington, D.C., 1987.

Code of Federal Regulations, 40, Part 761.

Hazardous Waste Consultant, January/February 1984.

OSHA: *Handbook for Small Business,* U.S. Department of Labor, Washington, D.C.

OSHA: "Electrical Hazard Fact Sheets," U.S. Department of Labor, Washington, D.C., January 1987.

OSHA: "Occupational Injuries and Illnesses in the United States by Industry," OSHA Bulletin 2278, U.S. Department of Labor, Washington, D.C., 1985.

SBE Newsletter, Los Angeles chapter, Society of Broadcast Engineers, Indianapolis, Ind.

Sprague Aluminum Electrolytic Capacitor Handbook, Sprague Corp., 1989.

"Current Intelligence Bulletin no. 45," National Institute for Occupational Safety and Health Division of Standards Development and Technology Transfer, Feb. 24, 1986.

"Electrical Standards Reference Manual," U.S. Department of Labor, Washington, D.C.

"Toxics Information Series," Office of Toxic Substances, July 1983.

15.8   ACKNOWLEDGMENT

Appreciation is expressed to Charles J. Fuhrman, Fuhrman Investigations, Inc.; and William S. Watkins, P.E., for their assistance in the preparation of portions of this chapter.

16

TABLES AND
REFERENCE DATA

16.1 STANDARD UNITS

NAME	SYMBOL	QUANTITY
ampere	A	electric current
ampere per meter	A/m	magnetic field strength
ampere per square meter	A/m^2	current density
becquerel	Bg	activity (of a radionuclide)
candela	cd	luminous intensity
coulomb	C	electric charge
coulomb per kilogram	C/kg	exposure (x and gamma rays)
coulomb per sq. meter	C/m^2	electric flux density
cubic meter	m^3	volume
cubic meter per kilogram	m^3/kg	specific volume
degree Celsius	xC	Celsius temperature
farad	F	capacitance
farad per meter	F/m	permittivity
henry	H	inductance
henry per meter	H/m	permeability
hertz	Hz	frequency
joule	J	energy, work, quantity of heat
joule per cubic meter	J/m^3	energy density
joule per kelvin	J/K	heat capacity
joule per kilogram K	J/(kgyK)	specific heat capacity
joule per mole	J/mol	molar energy
kelvin	K	thermodynamic temperature
kilogram	kg	mass
kilogram per cubic meter	kg/m^3	density, mass density

NAME	SYMBOL	QUANTITY
lumen	lm	luminous flux
lux	lx	luminance
meter	m	length
meter per second	m/s	speed, velocity
meter per second sq.	m/s^2	acceleration
mole	mol	amount of substance
newton	N	force
newton per meter	N/m	surface tension
ohm	Ω	electrical resistance
pascal	Pa	pressure, stress
pascal second	Pays	dynamic viscosity
radian	rad	plane angle
radian per second	rad/s	angular velocity
radian per second sq.	rad/s^2	angular acceleration
second	s	time
siemens	S	electrical conductance
square meter	m^2	area
steradian	sr	solid angle
tesla	T	magnetic flux density
volt	V	electrical potential
volt per meter	V/m	electric field strength
watt	W	power, radiant flux
watt per meter kelvin	W/(m*K)	thermal conductivity
watt per square meter	W/m^2	heat (power) flux density
weber	Wb	magnetic flux

16.2 STANDARD PREFIXES

MULTIPLE	PREFIX	SYMBOL
10^{18}	exa	E
10^{15}	peta	P
10^{12}	tera	T
10^9	giga	G
10^6	mega	M
10^3	kilo	k
10^2	hecto	h
10	deka	da
10^{-1}	deci	d
10^{-2}	centi	c
10^{-3}	milli	m

MULTIPLE	PREFIX	SYMBOL
10^{-6}	micro	μ
10^{-9}	nano	n
10^{-12}	pico	p
10^{-15}.	femto	f
10^{-18}	atto	a

16.3   COMMON STANDARD UNITS

UNIT	SYMBOL
centimeter	cm
cubic centimeter	cm^3
cubic meter per second	m^3/s
gigahertz	GHz
gram	g
kilohertz	kHz
kilohm	kΩ
kilojoule	kj
kilometer	km
kilovolt	kV
kilovoltampere	kVA
kilowatt	kW
megahertz	MHz
megavolt	MV
megawatt	MW
megohm	MΩ
microampere	μA
microfarad	μF
microgram	μg
microhenry	μH
microsecond	μs
microwatt	μW
milliampere	mA
milligram	mg
millihenry	mH
millimeter	mm
millisecond	ms
millivolt	mV
milliwatt	mW
nanoampere	nA
nanofarad	nF

UNIT	SYMBOL
nanometer	nm
nanosecond	ns
nanowatt	nW
picoampere	pA
picofarad	pF
picosecond	ps
picowatt	pW

16.4   CONVERSION REFERENCE DATA

TO CONVERT	INTO	MULTIPLY BY
abcoulomb	statcoulombs	2.998×10^{10}
acre	sq. chain (Gunters)	10
acre	rods	160
acre	square links (Gunters)	1×10^5
acre	Hectare or sq. hectometer	0.4047
acres	sq. feet	43,560.0
acres	sq. meters	4,047
acres	sq. miles	1.562×10^{-3}
acres	sq. yards	4,840
acre-feet	cu. feet	43,560.0
acre-feet	gallons	3.259×10^5
amperes/sq. cm.	amps/sq. in.	6.452
amperes/sq. cm.	amps/sq. meter	10^4
amperes/sq. in.	amps/sq. cm.	0.1550
amperes/sq. in.	amps/sq. meter	1,550.0
amperes/sq. meter	amps/sq. cm.	10^{-4}
amperes/sq. meter	amps/sq. in.	6.452×10^{-4}
ampere-hours	coulombs	3,600.0
ampere-hours	faradays	0.03731
ampere-turns	gilberts	1.257
ampere-turns/cm.	amp-turns/in.	2.540
ampere-turns/cm.	amp-turns/meter	100.0
ampere-turns/cm.	gilberts/cm.	1.257
ampere-turns/in.	amp-turns/cm.	0.3937
ampere-turns/in.	amp-turns/meter	39.37
ampere-turns/in.	gilberts/cm.	0.4950
ampere-turns/meter	amp-turns/cm.	0.01
ampere-turns/meter	amp-turns/in.	0.0254
ampere-turns/meter	gilberts/cm.	0.01257

TO CONVERT	INTO	MULTIPLY BY
Angstrom unit	inch	3937×10^{-9}
Angstrom unit	meter	1×10^{-10}
Angstrom unit	micron or (Mu)	1×10^{-4}
are	acre (U.S.)	0.02471
ares	sq. yards	119.60
ares	acres	0.02471
ares	sq. meters	100.0
astronomical unit	kilometers	1.495×10^{8}
atmospheres	ton/sq. inch	0.007348
atmospheres	cms. of mercury	76.0
atmospheres	ft. of water (at 4°C)	33.90
atmospheres	in. of mercury (at 0°C)	29.92
atmospheres	kgs./sq. cm.	1.0333
atmospheres	kgs./sq. meter	10,332
atmospheres	pounds/sq. in.	14.70
atmospheres	tons/sq. ft.	1.058
barrels (U.S., dry)	cu. inches	7056
barrels (U.S., dry)	quarts (dry)	105.0
barrels (U.S., liquid)	gallons	31.5
barrels (oil)	gallons (oil)	42.0
bars	atmospheres	0.9869
bars	dynes/sq. cm.	10^{4}
bars	kgs./sq. meter	1.020×10^{4}
bars	pounds/sq. ft.	2,089
bars	pounds/sq. in.	14.50
Baryl	Dyne/sq. cm.	1.000
bolt (U.S. cloth)	meters	36.576
Btu	liter-atmosphere	10.409
Btu	ergs	1.0550×10^{10}
Btu	foot-lbs.	778.3
Btu	gram-calories	252.0
Btu	horsepower-hrs.	3.931×10^{-4}
Btu	joules	1,054.8
Btu	kilogram-calories	0.2520
Btu	kilogram-meters	107.5
Btu	kilowatt-hrs.	2.928×10^{-4}
Btu/hr.	foot-pounds/sec.	0.2162
Btu/hr.	gram-cal./sec.	0.0700
Btu/hr.	horsepower-hrs.	3.929×10^{-4}
Btu/hr.	watts	0.2931
Btu/min.	foot-lbs./sec.	12.96

TO CONVERT	INTO	MULTIPLY BY
Btu/min.	horsepower	0.02356
Btu/min.	kilowatts	0.01757
Btu/min.	watts	17.57
Btu/sq. ft./min.	watts/sq. in.	0.1221
bucket (Br. dry)	cubic cm.	1.818×10^4
bushels	cu. ft.	1.2445
bushels	cu. in.	2,150.4
bushels	cu. meters	0.03524
bushels	liters	35.24
bushels	pecks	4.0
bushels	pints (dry)	64.0
bushels	quarts (dry)	32.0
calories, gram (mean)	B.T.U. (mean)	3.9685×10^{-3}
candle/sq. cm.	Lamberts	3.142
candle/sq. inch	Lamberts	0.4870
centares (centiares)	sq. meters	1.0
Centigrade	Fahrenheit	$(C° \times 9/5) + 32$
centigrams	grams	0.01
centiliter	ounce fluid (U.S.)	0.3382
centiliter	cubic inch	0.6103
centiliter	drams	2.705
centiliters	liters	0.01
centimeters	feet	3.281×10^{-2}
centimeters	inches	0.3937
centimeters	kilometers	10^{-5}
centimeters	meters	0.01
centimeters	miles	6.214×10^{-6}
centimeters	millimeters	10.0
centimeters	mils	393.7
centimeters	yards	1.094×10^{-2}
centimeter-dynes	cm.-grams	1.020×10^{-3}
centimeter-dynes	meter-kgs.	1.020×10^{-8}
centimeter-dynes	pound-feet	7.376×10^{-8}
centimeter-grams	cm.-dynes	980.7
centimeter-grams	meter-kgs.	10^{-5}
centimeter-grams	pound-feet	7.233×10^{-5}
centimeters of mercury	atmospheres	0.01316
centimeters of mercury	feet of water	0.4461
centimeters of mercury	kgs./sq. meter	136.0
centimeters of mercury	pounds/sq. ft.	27.85

TO CONVERT	INTO	MULTIPLY BY
centimeters of mercury	pounds/sq. in.	0.1934
centimeters/sec.	feet/min.	1.9686
centimeters/sec.	feet/sec.	0.03281
centimeters/sec.	kilometers/hr.	0.036
centimeters/sec.	knots	0.1943
centimeters/sec.	meters/min.	0.6
centimeters/sec.	miles/hr.	0.02237
centimeters/sec.	miles/min.	3.728×10^{-4}
centimeters/sec./sec.	feet/sec./sec.	0.03281
centimeters/sec./sec.	kms./hr./sec.	0.036
centimeters/sec./sec.	meters/sec./sec.	0.01
centimeters/sec./sec.	miles/hr./sec.	0.02237
chain	inches	792.00
chain	meters	20.12
chains (surveyors' or Gunter's)	yards	22.00
circular mils	sq. cms.	5.067×10^{-6}
circular mils	sq. mils	0.7854
circular mils	sq. inches	7.854×10^{-7}
circumference	Radians	6.283
cords	cord feet	8
cord feet	cu. feet	16
coulomb	statcoulombs	2.998×10^{9}
coulombs	faradays	1.036×10^{-5}
coulombs/sq. cm.	coulombs/sq. in.	64.52
coulombs/sq. cm.	coulombs/sq. meter	10^{4}
coulombs/sq. in.	coulombs/sq. cm.	0.1550
coulombs/sq. in.	coulombs/sq. meter	1,550
coulombs/sq. meter	coulombs/sq. cm.	10^{-4}
coulombs/sq. meter	coulombs/sq. in.	6.452×10^{-4}
cubic centimeters	cu. feet	3.531×10^{-5}
cubic centimeters	cu. inches	0.06102
cubic centimeters	cu. meters	10^{-6}
cubic centimeters	cu. yards	1.308×10^{-6}
cubic centimeters	gallons (U.S. liq.)	2.642×10^{-4}
cubic centimeters	liters	0.001
cubic centimeters	pints (U.S. liq.)	2.113×10^{-3}
cubic centimeters	quarts (U.S. liq.)	1.057×10^{-3}
cubic feet	bushels (dry)	0.8036
cubic feet	cu. cms.	28,320.0

TO CONVERT	INTO	MULTIPLY BY
cubic feet	cu. inches	1,728.0
cubic feet	cu. meters	0.02832
cubic feet	cu. yards	0.03704
cubic feet	gallons (U.S. liq.)	7.48052
cubic feet	liters	28.32
cubic feet	pints (U.S. liq.)	59.84
cubic feet	quarts (U.S. liq.)	29.92
cubic feet/min.	cu. cms./sec.	472.0
cubic feet/min.	gallons/sec.	0.1247
cubic feet/min.	liters/sec.	0.4720
cubic feet/min.	pounds of water/min.	62.43
cubic feet/sec.	million gals./day	0.646317
cubic feet/sec.	gallons/min.	448.831
cubic inches	cu. cms.	16.39
cubic inches	cu. feet	5.787×10^{-4}
cubic inches	cu. meters	1.639×10^{-5}
cubic inches	cu. yards	2.143×10^{-5}
cubic inches	gallons	4.329×10^{-3}
cubic inches	liters	0.01639
cubic inches	mil-feet	1.061×10^{5}
cubic inches	pints (U.S. liq.)	0.03463
cubic inches	quarts (U.S. liq.)	0.01732
cubic meters	bushels (dry)	28.38
cubic meters	cu. cms.	10^{6}
cubic meters	cu. feet	35.31
cubic meters	cu. inches	61,023.0
cubic meters	cu. yards	1.308
cubic meters	gallons (U.S. liq.)	264.2
cubic meters	liters	1,000.0
cubic meters	pints (U.S. liq.)	2,113.0
cubic meters	quarts (U.S. liq.)	1,057.
cubic yards	cu. cms.	7.646×10^{5}
cubic yards	cu. feet	27.0
cubic yards	cu. inches	46,656.0
cubic yards	cu. meters	0.7646
cubic yards	gallons (U.S. liq.)	202.0
cubic yards	liters	764.6
cubic yards	pints (U.S. liq.)	1,615.9
cubic yards	quarts (U.S. liq.)	807.9
cubic yards/min.	cubic ft./sec.	0.45
cubic yards/min.	gallons/sec.	3.367
cubic yards/min.	liters/sec.	12.74

TO CONVERT	INTO	MULTIPLY BY
Dalton	gram	1.650×10^{-24}
days	seconds	86,400.0
decigrams	grams	0.1
deciliters	liters	0.1
decimeters	meters	0.1
degrees (angle)	quadrants	0.01111
degrees (angle)	radians	0.01745
degrees (angle)	seconds	3,600.0
degrees/sec.	radians/sec.	0.01745
degrees/sec.	revolutions/min.	0.1667
degrees/sec.	revolutions/sec.	2.778×10^{-3}
dekagrams	grams	10.0
dekaliters	liters	10.0
dekameters	meters	10.0
Drams (apothecaries' or troy)	ounces (avoirdupois)	0.1371429
Drams (apothecaries' or troy)	ounces (troy)	0.125
Drams (U.S., fluid or apoth.)	cubic cm.	3.6967
drams	grams	1.7718
drams	grains	27.3437
drams	ounces	0.0625
dyne/cm.	erg/sq. millimeter	.01
dyne/sq. cm.	atmospheres	9.869×10^{-7}
dyne/sq. cm.	inch of Mercury at 0°C	2.953×10^{-5}
dyne/sq. cm.	inch of Water at 4°C	4.015×10^{-4}
dynes	grams	1.020×10^{-3}
dynes	joules/cm.	10^{-7}
dynes	joules/meter (newtons)	10^{-5}
dynes	kilograms	1.020×10^{-6}
dynes	poundals	7.233×10^{-5}
dynes	pounds	2.248×10^{-6}
dynes/sq. cm.	bars	10^{-6}
ell	inches	45
em, pica	inch	0.167
em, pica	cm.	0.4233
erg/sec.	Dyne--cm./sec.	1.000
ergs	Btu	9.480×10^{-11}

TO CONVERT	INTO	MULTIPLY BY
ell	cm.	114.30
ergs	dyne-centimeters	1.0
ergs	foot-pounds	7.367×10^{-8}
ergs	gram-calories	0.2389×10^{-7}
ergs	gram-cms.	1.020×10^{-3}
ergs	horsepower-hrs.	3.7250×10^{-14}
ergs	joules	10^{-7}
ergs	kg.-calories	2.389×10^{-11}
ergs	kg.-meters	1.020×10^{-8}
ergs	kilowatt-hrs.	0.2778×10^{-13}
ergs	watt-hours	0.2778×10^{-10}
ergs/sec.	Btu/min.	$5,688 \times 10^{-9}$
ergs/sec.	ft.-lbs./min.	4.427×10^{-6}
ergs/sec.	ft.-lbs./sec.	7.3756×10^{-8}
ergs/sec.	horsepower	1.341×10^{-10}
ergs/sec.	kg.-calories/min.	1.433×10^{-9}
ergs/sec.	kilowatts	10^{-10}
farads	microfarads	10^6
Faraday/sec.	ampere (absolute)	9.6500×10^4
faradays	ampere-hours	26.80
faradays	coulombs	9.649×10^4
fathom	meter	1.828804
fathoms	feet	6.0
feet	centimeters	30.48
feet	kilometers	3.048×10^{-4}
feet	meters	0.3048
feet	miles (naut.)	1.645×10^{-4}
feet	miles (stat.)	1.894×10^{-4}
feet	millimeters	304.8
feet	mils	1.2×10^4
feet of water	atmospheres	0.02950
feet of water	in. of mercury	0.8826
feet of water	kgs./sq. cm.	0.03048
feet of water	kgs./sq. meter	304.8
feet of water	pounds/sq. ft.	62.43
feet of water	pounds/sq. in.	0.4335
feet/min.	cms./sec.	0.5080
feet/min.	feet/sec.	0.01667
feet/min.	kms./hr.	0.01829

TO CONVERT	INTO	MULTIPLY BY
feet/min.	meters/min.	0.3048
feet/min.	miles/hr.	0.01136
feet/sec.	cms./sec	30.48
feet/sec.	kms./hr.	1.097
feet/sec.	knots	0.5921
feet/sec.	meters/min.	18.29
feet/sec.	miles/hr.	0.6818
feet/sec.	miles/min.	0.01136
feet/sec./sec.	cms./sec./sec.	30.48
feet/sec./sec.	kms./hr./sec.	1.097
feet/sec./sec.	meters/sec./sec.	0.3048
feet/sec./sec.	miles/hr./sec.	0.6818
feet/100 feet	per centigrade	1.0
foot-candle	lumen/sq. meter	10.764
foot-pounds	Btu	1.286×10^{-3}
foot-pounds	ergs	1.356×10^{7}
foot-pounds	gram-calories	0.3238
foot-pounds	hp.-hrs.	5.050×10^{-7}
foot-pounds	joules	1.356
foot-pounds	kg.-calories	3.24×10^{-4}
foot-pounds	kg.-meters	0.1383
foot-pounds	kilowatt-hrs.	3.766×10^{-7}
foot-pounds/min.	Btu/min.	1.286×10^{-3}
foot-pounds/min.	foot-pounds/sec.	0.01667
foot-pounds/min.	horsepower	3.030×10^{-5}
foot-pounds/min.	kg.-calories/min.	3.24×10^{-4}
foot-pounds/min.	kilowatts	2.260×10^{-5}
foot-pounds/sec.	Btu/hr.	4.6263
foot-pounds/sec.	Btu/min.	0.07717
foot-pounds/sec.	horsepower	1.818×10^{-3}
foot-pounds/sec.	kg.-calories/min.	0.01945
foot-pounds/sec.	kilowatts	1.356×10^{-3}
Furlongs	miles (U.S.)	0.125
furlongs	rods	40.0
furlongs	feet	660.0
gallons	cu. cms.	3,785.0
gallons	cu. feet	0.1337
gallons	cu. inches	231.0
gallons	cu. meters	3.785×10^{-3}

TO CONVERT	INTO	MULTIPLY BY
gallons	cu. yards	4.951×10^{-3}
gallons	liters	3.785
gallons (liq. Br. Imp.)	gallons (U.S. liq.)	1.20095
gallons (U.S.)	gallons (Imp.)	0.83267
gallons of water	pounds of water	8.3453
gallons/min.	cu. ft./sec.	2.228×10^{-3}
gallons/min.	liters/sec.	0.06308
gallons/min.	cu. ft./hr.	8.0208
gausses	lines/sq. in.	6.452
gausses	webers/sq. cm.	10^{-8}
gausses	webers/sq. in.	6.452×10^{-8}
gausses	webers/sq. meter	10^{-4}
gilberts	ampere-turns	0.7958
gilberts/cm.	amp-turns/cm.	0.7958
gilberts/cm.	amp-turns/in.	2.021
gilberts/cm.	amp-turns/meter	79.58
gills (British)	cubic cm.	142.07
gills	liters	0.1183
gills	pints (liq.)	0.25
grade	radian	0.01571
grains	drams (avoirdupois)	0.03657143
grains (troy)	grains (avdp.)	1.0
grains (troy)	grams	0.06480
grains (troy)	ounces (avdp.)	2.0833×10^{-3}
grains (troy)	pennyweight (troy)	0.04167
grains/U.S. gal.	parts/million	17.118
grains/U.S. gal.	pounds/million gal.	142.86
grains/Imp. gal.	parts/million	14.286
grams	dynes	980.7
grams	grains	15.43
grams	joules/cm.	9.807×10^{-5}
grams	joules/meter (newtons)	9.807×10^{-3}
grams	kilograms	0.001
grams	milligrams	1,000
grams	ounces (avdp.)	0.03527
grams	ounces (troy)	0.03215
grams	poundals	0.07093
grams	pounds	2.205×10^{-3}
grams/cm.	pounds/inch	5.600×10^{-3}
grams/cu. cm.	pounds/cu. ft.	62.43

TO CONVERT	INTO	MULTIPLY BY
grams/cu. cm.	pounds/cu. in.	0.03613
grams/cu. cm.	pounds/mil-foot	3.405×10^{-7}
grams/liter	grains/gal.	58.417
grams/liter	pounds/1,000 gal.	8.345
grams/liter	pounds/cu. ft.	0.062427
grams/liter	parts/million	1,000.0
grams/sq. cm.	pounds/sq. ft.	2.0481
gram-calories	Btu	3.9683×10^{-3}
gram-calories	ergs	4.1868×10^{7}
gram-calories	foot-pounds	3.0880
gram-calories	horsepower-hrs.	1.5596×10^{-6}
gram-calories	kilowatt-hrs.	1.1630×10^{-6}
gram-calories	watt-hrs.	1.1630×10^{-3}
gram-calories/sec.	Btu/hr.	14.286
gram-centimeters	Btu	9.297×10^{-8}
gram-centimeters	ergs	980.7
gram-centimeters	joules	9.807×10^{-5}
gram-centimeters	kg.-cal.	2.343×10^{-8}
gram-centimeters	kg.-meters	10^{-5}
hand	cm.	10.16
hectares	acres	2.471
hectares	sq. feet	1.076×10^{5}
hectograms	grams	100.0
hectoliters	liters	100.0
hectometers	meters	100.0
hectowatts	watts	100.0
henries	millihenries	1,000.0
horsepower	Btu/min.	42.44
horsepower	foot-lbs./min.	33,000
horsepower	foot-lbs./sec.	550.0
horsepower (metric) (542.5 ft. lb./sec.)	horsepower (550 ft. lb./sec.)	0.9863
horsepower (550 ft. lb./sec.)	horsepower (metric) (542.5 ft. lb./sec.)	1.014
horsepower	kg.-calories/min.	10.68
horsepower	kilowatts	0.7457
horsepower	watts	745.7
horsepower (boiler)	Btu/hr.	33.479
horsepower (boiler)	kilowatts	9.803

TO CONVERT	INTO	MULTIPLY BY
horsepower-hrs.	Btu	2,547
horsepower-hrs.	ergs	2.6845×10^{13}
horsepower-hrs.	foot-lbs.	1.98×10^{6}
horsepower-hrs.	gram-calories	641,190
horsepower-hrs.	joules	2.684×10^{6}
horsepower-hrs.	kg.-calories	641.1
horsepower-hrs.	kg.-meters	2.737×10^{5}
horsepower-hrs.	kilowatt-hrs.	0.7457
hours	days	4.167×10^{-2}
hours	weeks	5.952×10^{-3}
hundredweights (long)	pounds	112
hundredweights (long)	tons (long)	0.05
hundredweights (short)	ounces (avoirdupois)	1,600
hundredweights (short)	pounds	100
hundredweights (short)	tons (metric)	0.0453592
hundredweights (short)	tons (long)	0.0446429
inches	centimeters	2.540
inches	meters	2.540×10^{-2}
inches	miles	1.578×10^{-5}
inches	millimeters	25.40
inches	mils	1,000.0
inches	yards	2.778×10^{-2}
inches of mercury	atmospheres	0.03342
inches of mercury	feet of water	1.133
inches of mercury	kgs./sq. cm.	0.03453
inches of mercury	kgs./sq. meter	345.3
inches of mercury	pounds/sq. ft.	70.73
inches of mercury	pounds/sq. in.	0.4912
inches of water (at 4°C)	atmospheres	2.458×10^{-3}
inches of water (at 4°C)	inches of mercury	0.07355
inches of water (at 4°C)	kgs./sq. cm.	2.540×10^{-3}
inches of water (at 4°C)	ounces/sq. in.	0.5781
inches of water (at 4°C)	pounds/sq. ft.	5.204
inches of water (at 4°C)	pounds/sq. in.	0.03613
international ampere	ampere (absolute)	0.9998
international Volt	volts (absolute)	1.0003
international volt	joules (absolute)	1.593×10^{-19}
international volt	joules	9.654×10^{4}

TO CONVERT	INTO	MULTIPLY BY
joules	Btu	9.480×10^{-4}
joules	ergs	10^7
joules	foot-pounds	0.7376
joules	kg.-calories	2.389×10^{-4}
joules	kg.-meters	0.1020
joules	watt-hrs,	2.778×10^{-4}
joules/cm.	grams	1.020×10^4
joules/cm.	dynes	10^7
joules/cm.	joules/meter (newtons)	100.0
joules/cm.	poundals	723.3
joules/cm.	pounds	22.48
kilograms	dynes	980,665
kilograms	grams	1,000.0
kilograms	joules/cm.	0.09807
kilograms	joules/meter (newtons)	9.807
kilograms	poundals	70.93
kilograms	pounds	2.205
kilograms	tons (long)	9.842×10^{-4}
kilograms	tons (short)	1.102×10^{-3}
kilograms/cu. meter	grams/cu. cm.	0.001
kilograms/cu. meter	pounds/cu. ft.	0.06243
kilograms/cu. meter	pounds/cu. in.	3.613×10^{-5}
kilograms/cu. meter	pounds/mil-foot	3.405×10^{-10}
kilograms/meter	pounds/ft.	0.6720
kilogram/sq. cm.	dynes	980,665
kilograms/sq. cm.	atmospheres	0.9678
kilograms/sq. cm.	feet of water	32.81
kilograms/sq. cm.	inches of mercury	28.96
kilograms/sq. cm.	pounds/sq. ft.	2,048
kilograms/sq. cm.	pounds/sq. in.	14.22
kilograms/sq. meter	atmospheres	9.678×10^{-5}
kilograms/sq. meter	bars	98.07×10^{-6}
kilograms/sq. meter	feet of water	3.281×10^{-3}
kilograms/sq. meter	inches of mercury	2.896×10^{-3}
kilograms/sq. meter	pounds/sq. ft.	0.2048
kilograms/sq. meter	pounds/sq. in.	1.422×10^{-3}
kilograms/sq. mm.	kgs./sq. meter	10^6
kilogram-calories	Btu	3.968
kilogram-calories	foot-pounds	3,088

TO CONVERT	INTO	MULTIPLY BY
kilogram-calories	hp.-hrs.	1.560×10^{-3}
kilogram-calories	joules	4,186
kilogram-calories	kg.-meters	426.9
kilogram-calories	kilojoules	4.186
kilogram-calories	kilowatt-hrs.	1.163×10^{-3}
kilogram meters	Btu	9.294×10^{-3}
kilogram meters	ergs	9.804×10^{7}
kilogram meters	foot-pounds	7.233
kilogram meters	joules	9.804
kilogram meters	kg.-calories	2.342×10^{-3}
kilogram meters	kilowatt-hrs.	2.723×10^{-6}
kilolines	maxwells	1,000.0
kiloliters	liters	1,000.0
kilometers	centimeters	10^{5}
kilometers	feet	3,281
kilometers	inches	3.937×10^{4}
kilometers	meters	1,000.0
kilometers	miles	0.6214
kilometers	millimeters	10^{4}
kilometers	yards	1,094
kilometers/hr.	cms./sec.	27.78
kilometers/hr.	feet/min.	54.68
kilometers/hr.	feet/sec.	0.9113
kilometers/hr.	knots	0.5396
kilometers/hr.	meters/min.	16.67
kilometers/hr.	miles/hr.	0.6214
kilometers/hr./sec.	cms./sec./sec.	27.78
kilometers/hr./sec.	feet/sec./sec.	0.9113
kilometers/hr./sec.	meters/sec./sec.	0.2778
kilometers/hr./sec.	miles/hr./sec.	0.6214
kilowatts	Btu/min.	56.92
kilowatts	foot-lbs./min.	4.426×10^{4}
kilowatts	foot-lbs./sec.	737.6
kilowatts	horsepower	1.341
kilowatts	kg.-calories/min.	14.34
kilowatts	watts	1,000.0
kilowatt-hrs.	Btu	3,413
kilowatt-hrs.	ergs	3.600×10^{13}
kilowatt-hrs.	foot-lbs.	2.655×10^{6}
kilowatt-hrs.	gram-calories	859,850

TO CONVERT	INTO	MULTIPLY BY
kilowatt-hrs.	horsepower-hrs.	1.341
kilowatt-hrs.	joules	3.6×10^6
kilowatt-hrs.	kg.-calories	860.5
kilowatt-hrs.	kg.-meters	3.671×10^5
kilowatt-hrs.	pounds of water raised from 62° to 212°F	22.75
knots	feet/hr.	6,080
knots	kilometers/hr.	1.8532
knots	nautical miles/hr.	1.0
knots	statute miles/hr.	1.151
knots	yards/hr.	2,027
knots	feet/sec.	1.689
league	miles (approx.)	3.0
light year	miles	5.9×10^{12}
light year	kilometers	9.4637×10^{12}
lines/sq. cm.	gausses	1.0
lines/sq. in.	gausses	0.1550
lines/sq. in.	webers/sq. cm.	1.550×10^{-9}
lines/sq. in.	webers/sq. in.	10^{-8}
lines/sq. in.	webers/sq. meter	1.550×10^{-5}
links (engineer's)	inches	12.0
links (surveyor's)	inches	7.92
liters	bushels (U.S. dry)	0.02838
liters	cu. cm.	1,000.0
liters	cu. feet	0.03531
liters	cu. inches	61.02
liters	cu. meters	0.001
liters	cu. yards	1.308×10^{-3}
liters	gallons (U.S. liq.)	0.2642
liters	pints (U.S. liq.)	2.113
liters	quarts (U.S. liq.)	1.057
liters/min.	cu. ft./sec.	5.886×10^{-4}
liters/min.	gals./sec.	4.403×10^{-3}
lumens/sq. ft.	foot-candles	1.0
lumen	spherical candle power	0.07958
lumen	watt	0.001496
lumen/sq. ft.	lumen/sq. meter	10.76
lux	foot-candles	0.0929

TO CONVERT	INTO	MULTIPLY BY
maxwells	kilolines	0.001
maxwells	webers	10^{-8}
megalines	maxwells	10^6
megohms	microhms	10^{12}
megohms	ohms	10^6
meters	centimeters	100.0
meters	feet	3.281
meters	inches	39.37
meters	kilometers	0.001
meters	miles (naut.)	5.396×10^{-4}
meters	miles (stat.)	6.214×10^{-4}
meters	millimeters	1,000.0
meters	yards	1.094
meters	varas	1.179
meters/min.	cms./sec.	1,667
meters/min.	feet/min.	3.281
meters/min.	feet/sec.	0.05468
meters/min.	kms./hr.	0.06
meters/min.	knots	0.03238
meters/min.	miles/hr.	0.03728
meters/sec.	feet/min.	196.8
meters/sec.	feet/sec.	3.281
meters/sec.	kilometers/hr.	3.6
meters/sec.	kilometers/min.	0.06
meters/sec.	miles/hr.	2.237
meters/sec.	miles/min.	0.03728
meters/sec./sec.	cms./sec./sec.	100.0
meters/sec./sec.	ft./sec./sec.	3.281
meters/sec./sec.	kms./hr./sec.	3.6
meters/sec./sec.	miles/hr./sec.	2.237
meter-kilograms	cm.-dynes	9.807×10^7
meter-kilograms	cm.-grams	10^5
meter-kilograms	pound-feet	7.233
microfarad	farads	10^{-6}
micrograms	grams	10^{-6}
microhms	megohms	10^{-12}
microhms	ohms	10^{-6}
microliters	liters	10^{-6}
microns	meters	1×10^{-6}
miles (naut.)	feet	6,080.27

TO CONVERT	INTO	MULTIPLY BY
miles (naut.)	kilometers	1.853
miles (naut.)	meters	1,853
miles (naut.)	miles (statute)	1.1516
miles (naut.)	yards	2,027
miles (statute)	centimeters	1.609×10^5
miles (statute)	feet	5,280
miles (statute)	inches	6.336×10^4
miles (statute)	kilometers	1.609
miles (statute)	meters	1,609
miles (statute)	miles (naut.)	0.8684
miles (statute)	yards	1,760
miles/hr.	cms./sec.	44.70
miles/hr.	feet/min.	88
miles/hr.	feet/sec.	1.467
miles/hr.	kms./hr.	1.609
miles/hr.	kms./min.	0.02682
miles/hr.	knots	0.8684
miles/hr.	meters/min.	26.82
miles/hr.	miles/min.	0.1667
miles/hr./sec.	cms./sec./sec.	44.70
miles/hr./sec.	feet/sec./sec.	1.467
miles/hr./sec.	kms./hr./sec.	1.609
miles/hr./sec.	meters/sec./sec.	0.4470
miles/min.	cms./sec.	2,682
miles/min.	feet/sec.	88
miles/min.	kms./min.	1.609
miles/min.	knots/min.	0.8684
miles/min.	miles/hr.	60
mil-feet	cu. inches	9.425×10^{-6}
milliers	kilograms	1,000
millimicrons	meters	1×10^{-9}
milligrams	grains	0.01543236
milligrams	grams	0.001
milligrams/liter	parts/million	1.0
millihenries	henries	0.001
milliliters	liters	0.001
millimeters	centimeters	0.1
millimeters	feet	3.281×10^{-3}
millimeters	inches	0.03937
millimeters	kilometers	10^{-6}

TO CONVERT	INTO	MULTIPLY BY
millimeters	meters	0.001
millimeters	miles	6.214×10^{-7}
millimeters	mils	39.37
millimeters	yards	1.094×10^{-3}
million gals./day	cu. ft./sec.	1.54723
mils	centimeters	2.540×10^{-3}
mils	feet	8.333×10^{-5}
mils	inches	0.001
mils	kilometers	2.540×10^{-8}
mils	yards	2.778×10^{-5}
miner's inches	cu. ft./min.	1.5
minims (British)	cubic cm.	0.059192
minims (U.S., fluid)	cubic cm.	0.061612
minutes (angles)	degrees	0.01667
minutes (angles)	quadrants	1.852×10^{-4}
minutes (angles)	radians	2.909×10^{-4}
minutes (angles)	seconds	60.0
myriagrams	kilograms	10.0
myriameters	kilometers	10.0
myriawatts	kilowatts	10.0
nepers	decibels	8.686
Newton	dynes	1×105
ohm (international)	ohm (absolute)	1.0005
ohms	megohms	10^{-6}
ohms	microhms	10^{6}
ounces	drams	16.0
ounces	grains	437.5
ounces	grams	28.349527
ounces	pounds	0.0625
ounces	ounces (troy)	0.9115
ounces	tons (long)	2.790×10^{-5}
ounces	tons (metric)	2.835×10^{-5}
ounces (fluid)	cu. inches	1.805
ounces (fluid)	liters	0.02957
ounces (troy)	grains	480.0
ounces (troy)	grams	31.103481
ounces (troy)	ounces (avdp.)	1.09714
ounces (troy)	pennyweights (troy)	20.0

TO CONVERT	INTO	MULTIPLY BY
ounces (troy)	pounds (troy)	0.08333
ounce/sq. inch	dynes/sq. cm.	4,309
ounces/sq. in.	pounds/sq. in.	0.0625
parsec	miles	19×10^{12}
parsec	kilometers	3.084×10^{13}
parts/million	grains/U.S. gal.	0.0584
parts/million	grains/Imp. gal.	0.07016
parts/million	pounds/million gal.	8.345
pecks (British)	cubic inches	554.6
pecks (British)	liters	9.091901
pecks (U.S.)	bushels	0.25
pecks (U.S.)	cubic inches	537.605
pecks (U.S.)	liters	8.809582
pecks (U.S.)	quarts (dry)	8
pennyweights (troy)	grains	24.0
pennyweights (troy)	ounces (troy)	0.05
pennyweights (troy)	grams	1.55517
pennyweights (troy)	pounds (troy)	4.1667×10^{-3}
pints (dry)	cu. inches	33.60
pints (liq.)	cu. cms.	473.2
pints (liq.)	cu. feet	0.01671
pints (liq.)	cu. inches	28.87
pints (liq.)	cu. meters	4.732×10^{-4}
pints (liq.)	cu. yards	6.189×10^{-4}
pints (liq.)	gallons	0.125
pints (liq.)	liters	0.4732
pints (liq.)	quarts (liq.)	0.5
Planck's quantum	erg - second	6.624×10^{-27}
poise	gram/cm. sec.	1.00
pounds (avoirdupois)	ounces (troy)	14.5833
poundals	dynes	13,826
poundals	grams	14.10
poundals	joules/cm.	1.383×10^{-3}
poundals	joules/meter (newtons)	0.1383
poundals	kilograms	0.01410
poundals	pounds	0.03108
pounds	drams	256
pounds	dynes	44.4823×10^{4}
pounds	grains	7,000

TO CONVERT	INTO	MULTIPLY BY
pounds	grams	453.5924
pounds	joules/cm.	0.04448
pounds	joules/meter (newtons)	4.448
pounds	kilograms	0.4536
pounds	ounces	16.0
pounds	ounces (troy)	14.5833
pounds	poundals	32.17
pounds	pounds (troy)	1.21528
pounds	tons (short)	0.0005
pounds (troy)	grains	5,760
pounds (troy)	grams	373.24177
pounds (troy)	ounces (avdp.)	13.1657
pounds (troy)	ounces (troy)	12.0
pounds (troy)	pennyweights (troy)	240.0
pounds (troy)	pounds (avdp.)	0.822857
pounds (troy)	tons (long)	3.6735×10^{-4}
pounds (troy)	tons (metric)	3.7324×10^{-4}
pounds (troy)	tons (short)	4.1143×10^{-4}
pounds of water	cu. ft.	0.01602
pounds of water	cu. inches	27.68
pounds of water	gallons	0.1198
pounds of water/min.	cu. ft./sec.	2.670×10^{-4}
pound-feet	cm.-dynes	1.356×10^{7}
pound-feet	cm.-grams	13,825
pound-feet	meter-kgs.	0.1383
pounds/cu. ft.	grams/cu. cm.	0.01602
pounds/cu. ft.	kgs./cu. meter	16.02
pounds/cu. ft.	pounds/cu. in.	5.787×10^{-4}
pounds/cu. ft.	pounds/mil-foot	5.456×10^{-9}
pounds/cu. in.	gms./cu. cm.	27.68
pounds/cu. in.	kgs./cu. meter	2.768×10^{4}
pounds/cu. in.	pounds/cu. ft.	1,728
pounds/cu. in.	pounds/mil-foot	9.425×10^{-6}
pounds/ft.	kgs./meter	1.488
pounds/in.	gms./cm.	178.6
pounds/mil-foot	gms./cu. cm.	2.306×10^{6}
pounds/sq. ft.	atmospheres	4.725×10^{-4}
pounds/sq. ft.	feet of water	0.01602
pounds/sq. ft.	inches of mercury	0.01414
pounds/sq. ft.	kgs./sq. meter	4.882

TO CONVERT	INTO	MULTIPLY BY
pounds/sq. ft.	pounds/sq. in.	6.944×10^{-3}
pounds/sq. in.	atmospheres	0.06804
pounds/sq. in.	feet of water	2.307
pounds/sq. in.	inches of mercury	2.036
pounds/sq. in.	kgs./sq. meter	703.1
pounds/sq. in.	pounds/sq. ft.	144.0
quadrants (angle)	degrees	90.0
quadrants (angle)	minutes	5,400.0
quadrants (angle)	radians	1.571
quadrants (angle)	seconds	3.24×10^{5}
quarts (dry)	cu. inches	67.20
quarts (liq.)	cu. cms.	946.4
quarts (liq.)	cu. feet	0.03342
quarts (liq.)	cu. inches	57.75
quarts (liq.)	cu. meters	9.464×10^{-4}
quarts (liq.)	cu. yards	1.238×10^{-3}
quarts (liq.)	gallons	0.25
quarts (liq.)	liters	0.9463
radians	degrees	57.30
radians	minutes	3,438
radians	quadrants	0.6366
radians	seconds	2.063×10^{5}
radians/sec.	degrees/sec.	57.30
radians/sec.	revolutions/min.	9.549
radians/sec.	revolutions/sec.	0.1592
radians/sec./sec.	revs./min./min.	573.0
radians/sec./sec.	revs./min./sec.	9.549
radians/sec./sec.	revs./sec./sec.	0.1592
revolutions	degrees	360.0
revolutions	quadrants	4.0
revolutions	radians	6.283
revolutions/min.	degrees/sec.	6.0
revolutions/min.	radians/sec.	0.1047
revolutions/min.	revs./sec.	0.01667
revolutions/min./min.	radians/sec./sec.	1.745×10^{-3}
revolutions/min./min.	revs./min./sec.	0.01667
revolutions/min./min.	revs./sec./sec.	2.778×10^{-4}
revolutions/sec.	degrees/sec.	360.0

TO CONVERT	INTO	MULTIPLY BY
revolutions/sec.	radians/sec.	6.283
revolutions/sec.	revs./min.	60.0
revolutions/sec./sec.	radians/sec./sec.	6.283
revolutions/sec./sec.	revs./min./min.	3,600.0
revolutions/sec./sec.	revs./min./sec.	60.0
rod	chain (Gunters)	0.25
rod	meters	5.029
rods (surveyors' meas.)	yards	5.5
rods	feet	16.5
scruples	grains	20
seconds (angle)	degrees	2.778×10^{-4}
seconds (angle)	minutes	0.01667
seconds (angle)	quadrants	3.087×10^{-6}
seconds (angle)	radians	4.848×10^{-6}
slug	kilogram	14.59
slug	pounds	32.17
sphere	steradians	12.57
square centimeters	circular mils	1.973×10^{5}
square centimeters	sq. feet	1.076×10^{-3}
square centimeters	sq. inches	0.1550
square centimeters	sq. meters	0.0001
square centimeters	sq. miles	3.861×10^{-11}
square centimeters	sq. millimeters	100.0
square centimeters	sq. yards	1.196×10^{-4}
square feet	acres	2.296×10^{-5}
square feet	circular mils	1.833×10^{8}
square feet	sq. cms.	929.0
square feet	sq. inches	144.0
square feet	sq. meters	0.09290
square feet	sq. miles	3.587×10^{-8}
square feet	sq. millimeters	9.290×10^{4}
square feet	sq. yards	0.1111
square inches	circular mils	1.273×10^{6}
square inches	sq. cms.	6.452
square inches	sq. feet	6.944×10^{-3}
square inches	sq. millimeters	645.2
square inches	sq. mils	10^{6}
square inches	sq. yards	7.716×10^{-4}
square kilometers	acres	247.1

TO CONVERT	INTO	MULTIPLY BY
square kilometers	sq. cms.	10^{10}
square kilometers	sq. ft.	10.76×10^6
square kilometers	sq. inches	1.550×10^9
square kilometers	sq. meters	10^6
square kilometers	sq. miles	0.3861
square kilometers	sq. yards	1.196×10^6
square meters	acres	2.471×10^{-4}
square meters	sq. cms.	10^4
square meters	sq. feet	10.76
square meters	sq. inches	1,550
square meters	sq. miles	3.861×10^{-7}
square meters	sq. millimeters	10^6
square meters	sq. yards	1.196
square miles	acres	640.0
square miles	sq. feet	27.88×10^6
square miles	sq. kms.	2.590
square miles	sq. meters	2.590×10^6
square miles	sq. yards	3.098×10^6
square millimeters	circular mils	1,973
square millimeters	sq. cms.	0.01
square millimeters	sq. feet	1.076×10^{-5}
square millimeters	sq. inches	1.550×10^{-3}
square mils	circular mils	1.273
square mils	sq. cms.	6.452×10^{-6}
square mils	sq. inches	10^{-6}
square yards	acres	2.066×10^{-4}
square yards	sq. cms.	8,361
square yards	sq. feet	9.0
square yards	sq. inches	1,296
square yards	sq. meters	0.8361
square yards	sq. miles	3.228×10^{-7}
square yards	sq. millimeters	8.361×10^5
temperature (°C) + 273	absolute temp. (°C)	1.0
temperature (°C) + 17.78	temperature (°F)	1.8
temperature (°F) + 460	absolute temp. (°F)	1.0
temperature (°F) - 32	temperature (°C)	5/9
tons (long)	kilograms	1,016
tons (long)	pounds	2,240
tons (long)	tons (short)	1.120

TO CONVERT	INTO	MULTIPLY BY
tons (metric)	kilograms	1,000
tons (metric)	pounds	2,205
tons (short)	kilograms	907.1848
tons (short)	ounces	32,000
tons (short)	ounces (troy)	29,166.66
tons (short)	pounds	2,000
tons (short)	pounds (troy)	2,430.56
tons (short)	tons (long)	0.89287
tons (short)	tons (metric)	0.9078
tons (short)/sq. ft.	kgs./sq. meter	9,765
tons (short)/sq. ft.	pounds/sq. in.	2,000
tons of water/24 hrs.	pounds of water/hr.	83.333
tons of water/24 hrs.	gallons/min.	0.16643
tons of water/24 hrs.	cu. ft./hr.	1.3349
volt/inch	volt/cm.	0.39370
volt (absolute)	statvolts	0.003336
watts	Btu/hr.	3.4129
watts	Btu/min.	0.05688
watts	ergs/sec.	107
watts	foot-lbs./min.	44.27
watts	foot-lbs./sec.	0.7378
watts	horsepower	1.341×10^{-3}
watts	horsepower (metric)	1.360×10^{-3}
watts	kg.-calories/min.	0.01433
watts	kilowatts	0.001
watts (Abs.)	B.T.U. (mean)/min.	0.056884
watts (Abs.)	joules/sec.	1
watt-hours	Btu	3.413
watt-hours	ergs	3.60×10^{10}
watt-hours	foot-pounds	2,656
watt-hours	gram-calories	859.85
watt-hours	horsepower-hrs.	1.341×10^{-3}
watt-hours	kilogram-calories	0.8605
watt-hours	kilogram-meters	367.2
watt-hours	kilowatt-hrs.	0.001
watt (international)	watt (absolute)	1.0002
webers	maxwells	10^8
webers	kilolines	10^5
webers/sq. in.	gausses	1.550×10^7
webers/sq. in.	lines/sq. in.	10^8

TO CONVERT	INTO	MULTIPLY BY
webers/sq. in.	webers/sq. cm.	0.1550
webers/sq. in.	webers/sq. meter	1,550
webers/sq. meter	gausses	10^4
webers/sq. meter	lines/sq. in.	6.452×10^4
webers/sq. meter	webers/sq. cm.	10^{-4}
webers/sq. meter	webers/sq. in.	6.452×10^{-4}
yards	centimeters	91.44
yards	kilometers	9.144×10^{-4}
yards	meters	0.9144
yards	miles (naut.)	4.934×10^{-4}
yards	miles (stat.)	5.682×10^{-4}
yards	millimeters	914.4

16.5   ELECTROMAGNETIC-RADIATION SPECTRUM

The usable spectrum of electromagnetic-radiation frequencies extends over a range from below 100 Hz for power distribution to 10^{20} Hz for the shortest X-rays. The lower frequencies are used primarily for terrestrial broadcasting and communications. The higher frequencies include visible and near-visible infrared and ultraviolet light, and X-rays. The primary frequency bands used for communications purposes range from 30 kHz and up.

16.5.1   Low Frequency (LF)

The 30 kHz to 300 kHz band is used for around-the-clock communications services over long distances, and where adequate power is available to overcome high levels of atmospheric noise. Applications include:

- Radionavigation
- Fixed/maritime communications and navigation
- Aeronautical radionavigation
- Low frequency broadcasting (Europe)
- Underwater submarine communications (10-30 kHz)

16.5.2   Medium Frequency (MF)

The 300 kHz to 3 MHz low frequency portion of this band is used for around-the-clock communication services over moderately long distances. The upper portion of the MF band is used principally for moderate distance voice communications. Applications include:

- AM radio broadcasting (535.5-1605.5 kHz)
- Radionavigation
- Fixed/maritime communications
- Aeronautical radionavigation
- Fixed and mobile commercial communications
- Amateur radio
- Standard time and frequency services

16.5.3   High Frequency (HF)

The 3 MHz to 30 MHz band provides reliable medium-range coverage during daylight and, when the transmission path is in total darkness, worldwide long-distance service. Applications include

- Shortwave broadcasting
- Fixed and mobile service
- Telemetry
- Amateur radio
- Fixed/maritime mobile
- Standard time and frequency services
- Radio astronomy
- Aeronautical fixed and mobile

16.5.4   Very High Frequency (VHF)

The 30 MHz to 300 MHz band is characterized by reliable transmission over medium distances. At the higher portion of the VHF band, communications is limited by the horizon. Applications include:

- FM radio broadcasting (88-108 MHz)
- Low band VHF television broadcasting (54-72 MHz and 76-88 MHz)
- High band VHF television broadcasting (174-216 MHz)
- Commercial fixed and mobile radio
- Aeronautical radionavigation
- Space research
- Fixed/maritime mobile
- Amateur radio
- Radiolocation

16.5.5   Ultra High Frequency (UHF)

The 300 MHz to 3 GHz band is typically used for line-of-sight communications. Short wavelengths at the upper end of the band permit the use of highly directional parabolic or multielement antennas. Applications include:

- UHF terrestrial television (470-806 MHz)
- Fixed and mobile communications
- Telemetry
- Meteorological aids
- Space operations
- Radio astronomy
- Radionavigation
- Satellite communications
- Point-to-point microwave relay

16.5.6   Super High Frequency (SHF)

The 3 GHz to 30 GHz band provides strictly line-of-sight communications. Very short wavelengths permit the use of parabolic transmit and receive antennas of exceptional gain. Applications include:

- Satellite communications
- Point-to-point wideband relay
- Radar
- Specialized wideband communications
- Developmental research
- Military support systems
- Radiolocation
- Radionavigation
- Space research

16.6   REFERENCE TABLES

Table 16.1   Power conversion factors (decibels to watts)

dBm	dBW	Watts Whole Number or Decimal Number	Multiple or Submultiple	Prefix
+150	+120	1,000,000,000,000	10^{12}	1 Terawatt
+140	+110	100,000,000,000	10^{11}	100 Gigawatts
+130	+100	10,000,000,000	10^{10}	10 Gigawatts
+120	+90	1,000,000,000	10^{9}	1 Gigawatt
+110	+80	100,000,000	10^{8}	100 Megawatts
+100	+70	10,000,000	10^{7}	10 Megawatts
+90	+60	1,000,000	10^{6}	1 Megawatt
+80	+50	100,000	10^{5}	100 Kilowatts
+70	+40	10,000	10^{4}	10 Kilowatts
+60	+30	1,000	10^{3}	1 Kilowatt
+50	+20	100	10^{2}	1 Hectrowatt (100 w)
+40	+10	10	10^{1}	1 Decawatt (10 w)
+30	0	1	10^{0}	1 Watt
+20	−10	0.1	10^{-1}	1 Deciwatt (100 mw)
+10	−20	0.01	10^{-2}	1 Centiwatt (10 mw)
0	−30	0.001	10^{-3}	1 Milliwatt
−10	−40	0.0,001	10^{-4}	100 Microwatts
−20	−50	0.00,001	10^{-5}	10 Microwatts
−30	−60	0.000,001	10^{-6}	1 Microwatt
−40	−70	0.0,000,001	10^{-7}	100 Nanowatts
−50	−80	0.00,000,001	10^{-8}	10 Nanowatts
−60	−90	0.000,000,001	10^{-9}	1 Nanowatt
−70	−100	0.0,000,000,001	10^{-10}	100 Picowatts
−80	−110	0.00,000,000,001	10^{-11}	10 Picowatts
−90	−120	0.000,000,000,001	10^{-12}	1 Picowatt

Table 16.2   Voltage standing wave ratio (VSWR) relationships

VSWR	REFLECTION COEFFICIENT	RETURN LOSS	POWER RATIO	PERCENT REFLECTED
1.01 : 1	.0050	46.1 dB	.00002	.002%
1.02 : 1	.0099	40.1 dB	.00010	.010%
1.04 : 1	.0196	34.2 dB	.00038	.038%
1.06 : 1	.0291	30.7 dB	.00085	.085%
1.08 : 1	.0385	28.3 dB	.00148	.148%
1.10 : 1	.0476	26.4 dB	.00227	.227%
1.20 : 1	.0909	20.8 dB	.00826	.826%
1.30 : 1	.1304	17.7 dB	.01701	1.7%
1.40 : 1	.1667	15.6 dB	.02778	2.8%
1.50 : 1	.2000	14.0 dB	.04000	4.0%
1.60 : 1	.2308	12.7 dB	.05325	5.3%
1.70 : 1	.2593	11.7 dB	.06722	6.7%
1.80 : 1	.2857	10.9 dB	.08163	8.2%
1.90 : 1	.3103	10.2 dB	.09631	9.6%
2.00 : 1	.3333	9.5 dB	.11111	11.1%
2.20 : 1	.3750	8.5 dB	.14063	14.1%
2.40 : 1	.4118	7.7 dB	.16955	17.0%
2.60 : 1	.4444	7.0 dB	.19753	19.8%
2.80 : 1	.4737	6.5 dB	.22438	22.4%
3.00 : 1	.5000	6.0 dB	.25000	25.0%
3.50 : 1	.5556	5.1 dB	.30864	30.9%
4.00 : 1	.6000	4.4 dB	.36000	36.0%
4.50 : 1	.6364	3.9 dB	.40496	40.5%
5.00 : 1	.6667	3.5 dB	.44444	44.4%
6.00 : 1	.7143	2.9 dB	.51020	51.0%
7.00 : 1	.7500	2.5 dB	.56250	56.3%
8.00 : 1	.7778	2.2 dB	.60494	60.5%
9.00 : 1	.8000	1.9 dB	.64000	64.0%
10.00 : 1	.8182	1.7 dB	.66942	66.9%
15.00 : 1	.8750	1.2 dB	.76563	76.6%
20.00 : 1	.9048	.9 dB	.81859	81.9%
30.00 : 1	.9355	.6 dB	.87513	87.5%
40.00 : 1	.9512	.4 dB	.90482	90.5%
50.00 : 1	.9608	.3 dB	.92311	92.3%

$$\text{VSWR} = \frac{1 + |p|}{1 - |p|} = \frac{1 + \sqrt{(Prfl/Pfwd)}}{1 - \sqrt{(Prfl/Pfwd)}}$$

$$\text{POWER RATIO} = (Prfl/Pfwd)$$

$$p = \frac{\text{VSWR} - 1}{\text{VSWR} + 1} = \text{REFLECTION COEFFICIENT}$$

$$\text{RETURN LOSS} = -20 \log |p|$$

Table 16.3 Specifications of standard copper wire sizes

Wire Size A.W.G. (B&S)	Diam. in Mils	Circular Mil Area	Turns per Linear Inch (25.4 mm)			Ohms per 1000 ft. 20°C	Current Carrying Capacity at 700 C.M. per Amp	Diam. in mm.
			Enamel	S.C.E.	D.C.C.			
1	289.3	83690	—	—	—	.1239	119.6	7.348
2	257.6	66370	—	—	—	.1563	94.8	6.544
3	229.4	52640	—	—	—	.1970	75.2	5.827
4	204.3	41740	—	—	—	.2485	59.6	5.189
5	181.9	33100	—	—	—	.3133	47.3	4.621
6	162.0	26250	—	—	—	.3951	37.5	4.115
7	144.3	20820	—	—	—	.4982	29.7	3.665
8	128.5	16510	7.6	—	7.1	.6282	23.6	3.264
9	114.4	13090	8.6	—	7.8	.7921	18.7	2.906
10	101.9	10380	9.6	9.1	8.9	.9989	14.8	2.588
11	90.7	8234	10.7	—	9.8	1.26	11.8	2.305
12	80.8	6530	12.0	11.3	10.9	1.588	9.33	2.053
13	72.0	5178	13.5	—	12.8	2.003	7.40	1.828
14	64.1	4107	15.0	14.0	13.8	2.525	5.87	1.628
15	57.1	3257	16.8	—	14.7	3.184	4.65	1.450
16	50.8	2583	18.9	17.3	16.4	4.016	3.69	1.291
17	45.3	2048	21.2	—	18.1	5.064	2.93	1.150
18	40.3	1624	23.6	21.2	19.8	6.385	2.32	1.024
19	35.9	1288	26.4	—	21.8	8.051	1.84	.912
20	32.0	1022	29.4	25.8	23.8	10.15	1.46	.812
21	28.5	810	33.1	—	26.0	12.8	1.16	.723
22	25.3	642	37.0	31.3	30.0	16.14	.918	.644
23	22.6	510	41.3	—	37.6	20.36	.728	.573
24	20.1	404	46.3	37.6	35.6	25.67	.577	.511
25	17.9	320	51.7	—	38.6	32.37	.458	.455
26	15.9	254	58.0	46.1	41.8	40.81	.363	.405
27	14.2	202	64.9	—	45.0	51.47	.288	.361
28	12.6	160	72.7	54.6	48.5	64.9	.228	.321
29	11.3	127	81.6	—	51.8	81.83	.181	.286
30	10.0	101	90.5	64.1	55.5	103.2	.144	.255
31	8.9	80	101	—	59.2	130.1	.114	.227
32	8.0	63	113	74.1	61.6	164.1	.090	.202
33	7.1	50	127	—	66.3	206.9	.072	.180
34	6.3	40	143	86.2	70.0	260.9	.057	.160
35	5.6	32	158	—	73.5	329.0	.045	.143
36	5.0	25	175	103.1	77.0	414.8	.036	.127
37	4.5	20	198	—	80.3	523.1	.028	.113
38	4.0	16	224	116.3	83.6	659.6	.022	.101
39	3.5	12	248	—	86.6	831.8	.018	.090

Table 16.4 Pinout patterns for common semiconductor cases

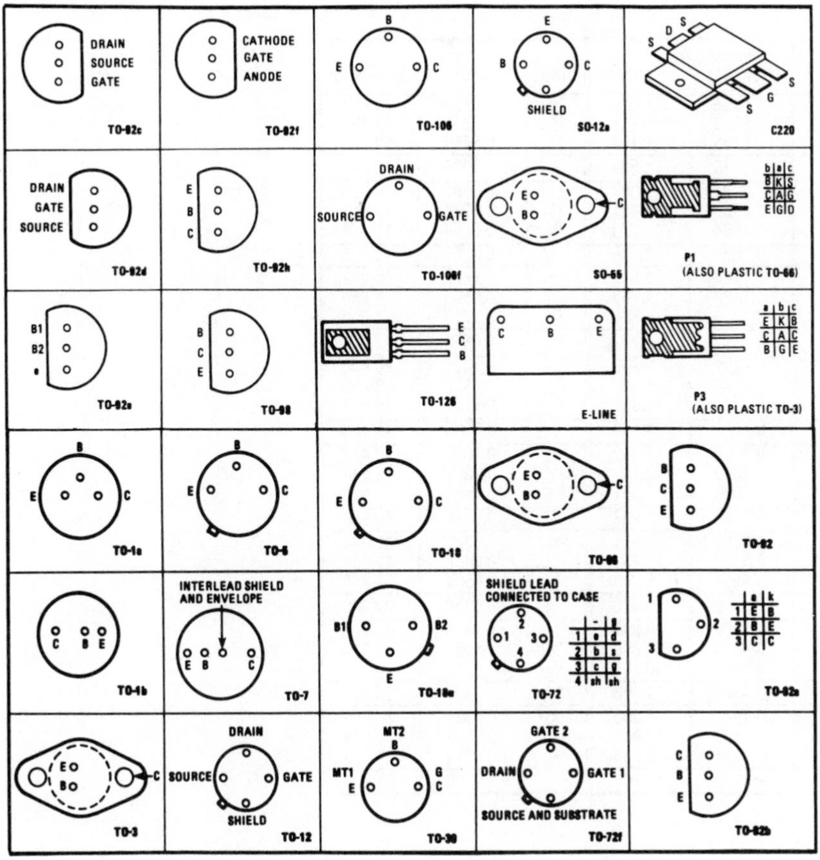

Table 16.5 Inch to millimeter conversion table

1 Inch = 25.4 mm

Inch	Decimal Inch	Millimeter	Inch	Decimal Inch	Millimeter
1/64	0.0156	0.397	33/64	0.5156	13.097
1/32	0.0313	0.794	17/32	0.5313	13.494
3/64	0.0469	1.191	35/64	0.5469	13.891
1/16	0.0625	1.588	9/16	0.5625	14.288
5/64	0.0781	1.984	37/64	0.5781	14.684
3/32	0.0938	2.381	19/32	0.5938	15.081
7/64	0.1094	2.778	39/64	0.6094	15.478
1/8	0.1250	3.175	5/8	0.6250	15.875
9/64	0.1406	3.572	41/64	0.6406	16.272
5/32	0.1563	3.969	21/32	0.6563	16.689
11/64	0.1719	4.366	43/64	0.6719	17.066
3/16	0.1875	4.763	11/16	0.6875	17.463
13/64	0.2031	5.159	45/64	0.7031	17.859
7/32	0.2188	5.556	23/32	0.7188	18.256
15/64	0.2344	5.953	47/64	0.7344	18.653
1/4	0.2500	6.350	3/4	0.7500	19.050
17/64	0.2656	6.747	49/64	0.7656	19.447
9/32	0.2813	7.144	25/32	0.7813	19.844
19/64	0.2969	7.541	51/64	0.7969	20.241
5/16	0.3125	7.938	13/16	0.8125	20.638
21/64	0.3281	8.334	53/64	0.8281	21.034
11/32	0.3438	8.731	27/32	0.8438	21.431
23/64	0.3594	9.128	55/64	0.8594	21.828
3/8	0.3750	9.525	7/8	0.8750	22.225
25/64	0.3906	9.922	57/64	0.8906	22.622
13/32	0.4063	10.319	29/32	0.9063	23.019
27/64	0.4219	10.716	59/64	0.9219	23.416
7/16	0.4375	11.113	15/16	0.9375	23.813
29/64	0.4531	11.509	61/64	0.9531	24.209
15/32	0.4688	11.906	31/32	0.9688	24.606
31/64	0.4844	12.303	63/64	0.9844	25.003
1/2	0.5000	12.700	1	1.0000	25.400

Table 16.6 Celsius to Fahrenheit conversion table

$$32 + \frac{9}{5}\,°C = °F \qquad \frac{5}{9}(°F - 32) = °C$$

°C	°F	°C	°F
− 50	− 58	125	257
− 45	− 49	130	266
− 40	− 40	135	275
− 35	− 31	140	284
− 30	− 22	145	293
− 25	− 13	150	302
− 20	− 4	155	311
− 15	5	160	320
− 10	14	165	329
− 5	23	170	338
0	32	175	347
5	41	180	356
10	50	185	365
15	59	190	374
20	68	195	383
25	77	200	392
30	86	205	401
35	95	210	410
40	104	215	419
45	113	220	428
50	122	225	437
55	131	230	446
60	140	235	455
65	149	240	464
70	158	245	473
75	167	250	482
80	176	255	491
85	185	260	500
90	194	265	509
95	203	270	518
100	212	275	527
105	221	280	536
110	230	285	545
115	239	290	554
120	248	295	563
		300	572

Table 16.7   Decimal equivalents of fractional inches

Inch	0	1/8	1/4	3/8	1/2	5/8	3/4	7/8	Inch
0	0.0	3.18	6.35	9.52	12.70	15.88	19.05	22.22	0
1	25.40	28.58	31.75	34.92	38.10	41.28	44.45	47.62	1
2	50.80	53.98	57.15	60.32	63.50	66.68	69.85	73.02	2
3	76.20	79.38	82.55	85.72	88.90	92.08	95.25	98.42	3
4	101.6	104.8	108.0	111.1	114.3	117.5	120.6	123.8	4
5	127.0	130.2	133.4	136.5	139.7	142.9	146.0	149.2	5
6	152.4	155.6	158.8	161.9	165.1	168.3	171.4	174.6	6
7	177.8	181.0	184.2	187.3	190.5	193.7	196.8	200.0	7
8	203.2	206.4	209.6	212.7	215.9	219.1	222.2	225.4	8
9	228.6	231.8	235.0	238.1	241.3	244.5	247.6	250.8	9
10	254.0	257.2	260.4	263.5	266.7	269.9	273.0	276.2	10
11	279	283	286	289	292	295	298	302	11
12	305	308	311	314	317	321	324	327	12
13	330	333	337	340	343	346	349	352	13
14	356	359	362	365	368	371	375	378	14
15	381	384	387	391	394	397	400	403	15
16	406	410	413	416	419	422	425	429	16
17	432	435	438	441	445	448	451	454	17
18	457	460	464	467	470	473	476	479	18
19	483	486	489	492	495	498	502	505	19
20	508	511	514	518	521	524	527	530	20
Inch	0	1/8	1/4	3/8	1/2	5/8	3/4	7/8	Inch

INDEX

ABOUT THE AUTHOR

Jerry Whitaker is Associate Publisher of *Broadcast Engineering* magazine and *Video Systems* magazine, published by Intertec Publishing Corp. of Overland Park, Kan. He is a Fellow of the Society of Broadcast Engineers, and an SBE-certified senior AM-FM engineer. He is also a member of the SMPTE, AES, ITVA, and IEEE (Broadcast Society, Power Electronics Society, and Reliability and Maintainability Society). He has written and lectured extensively on the topic of electronic systems maintenance. Mr. Whitaker is a former radio chief engineer and television news producer. He is the author of the McGraw-Hill publication, *Radio Frequency Transmission Systems: Design and Operation*, and co-author of McGraw-Hill's *Television and Audio Handbook for Technicians and Engineers*. Mr. Whitaker is a contributor to the McGraw-Hill *Audio Engineering Handbook*, and to the National Association of Broadcasters' *NAB Engineering Handbook, 7th and 8th editions*. He has twice received a Jesse H. Neal Award Certificate of Merit from the Association of Business Publishers for editorial excellence.